Probability and Statistics
A Didactic Introduction

Probability and Statistics
A Didactic Introduction

Editors
José I. Barragués
Polytechnical College of San Sebastian
University of the Basque Country
San Sebastian, Spain
Adolfo Morais
Polytechnical College of San Sebastian
University of the Basque Country
San Sebastian, Spain
Jenaro Guisasola
Polytechnical College of San Sebastian
University of the Basque Country
San Sebastian, Spain

CRC Press
Taylor & Francis Group
Boca Raton London New York

CRC Press is an imprint of the
Taylor & Francis Group, an **informa** business
A SCIENCE PUBLISHERS BOOK

CRC Press
Taylor & Francis Group
6000 Broken Sound Parkway NW, Suite 300
Boca Raton, FL 33487-2742

Printed in the United States of America on acid-free paper

International Standard Book Number: 978-1-4822-1977-7 (Hardback)

Visit the Taylor & Francis Web site at
http://www.taylorandfrancis.com

CRC Press Web site at
http://www.crcpress.com

Science Publishers Web site at
http://www.scipub.net

ADDITIONAL RESOURCES

A GoogleDocs folder has been provided for the reader comprising additional resources such as spreadsheets, data files, SPSS and R files, to name but a few. This toolkit is free and is contained in the zip file "Probability_and_Statistics_a_didactic_introduction.zip". You may download this file using the link provided below, or by using the QR code:

i) https://docs.google.com/file/d/0B-tSC_22acQjbVhuSTZFOFh6SDQ/edit?usp=sharing

ii)

Preface

Competence in statistics is considered essential for citizens to be able to deal with an environment of close social, political and economic interdependencies, in which decisions with a social impact are often taken on the basis of statistical analysis. The Development of statistics has broadened the range of processes in which users can use statistical methods. This may explain the observed growth in the demand for training in statistics in fields such as engineering, psychology, education, health sciences, natural sciences, and business, to name but a few.

This book attempts to provide a comprehensive introduction to the foundations of Probability and Statistics for which leading authors have been invited to write specific papers. Each of the chapters is dedicated to a major topic and offers an intuitive view of the subject matter, methodologies, concepts, terms and related applications. The book is suitable for entry level courses in first year university courses of science and engineering, higher level courses, postgraduate university studies and for the research community.

In writing the chapters, we sought to avoid the traditional definition-theorem-example exposition format, used in most university textbooks. Instead, a problem-oriented approach was adopted. That is, a problem situation is posed that highlights the need for a theoretical framework that can provide a satisfactory answer to the proposed situation. The theoretical framework is then reconstructed using intuitive approaches, discussions and activities for the reader.

The book has been organized as follows:

- Key elements of the theoretical framework have been developed, laying greater, emphasis on concepts than formal demonstrations. The author provides a personal perspective of the significance and the utility of this theoretical framework for the resolution of problems in sciences, engineering and other fields.
- The methodology for the development of the content focuses on mathematical meanings, and orienting it toward the acquisition of mathematical competence through problem-solving. In this sense, the

concepts are constructed as they are needed for the solving of new problems.
- Self-assessment activities have been proposed throughout the chapter, not just at the end. The aim of these activities is to involve the reader in actively participating in the construction of the theoretical framework, so that the reader reflects on the meanings that are being constructed, their utility and their practical applications.
- Examples of applications, solved problems and additional problems for readers have been provided.
- Paying attention to potential students' learning difficulties. Some of these have been widely studied by the research community in the field of Mathematics Education.
- Including activities that use the computer to explore the meaning of the concepts in greater depth, to experiment or to investigate problems.

We would like to thank the authors for the interest and care that they have shown in completing their work. They have brought not only their knowledge of the discipline, but also valuable experience in university teaching and current practical applications of Probability and Statistics.

José I. Barragués
Adolfo Morais
Jenaro Guisasola

Contents

CHAPTER 1

Descriptive Statistics

Nicholas Watier, Claude Lamontagne* and *Sylvain Chartier*

Dare to know!
—Horace

1. Introduction

The motto of this chapter was used by Immanuel Kant to describe the spirit of the enlightenment—a philosophical movement that championed mankind's intellect and encouraged the personal use of reason in all matters of inquiry. In describing the enlightenment, Kant explicitly challenges his readers to have the courage to use their own intellect without depending on the counsel or approval of an expert or authority.

In this chapter we propose a similar challenge to you, the reader. Throughout the chapter you will uncover and encounter problems. These problems are intended to promote speculation on key concepts in descriptive statistics. We encourage you to offer solutions to these problems before progressing onto the next section in the chapter. Your solutions may be partial, incomplete, or just plain incorrect, and odd as it may sound, we are actually hoping for this to be the case. Proposing solutions to problems affords you the opportunity to wonder, to speculate and to learn. And by exploring the effectiveness of your solutions and discovering their limitations, insights into statistical concepts can be obtained without the need for direct instruction or complete reliance on the teachings of an expert. In this way, you can discover key concepts in descriptive statistics on your own.

University of Ottawa, School of Psychology, Ottawa, Ontario, Canada.
*Corresponding author

In short, our aim in this chapter is to give you the opportunity to develop *your own understanding* of descriptive statistics and create *your own meaning* of a particular statistical concept through a speculative problem-solving process. At various points in the chapter, problems will be presented (or discovered by you) and popular solutions (or solutions that we think you'll propose) will be discussed. The effectiveness of these solutions will be explored and those solutions that solve the problem satisfactorily will be examined in more detail. Each solution will lead to further sub-problems, thus creating a problem-solution hierarchy. To gain the full benefits of this approach, we strongly encourage you to play along and give an honest effort in thinking about the problems and proposing solutions. Cues (e.g., Try it!) are included in the text to encourage you to think about the problem. You have nothing to lose and everything to gain by doing so. Heed Kant's challenge: have the courage to use your own intellect!

2. Descriptive Statistics

As a starting point, we should set discussion of descriptive statistics in a familiar context. Grades are often a concern for students. Students are not only interested in the value of their final grade or whether they have successfully completed a course, but also how well they have performed in relation to their peers. Grades can be represented in a variety of formats, from letter grades (e.g., A, B, C, D, F) and descriptive statements (e.g., Excellent, Good, Poor) to numerical values (e.g., 80,100), the last of which affords the opportunity of making quantitative comparisons among individuals. For these reasons, grades will be used as an entry point for discussing descriptive statistics.

Let us say, for the sake of example, that your final grade in an introductory statistics course is 75 out of 100. On the basis of this information, how would you describe your performance in the course? Try it!

You might have concluded that you passed the course because your final grade is above 50, but little can be concluded beyond that. Is there any additional information that could help you describe your performance?

Perhaps the final grades of your classmates might be helpful. The distribution[1] of final grades for the course (a box surrounds your final grade) is as follows:

$$x_i = \{80, 81, 65, 60, 62, \boxed{75}, 70, 72, 71, 75, 55\}$$

[1] A brief note on notation is in order. In general, the symbol x_i is used to represent scores in a distribution without specifying any particular score. The letter x is a stand-in for the thing being measured or observed. In the example above, x refers to final grades. The lower case italicized letter i (*i*) refers to a numerical index, where the value of *i* indicates the location of a score in a distribution. For example, the final grade of 80 is represented by x_1 because it is the first score in the distribution. Your final grade of 75 is represented by x_6.

This information forms our data set. A collection of observations is typically referred to as data (a single observation is a datum or a score). The present data set consists of 11 grades: your grade of 75 and the grades of ten classmates. With this new information how would you describe your performance in the course? Make a note of the strategy you are using. Try it!

Your strategy probably involved some sort of comparison. For the sake of exposition, let us assume that you decided to compare your grade with a classmate's grade. This comparison would reveal how well you performed in relation to someone else. The problem, however, is in selecting another grade to use for the comparison. For instance, if you compared your grade with the lowest grade in the distribution (i.e., 55), then you might conclude that you performed pretty well in the course. Conversely, if you used the highest grade in the distribution (i.e., 81), then you might conclude that your performance was not that good. The point to be made here is that comparing your grade with only one score from the data is an ineffective strategy. Any conclusion about your performance will depend on the particular grade you used in the comparison. Can you think of another strategy for describing your performance that is not dependent on only one score? Try it!

At this point, you might have considered comparing your grade with every other grade in the distribution and noting whether your grade was above or below. For instance, your grade of 75 is 5 points below the first grade in the distribution, 6 points below the second grade, 10 points above the third grade, and so on. By using this strategy, what can you conclude with regard to your performance?

The value of your grade is greater than seven of your classmates, which suggests that your performance was better than the majority of students in the course. So far you are off to a good start in your statistics career!

Comparing one grade with every other grade was easy to accomplish with a data set of only 11 scores. However, this strategy might not be ideal if the data set contained 100 scores. For example, the distribution of final grades in the course is now:

x_i =	{ 75	91	13	91	63	10	28	55	96	96	16	97
	96	49	80	14	42	92	79	96	66	40	85	93
	68	76	74	39	66	17	71	30	28	50	10	82
	69	32	95	30	44	38	77	80	19	49	45	65
	71	75	28	68	66	16	12	50	96	34	59	22
	75	26	51	70	89	96	55	14	15	26	84	25
	81	24	93	35	20	25	62	47	35	83	59	55
	92	29	76	75	38	57	80	50	53	78	93	13
	57	47	19	34}								

Obviously, the previous strategy of comparing one grade with all the others in the data set becomes tedious and cumbersome as the amount of data increases. Can you think of a method that might simplify the comparison process?

The comparison would be much easier if we used a single number to summarize the data set. Moreover, a single number would also allow us to describe the data set efficiently. The necessity of summarizing and describing data is common among all disciplines of science. Empirical investigations of a natural phenomenon involve some form of observation. In order to obtain a reliable measurement of the phenomenon, researchers often collect numerous observations rather than rely on a single observation. Interestingly, this scheme creates some sort of a paradox: numerous observations of a phenomenon are collected, but the sheer amount of observations makes it difficult to pinpoint the defining characteristics of the phenomenon. In the current example, you are presented with multiple observations (i.e., the final grades of your classmates). These observations provide a reliable measurement of the overall performance in the statistics course. If you were to describe your performance relative to your classmates, it would be ideal to use a single grade that represents the class as a whole. Otherwise, comparing your grade with every other grade in the course would be too laborious and difficult to interpret.

2.1 The Problem

The above line of questioning was intended to provide two insights. First, individual observations are somewhat meaningless on their own. The informative value of a single score begins to take shape only when that score can be *compared* to something else. Comparing plays a vital role in science and you will soon discover that it underlies many statistical concepts. The second insight is that it is difficult to describe a phenomenon precisely purely on the basis of raw data. At face value, data sets simply contain too much information for a researcher to make sense of it all, let alone communicate that information in meaningful way. On the one hand, numerous observations ensure a reliable measurement of a phenomenon and enable comparisons to be made, but on the other, numerous observations obfuscate any pattern or characteristics of a phenomenon and increase the amount of comparisons that can be made. This tension highlights the need for techniques that summarize and describe a data set. These techniques are known as descriptive statistics. The central problem of this chapter, and the problem addressed by descriptive statistics is:

P1. How can I summarize a data set accurately and efficiently?

As a starting point for solving this problem, we should speculate on how the data could be summarized. Take a minute to think about how a data set could be summarized accurately and efficiently. Try it!

First, it would be ideal to discover a single value that could represent an entire data set. Condensing the data into a single value would allow researchers to summarize and describe their observations efficiently. A single value would also be much easier to work with when making comparisons. Second, the degree to which a single value represents the data accurately could also prove to be informative. For instance, knowing whether the majority of scores cluster around a single value or whether they are spread apart would provide information concerning the variability of phenomenon. This would not only describe the degree in which observations differ from one another, but also the degree in which a single value represents the data accurately. With these ideas in mind, a tentative solution to P1 could be:

S1. Determine a single value that best represents my data and determine the degree to which that single value represents my data accurately.

3. Measures of Central Tendency

The tentative solution described in S1 opens up to further sub-problems. If our goal is to determine a single value that best represents the data, then we need to determine a method that will allow us to accomplish our goal. Thus, our first sub-problem is:

P2. How can I determine a single value that best represents my data?

Measures of central tendency are the class of statistics that were created to solve this problem. The purpose of a measure of central tendency is to determine a single value that best represents a distribution of data. Indeed, the term 'measure of *central* tendency' implies that the measure seeks to identify the *central* value in a data set, that is—the value in which the majority of scores tend to cluster around. This value could be considered representative of the data.

It might be beneficial to speculate on some ideal properties of a measure of central tendency before proposing tentative solutions to P2. If you could create your own measure of central tendency, what properties would it have? We encourage you to speculate wildly.

First, the measure of central tendency should result in a single unique value. A single value is desirable because it alleviates the need to distinguish among multiple values when describing data. Second, the measure should take into account all of the data. If a researcher went through the trouble of collecting numerous observations, all of them should be included when summarizing and describing the data set. Finally, the measure should be unaffected by aberrant or anomalous observations. Anomalous observations are typically referred to as outliers. If an outlier is present in the data, it should not have a strong influence on the value that is representative of the data as a whole.

3.1 The Mode

Armed with these ideal properties, we can begin to speculate on tentative solutions to P2. An exercise might be helpful in this regard. Three distributions of grades are presented below. For each distribution, our goal is to determine a single value that best represents the data. The strategy that we use to accomplish our goal could serve as a suitable answer to P2. Thus, for each of the following distributions, the question to be answered is: If you had to pick one value that best represents the distribution, which value would you pick? Make a note of the strategy that you are using. Try it!

1. $x_i = \{6,6,6,6,6,6,6,6,6\}$
2. $x_i = \{6,6,6,6,6,6,6,6,6,6,6,6,7\}$
3. $x_i = \{1,6,6,6,6,6,6,6,6,6,6,6,6,6,6,6,6,6,6,6,100\}$

For the first distribution, every score has a value of 6, and as a result, the most representative value for the distribution is 6. For the second distribution, a value of 6 is the most frequent and as a result it appears to be the most representative. Likewise for the third distribution, a value of 6 is the most frequent and once again appears be the most representative. In summary, the single value that best represents each of the three distributions above is: 6, 6, and 6,[2] respectively.

The strategy that was used to determine a single value that best represented the data involved selecting the most frequently occurring value. One might say that the most frequently occurring value in a data set is the most fashionable value: it is the value that is characteristic of all the others. Considering that the French phrase for fashionable is 'à la mode', the term that is given to the most frequently occurring value in a distribution of data is the mode. The mode is a measure of the central tendency. It is the most frequently occurring value in a data set and as such is considered to be representative of the data. Thus one tentative solution to P2 is:

S2-1. Identify the most frequently occurring value in the data set. The most frequently occurring value is called the mode.

To gain some insight into the effectiveness (or ineffectiveness) of the mode, use the mode to describe the following data sets. As you are working through each exercise, make a note of any advantages or disadvantages of the mode that come to mind. Try it!

---

[2]Anecdotally, the number 6 is a perfect number. Perfect numbers are those positive integers whose sum of their proper positive divisors equals that number. For example, the proper positive divisors of 6 are 1, 2, and 3, and their sum is equal to 6.

1. $x_i = \{6,6,6,6,6,6,6,6,100\}$
2. x_i = {male, male, male, male, female}
3. $x_i = \{2,2,2,2,6,6,6,6,7\}$
4. $x_i = \{1.23,1.24,4.56,3.21,7.65,4.55,4.54,1.34\}$

Using the mode as a measure of central tendency, one would conclude that the most representative value in the first data set is 6. Notice that outliers do not affect the value of the mode. Although the first data set contains an anomalous value (i.e., 100), the mode is not affected by it. The most representative value in the second data set is 'male' because it is the most frequently occurring one. The second data set reveals that the mode is a useful measure of central tendency for non-numerical scores. In cases where observations take the form of verbal categories, these observations are referred to as nominal or categorical data. Nominal data does not contain any quantitative information. For example, the first and last scores in the distribution reveal that two observations differ in respect to gender, but the difference between 'male' and 'female' does not contain any magnitude or direction. Another example of nominal data would be a student's program of enrollment. Although students can be differentiated on the basis of enrolling in chemistry, psychology, history, or biology, there is no quantitative distinction among these categories.

Problems arise when applying the mode to the remaining data sets. In the third data set, there are two modes: the values of 2 and 6 occur with the same frequency. Data sets with more than one mode are referred to as multimodal. Using the mode as a measure of central tendency for multimodal data sets can be troublesome. For instance, which value should we choose to represent the third data set, 2 or 6? The mode is ineffective at determining a *single* value that is representative of multimodal data.

The fourth data set is an example of continuous data. In this case, there is no single score whose value has a frequency greater than one. Continuous data contain real numbers, and as such, it is theoretically possible for an infinite amount of values to fall between two scores. For example, an infinite amount of values can fall between the first two scores in the distribution (i.e., 1.23 and 1.24), as long as one is willing to prohibit any limits on the number of reported decimal places. Examples of continuous data are measurements of time, space, and mass. Using the mode to represent continuous data is often problematic because a single value is rarely repeated in the data set. In contrast to continuous data, discrete data does not contain any intermediate values between scores. Discrete data is often represented by whole integers, which makes it impossible for an infinite amount of values to fall between two scores. For example, the number of goals scored in a football match or the number of dots displayed on the face of a dice are limited to whole numbers. It is impossible to score 2.5 goals in a match or roll a dice and obtain a value of 1.5.

After gaining some practice at applying the mode to a variety of data sets, can you list any limitations of the mode or describe the conditions in which the mode is an effective measure of central tendency? To help you with this question, try comparing the mode with the ideal properties of a measure of central tendency.

One limitation of the mode is that it is not guaranteed to result in a single value that represents the data. A data set may contain multiple modes, and in such circumstances it is not appropriate to use the mode as a measure of central tendency. Another limitation is that the mode only reflects the most frequently occurring value and ignores all the others. Consequently, the mode neglects a considerable amount of information in the data. Although this prevents the mode from taking into account all the scores, it also prevents outliers from having an influence on the mode. If a distribution contains anomalous values, the mode can still be used to represent that distribution accurately. Third, while the mode is useful for summarizing discrete and nominal data, it could be problematic for summarizing continuous data. Distributions of continuous data are unlikely to contain values that have a frequency greater than one. For example, if a researcher measures reaction time in milliseconds, it would be rare for two participants to obtain the exact same value of reaction time.

In short, the tentative solution described in S2-1, the mode, is useful for summarizing and describing some data sets but not others. The mode takes into account only the frequency of a particular score and ignores its rank or magnitude.

3.2 The Median

Perhaps we can discover an alternative solution to P2 that can account for some of the shortcomings of the mode. Once again an exercise might be helpful. Four distributions of grades are presented below. For each distribution, our goal is to determine a single value that best represents the data. The strategy that we use to accomplish our goal could serve as another suitable answer to P2. For each of the following distributions, the question to be answered is: If you had to pick one value that best represents the distribution, which one would you pick? Make a note of the strategy you are using. Try it!

1. x_i = {Excellent, Excellent, Good, Poor, Poor}

2. x_i = {A+, A, A−, B, C−, D, F}

3. $x_i = \{1,2,3,4,5,6,7\}$

4. $x_i = \{20,30,40,99,100\}$

The first distribution contains two students who performed very well, two students who performed poorly, and one student who performed somewhere in between. It does not seem reasonable to conclude that a grade of 'Excellent' best represents this distribution because some students performed poorly. Likewise, a grade of 'Poor' does not accurately represent the distribution because some students performed excellently. A grade of 'Good', however, is the middle ground between the high performing students and the low performing students. As a result, it is reasonable to suggest that a score of 'Good' best represents the first distribution.

Of the seven students in the second distribution, three received a grade greater than 'B' and three students received a grade less than 'B'. A grade of 'B' strikes a balance between the high performing and low performing students, and as a result, is arguably representative of the distribution.

Similar considerations apply to the third and fourth distributions. In the third distribution, three grades have a value less than 4 (i.e., 1, 2, 3) and three grades have a value greater than 4 (i.e., 5, 6, 7). The grade of 4 appears to be representative of the data because half of the grades are above it and half are below it. In the fourth distribution, a grade of 40 is directly in between the highest (i.e., 99, 100) and lowest (i.e., 20, 30) grades. Although the highest grades are considerably greater than 40, the *number* of grades greater than 40 is equal to the *number* of grades less than 40. Given that the value of 40 balances the number of grades above and below it, it is reasonable to suggest that is it representative of the distribution.

The representative value for each of these distributions belonged to the score that split the distribution in half. One can think of this value as the middle of the distribution because it separates the top 50% of scores from the bottom 50% of the scores. Considering that the Latin root for 'middle' is 'medi-', the term that is given to the score that separates upper and lower halves of a distribution is the median. The median is a measure of central tendency. Indeed, the median identifies *the* central value in a distribution such that it is directly in between the upper and lower halves of the distribution. For this reason, it is considered to be representative of the data. Thus another tentative solution to P2 is:

S2-2. Identify the score whose value splits the distribution in half such that 50% of all scores are greater than that value and 50% of all scores are less than that value. The score that separates the upper and lower halves of the distribution is called the median.

The median has the potential to overcome some of the shortcomings of the mode. The effectiveness (or ineffectiveness) of the median at representing a distribution can be explored through a few exercises. For the following data sets, use the median to identify a single value that best represents the data. Make a note of any limitations that come to mind when using the median as a measure of central tendency. Try it!

1. x_i = {69,70,87}
2. x_i = {A, A, C, F, D+}
3. x_i = {80,20,40,50,75}
4. x_i = {60,70,80,90}
5. x_i = {58,59,60,70,80 and above}
6. x_i = {Male, Male, Female}
7. x_i = {1.23, 1.24, 4.56, 3.21, 7.65, 4.55, 4.54, 1.34}

The median of the first distribution is 70. This becomes evident once we notice that of the three scores in the distribution, one has a value less than 70 and one has a value greater than 70. A value of 70 splits the distribution in half such that an equal number of scores are above and below it. The actual magnitude of the scores that are greater than and less than the median has no significance on the value of the median. It does not matter than the score below 70 (i.e., 69) is actually closer in magnitude to a value of 70 than the score above 70 (i.e., 87). It is the *rank* of a score that determines the value of the median rather than a score's magnitude. The value of the median for this distribution would remain at 70 if the first score changed from 69 to 1 or 2 or 34 or 69.99. As long as the value of the score is less than or equal to 70, the median of the distribution will be 70. In the eyes of the median, the actual magnitude of a score is somewhat arbitrary.

When the values of scores are only used to establish order or rank, the scores are referred to as ordinal data. The magnitude of difference among scores in ordinal data is meaningless. The magnitude of an ordinal score only establishes the rank of that score, not the degree to which that score differs from another. For example, let us assume that the scores in the first distribution represent ordinal data. The magnitude of the first score in the first distribution only establishes that the score is less than the others. We cannot conclude that the first score differs from the second score by 1 point; rather, we can only conclude that the first score is less than the second and third scores.

The letter grades in the second distribution are a prime example of ordinal data. In terms of a student's performance, a letter grade of 'A' is greater than a letter grade of 'B', which is greater than a letter grade of 'C', and so on. However, the magnitude of difference between 'A' and 'B' or 'C' is absent. A score of 'A' only indicates that a student's performance was greater than 'B' and 'C', it does not indicate the degree to which performance was greater than 'B' and 'C'. The median is an appropriate measure of central tendency for ordinal data because it only considers the rank of the data rather than the magnitude. With this in mind, the median for sixth distribution is a grade of 'C' because half of the grades in the distribution are below 'C' and half are above 'C'.

Admittedly the third distribution was intended to trick you. The median of this distribution is 50. The score that is located in the middle of the distribution is 40, but this value does not satisfy the requirements of the median. Three scores are greater than a value of 40 (i.e., 80, 50, 75), whereas only one score is less than a value of 40 (i.e., 20). Consequently, the median is not 40 even though the score of 40 happens to be located in the middle of the distribution. The score of 50, on the other hand, satisfies the requirements of the median. Two scores have a value greater than 50 (i.e., 75, 80) and two scores have a value less than 50 (i.e., 20, 40). It is important to realize that the median is not necessarily the score that is *located* in the middle of the distribution. Only if scores are ordered will the median be located in the middle of the distribution.

Complications arise when using the median for the fourth distribution. In this case there is an even number of scores, and as a result, a single score does not appear to separate the distribution in half. Fortunately there is a solution to this problem. When distributions contain an even number of scores, the median will be equal to the value that is directly in between the two middle scores. The two middle scores for this distribution are 70 and 80. A value of 75 is halfway between 70 and 80; consequently, a value of 75 satisfies the requirement for the median: it separates the upper and lower halves of the distribution. Although the value of 75 does not actually exist in the distribution, it is nonetheless representative of the distribution. This situation raises an interesting point: it is possible for a measure of central tendency to result in a value that does not have a parallel in the distribution.

Keeping in mind that the median focuses only on the rank of scores, it can be used to describe open-ended data. The last score in the fifth distribution is open-ended: it can refer to any value that is equal to or greater than 80. Nevertheless, any score that is equal to or greater than 80 will always be greater than a value of 79, 70, or 60. Consequently, the median in this distribution is 60 because the number of scores less than 60 is equal to the number of scores greater than 60.

The median cannot be determined for the sixth distribution. The scores in this distribution do not contain any inherent order, rank, or direction of difference; males are not greater than or less than females. This distribution reveals that the median is an inappropriate measure of central tendency for nominal data.

The seventh distribution contains continuous data and an even number of scores. The characteristics of this distribution pose no problems for the median. Once the scores are ordered according to their magnitude, the two middle scores are 3.21 and 4.56. The value of the median is directly in between these two scores. How can we determine the midpoint between 3.21 and 4.56?

One way of thinking about a midpoint is illustrated in Fig. 1. Adding two scores together and dividing their sum by two is equivalent to splitting their sum in half. Considering that the two scores contributed to the sum, halving the sum would be equivalent to finding the midpoint between the two scores. Thus, one way to determine the midpoint between 3.21 and 4.56 is to add the two scores together and divide by two. The resulting value is 3.89, which is the median for this distribution.

After using the median on a variety of data sets, how would you characterize its effectiveness at identifying a representative value? Are there any advantages or disadvantage of using the median instead of the mode as a measure of central tendency? To help you answer these questions, ask yourself if the median was able to summarize and describe distributions that the mode could not. Try it!

Unlike the mode, the median will always result in a single value that summarizes and describes a distribution. Regardless of whether the data is discrete or continuous, a distribution will always contain one median. Another advantageous property of the median is that it is insensitive to outliers. The value of the median is unaffected by the actual magnitude of a score; the ordered relationship among scores is the only factor that determines the value of the median. As a result, one score can be

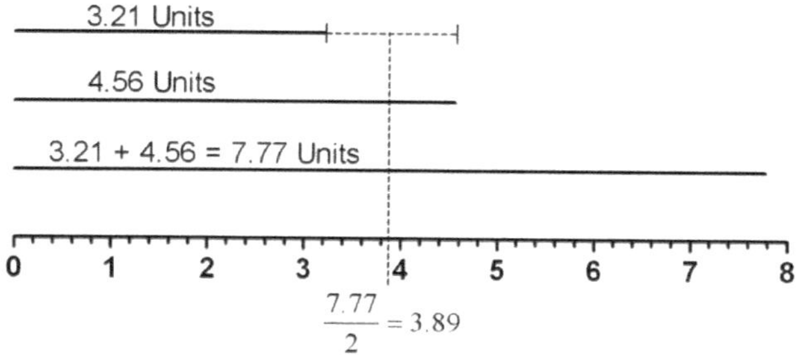

Figure 1. The midpoint between two scores is equivalent to halving the sum of the two scores. From top to bottom, the first and second lines represent two scores, whereas the third line represents the sum of the two scores. The length of a line corresponds to the magnitude of a score or sum. The dashed vertical line indicates the location of the midpoint between the two scores (i.e., 3.89). The location of the midpoint between the two scores is equivalent to the location of the halfway point of the sum of the two scores.

considerably greater than all the others and have no effect on the median. All that matters is that a score is *greater than* or *less than* the others. Although this property enables the median to be used in distributions with outliers and open-ended values, it does so at the cost of neglecting potentially useful information. Indeed, the value of the median does not take into account all the data as it only reflects the middle of the distribution.

3.3 The Mean

The median and the mode reflect the rank and frequency of a score, respectively. We have yet to propose a measure of central tendency that takes into account the magnitude of a score. Furthermore, we have yet to determine a method that summarizes a data set without neglecting any of the scores in the data set. Perhaps we can discover another solution to P2 that addresses these issues. With this in mind, try to determine a single value that best represents each of the following distributions.[3] Once again, make a note of the strategy you are using. Try it!

1. $x_i = \{1,1,1,3,3,6,7,10\}$
2. $x_i = \{1,1,1,3,3,6,7,26\}$
3. $x_i = \{1,1,1,3,3,6,7,42\}$

A crude way of summarizing a distribution without neglecting any of the scores is to simply add the scores. Indeed, the word 'sum' originates from the Latin noun 'summa', which means whole or essence. Hence the sum of a distribution can be thought of as capturing the whole or essence of the distribution. Surely the magnitude of a score is part of the "essence" of a distribution. It is not surprising then that the sum is sensitive to magnitude. To see this, calculate the sum for each of the distributions above and note what happens to the value of the sum as the magnitude of the last score in the distribution increases. Try it!

Although the relative rank of the last score in each distribution remains the same,[4] the magnitude of the last score becomes increasingly larger. Accordingly, the sums for each of the distributions become increasingly larger: the sum of the first distribution is 32, the sum of the second distribution is 48, and the sum of the third distribution is 64. The value of the sum increases as the magnitude of a score increases. This reveals that, unlike the median or the mode, the sum is sensitive to magnitude.

---

[3] Notice that the value of the median or the mode does not differ among these distributions, even though these distributions differ from one another.

[4] 10, 26, and 42 are all greater than the remaining scores in each of their respective distributions.

The sum seems to satisfy our requirements: it incorporates all of the scores in the data and is sensitive to magnitude. However, the value of the sum does not appear to be representative of the scores in the distribution. For example, none of the scores in the third distribution have a value similar to 64. Although the sum is an efficient summarizer, it is not an accurate descriptor. To highlight this point, consider the following three distributions of grades:

1. $x_i = \{1,1,1,3,3,6,7,10\}$
2. $x_i = \{1,1,1,1,1,1,1,1,1,1,1,1,1,1,1,1,1,1,1,1,1,1,1,1,1,1,1,1,1,1,1,1\}$
3. $x_i = \{10,10,10,2\}$

The sum of each of these distributions is 32. A value of 32 does not appear to be representative of the first distribution, nor the second distribution, nor the third distribution. Because the sum is useful for summarizing, it should not be abandoned in our search for a measure of central tendency, but its poverty as a descriptive statistic needs to be resolved. Is there any additional information about the distribution that we should consider when interpreting the sum? To help you with this question, compare the aforementioned second and third distributions. Although both distributions have identical sums, they differ in some respects. How do they differ?

By comparing the second and third distributions, you might have noticed that the *number* of scores in a distribution influences the value of the sum. As the *number* of (non-zero positive) scores in a distribution increases, the value of the sum also increases. Consequently, the value of the sum might not reflect the magnitude of the scores but rather the number of scores. This point is highlighted in the distributions above. A sum of 32 does not necessarily mean that the scores in the distribution are comparable in magnitude to 32. A sum of 32, or any sum for that matter, can be obtained with numerous scores of a lesser magnitude, as in the case of the second distribution. How can we reconcile the sum of a distribution with the number of scores in a distribution? Wild speculation is a must.

One solution is to relativize the sum with the number of scores. This can be accomplished by dividing the sum by the number of scores. By doing so, the distribution would be represented by a single value that is sensitive to both the magnitude and number of scores. Try using this strategy on the aforementioned distributions. For each distribution, calculate the sum and divide the sum by the number of scores. Is the resulting value representative of the distribution?

The sum of the first distribution is 32 and there are 8 grades in the distribution. According to our tentative strategy, the single value that best represents this distribution is equal to 32/8 or 4, which is reasonably representative of the grades in this distribution. The sum of the second distribution is also 32 and there are 32 grades in the distribution. According to our strategy, the single value that would best represent this distribution is equal to 32/32 or 1. In this case a value of 1 is perfectly representative. Finally, the representative value for the third distribution would be equal to 32/4 or 8. A value of 8 is comparable in magnitude to the majority of grades in this distribution and it is within reason to suggest it is representative.

To review, our strategy for determining a single value that best represents a distribution involves summing the scores in a distribution and dividing the sum by the number of scores in a distribution. Algebraically, this strategy is represented by the following equation:[5]

$$\mu = \frac{\sum x_i}{N}$$

The resulting value of this strategy is called the mean and is symbolized by the Greek letter mu (μ). The mean is a measure of central tendency. It takes into account: 1) all of the scores in a distribution, via the sum; 2) the magnitude of the scores in a distribution, via the sum; and 3) the number of scores in a distribution, symbolized by N.

The mean can be playfully thought of as a "socialist" measure of central tendency. Socialism is an ideology that is characterized by the equal distribution of wealth among all members of a society. Considering that the mean involves summing the scores and dividing the sum by the number of scores, the mean is equivalent to determining how the sum could be equally proportioned to all the scores in the distribution. From this point of view, the mean is an intuitively appealing measure of central tendency because it is the single value that is common to all of the scores. The etymology of mean and sum illustrates this point well. Recall that the word 'sum' originates from the Latin noun 'summa', which refers to a whole or essence. Dividing the sum by N is equivalent to partitioning the whole of the distribution equally to all the members of the distribution. Accordingly, the value that each score would receive is equal to the value of the mean. It is not surprising then that etymology of mean contains a "shared by all" connotation. It is the single value that is common to all the scores. For this reason the mean is considered to be a representative value of a distribution. Thus another tentative solution to P2 is:

S2-3. Sum all of the scores in the distribution and divide the sum by the number of scores in the distribution. The resulting value is called the mean.

Once again, the effectiveness (or ineffectiveness) of the mean at representing a distribution can be explored through a few exercises. For each of following

[5] The capital Greek letter sigma (Σ) is used to represent the summation function. Σx_i refers to the sum of all the scores in a distribution.

data sets, use the mean to determine a representative value. Make a note of any drawback that come to mind when using the mean as a measure of central tendency. Try it!

1. x_i = {1,1,1,3,3,6,7,2}
2. x_i = {1,1,1,3,3,6,7,106}
3. x_i = {58,59,60,70,80 and above}
4. x_i = {Male,Male,Female}
5. x_i = {A, A, C, F, D+}
6. x_i = {1.23, 1.24, 4.56, 3.21, 7.65, 4.55, 4.54, 1.34}
7. x_i = {2,2,2,2,6,6,6,6,7}

The means for the first and second distribution are 3 and 16, respectively. Although a value of 3 is representative of the first distribution, a value of 16 does not appear to be representative of the second distribution. The only difference between these two distributions is the magnitude of the final score. Why does the value of the mean change drastically from the first distribution to the second distribution?

In the second distribution, the score of 106 is an outlier. Considering that the mean relies on the sum, and the sum incorporates the magnitude of all of the scores, increasing or decreasing the magnitude of one score will have an impact on the value of the mean. In particular, the value of the mean will be pulled in the direction of an outlier, which is what occurred in the second distribution. The value of the mean increased because the value of the outlier was considerably greater than the rest of the distribution. Unfortunately, the presence of outliers distorts the representativeness of the mean.

The mean cannot be used to represent the third, fourth, and fifth distributions. The reason is that a sum cannot be calculated if a distribution contains open-ended scores, nominal data, or ordinal data. The magnitude of open-ended scores is ambiguous and cannot be included in the sum. It would not make sense to sum the magnitude of ordinal data because the magnitude of ordinal data is arbitrary. Finally, nominal data does not contain quantitative information to be summed.

The means for the sixth and seventh distributions are 3.54 and 4.33, respectively. For each of these distributions the mean appears to be a representative value. As the sixth distribution reveals, the mean can be used to adequately represent continuous data. Even though a value of 3.54 does not actually exist in the distribution, it still appears to be representative of the distribution. This reveals that the mean might not have a parallel in the data, and like the median, it can be considered an abstraction from the data.

Using the mean to represent discrete data could pose some problems. Discrete data places limits on the existence of intermediate values between scores. If the mean for a discrete data set results in a fractional value then it might not be useful for representing that data set. For example, if the data in the seventh distribution represents the number of children per household, it would be inappropriate to use the value of the mean to best represent this distribution. It is impossible to have 4.33 children in a household. Ultimately the mean is useful for representing continuous data, but might not be effective at representing discrete data.

Although the mean is not perfect, it is the most commonly used measure of central tendency. Can you think of any advantages that the mean has over the median and mode? Try it!

First, the mean does not neglect any of the scores in the distribution. By using the sum, the mean incorporates all of the scores to ultimately arrive at representative value rather than focusing on only the most frequent score or the middle score. Second, the mean's sensitivity to magnitude affords a degree of precision that is not present in the mode or median. Changing the value of a score will not affect the median or the mode as long as a score's frequency or rank is unaffected. However, changing the value of a score will affect the value of the mean. Finally, the mean can be expressed algebraically whereas the mode and median cannot. This allows for the mean to be included in more complex equations.

We have previously discussed the median in terms of a balancing point. A similar conceptualization can be applied to the mean. Recall that the median separates the upper and lower halves of a distribution such that the *number* of scores that is greater than the value of the median is equal to the *number* of scores that is less than the value of the median. For instance, consider the following distribution:

1. $x_i = \{1,1,1,3,3,6,7,10\}$

As illustrated in Fig. 2, the scores in this distribution can be represented on a number line. Each gray box represents the existence of a score in the distribution. The median for the distribution is equal to 3, and as depicted in Fig. 2, the number of scores (grey boxes) greater than the median is equal to the number of scores less than the median.

The mean can also be thought of as a balancing point. Before we present this conceptualization, take a few minutes the think about how the mean might act as a balancing point. Try it!

Unlike the median, the mean does not balance the *number of scores;* rather, the mean balances the *deviations of scores*. A deviation is the difference between the value of a score and the value of a reference point in the distribution. Mean deviations measure the difference between a score and the mean. As illustrated in Fig. 3, the value of the mean is the location in

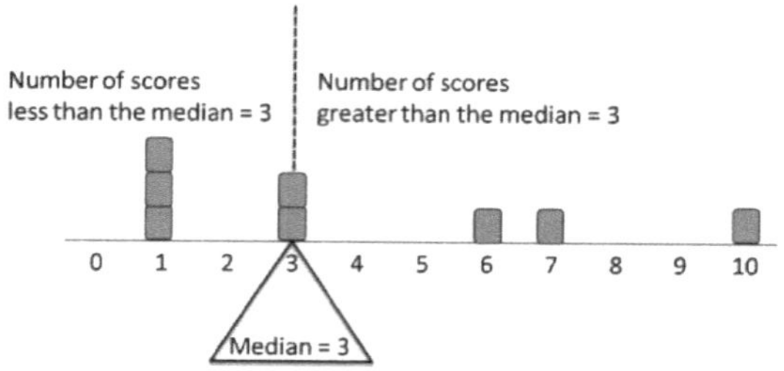

Figure 2. The median depicted as a fulcrum. The distribution {1,1,1,3,3,6,7,10} is represented on a number line. Each grey box represents the existence of a score in the distribution. The number of scores greater than the median is equal to the number of scores less than the median. Thus, the median acts as a balancing point in the distribution.

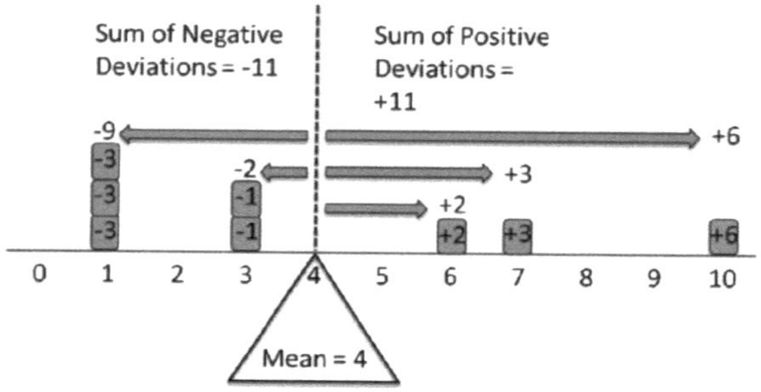

Figure 3. The mean depicted as a fulcrum. The distribution {1,1,1,3,3,6,7,10} is represented on a number line. Each grey box represents the existence of a score in the distribution. The numbers inside each box represents the deviation from the mean. The sum of negative deviations from the mean is equal to the sum of positive deviations from the mean. Thus, the mean acts as a balancing point in the distribution.

the distribution where the sum of negative deviations from that location is equal to the sum of positive deviations from that location. Thus the degree to which scores differ from the mean is balanced—the total of deviations above the mean is equal to the total of deviations below the mean. The mean is the only reference point in a distribution[6] that satisfies this property. If you select any other value in a distribution and calculate the deviation that each score is from that value, the sum of positive deviations will not be equal to the sum of negative deviations. Try it and see for yourself!

The fact that the mean balances the deviations of scores whereas the median balances the number of scores might have been anticipated if you remembered that the mean makes use of magnitude whereas the median makes use of rank. The median does not take into account the magnitude of difference between scores; it only relies on whether one score is greater or less

[6] The value of the mean might not actually exist in the distribution, but it nonetheless belongs to the distribution.

than another. The median cannot balance the deviations of scores because deviations require magnitude. The sensitivity of the mean to magnitude, on the other hand, affords the opportunity to balance deviations.

Conceptualizing the mean as a fulcrum brings to light some interesting properties. If the mean is a balancing point then it must be located within the confines of the distribution. As a result, it is impossible for the value of the mean to be greater than the maximum score or less than the minimum score. Second, the sum of *negative* deviations will always be equal to the sum of *positive* deviations. It follows that the sum of deviations from the mean will always equal zero.

3.3 Summary

The mean, median, and mode are three of the most common measures of central tendency. Each measure attempts to summarize and describe a distribution by determining a single value that best[7] represents the distribution. The choice of a measure of central tendency is dependent on the characteristics of the to-be-described data. The mode is the most frequently occurring value in a distribution and is appropriate for summarizing nominal and discrete data. It does not make use of any quantitative distinctions among scores and will not result in fractional values. The median is the value that separates the upper and lower halves of a distribution and is appropriate for data that contains an ordered relationship among scores, be it discrete or continuous. The mean is the sum of scores divided by the number of scores and is ideal for data where the magnitude of difference among scores is unambiguous, quantifiable, and meaningful. Although the mean can be distorted by the presence of outliers, it is the only measure that uses all the scores in a distribution and is the only measure that is sensitive to the magnitude of scores. Furthermore, the mean can be expressed algebraically whereas the other measures cannot. For these and other reasons to be discussed, the mean is often used to summarize and describe a data set.

3.4 Local-Global Relationships in Measures of Central Tendency

The preceding section highlighted the differences among the measures of central tendency. In this section we would like to adopt a different point of view. Can you think of any similarities among the measures of central

[7] The word best is used in the sense of the most currently advantageous. We do not wish to imply that there is an ultimate or final measure of central tendency that is always effective at representing a distribution.

tendency? Is there a common theme among them? To stimulate your speculation on this question, consider the purpose of measures of central tendency and how each measure accomplishes that purpose. Try it!

All the measures of central tendency act as a globalizing function. Each measure of central tendency takes as its input the scores in a distribution and outputs a value that is thought to be representative of that distribution. The meaning of the resulting value is different from, but at the same time reliant on, the scores that contributed to it. For example, the value of the mode has a different meaning than the value of a score, even though the values of the scores determine the value of the mode. Within this context, the scores in a distribution can be thought of as local elements which, according to some rule, form a global element. The rule specifies the scores that will be used and how they will be used to form the globality. As a result, the rule determines the meaning of the resulting globality.

We consider scores to be local elements of a measure of central tendency because the scores are *part of* a measure of central tendency, much like leaves are local elements of a tree because they are *part of* a tree. Although measures of central tendency are more global compared to scores, they might nonetheless act as local elements in higher-level statistical techniques, allowing for the possibility of local-global hierarchies. The notion of a local-global hierarchy might seem superfluous and unrelated to statistics. However, local-global relationships are prevalent among all statistical techniques, and are useful for establishing conceptual links among statistical topics. We encourage you to keep this local-global distinction in mind throughout the remainder of the chapter.

4. Measures of Variability

Thus far we have examined several techniques that summarize and describe a data set. Given a distribution of data, we can use the mean, for example, to represent that distribution. But does the mean, or any measure of central tendency for that matter, *fully describe* a distribution? The following exercise may help to stimulate your intuition regarding this question.

The final grades for an introductory statistics course have a $\mu = 70$ and you received a final grade of 75. How well did you perform in the course?

Although your grade is greater than the mean, does this suggest that you performed better than most of your classmates? How do you know whether your classmates were close to the mean or far from the mean? How do you even know whether the mean accurately represents the distribution?

The answer to these questions is: I do not know. The value of the mean, on its own, is not entirely descriptive of a distribution. Although the mean reveals the center of the distribution, it does not reveal *variability* in the distribution. By itself, the mean does not provide any information concerning its accuracy at representing a distribution or provide any information concerning how the scores are spread apart in a distribution.

Let us consider another exercise. For each of the following distributions, describe the accuracy of the mean at representing the distribution. Try it!

1. $x_i = \{50,50,50,50,50,50,50\}$, $\mu = 50$
2. $x_i = \{47,48,49,50,51,52,53\}$, $\mu = 50$
3. $x_i = \{35,40,45,50,55,60,65\}$, $\mu = 50$
4. $x_i = \{10,30,40,50,60,70,90\}$, $\mu = 50$

The degree to which the mean accurately represents each of these distributions varies. While the mean represents the first distribution perfectly, its accuracy declines for the remaining distributions, especially the fourth. By knowing only the value of the mean, however, you cannot anticipate whether the scores in the distribution are similar to each other and to the mean (e.g., the first distribution) or whether they differ from each other and from the mean (e.g., the fourth distribution).

The previous exercises were intended to prompt you to question the adequacy of a measure of central tendency at describing a data set. Recall that the problem addressed by descriptive statistics is:

P1. How can I summarize a data set accurately and efficiently?

Our tentative solution to this problem is:

S1. Determine a single value that best represents my data and determine the degree to which that single value accurately represents my data.

Measures of central tendency address the first aspect of this solution. We now need address the second aspect:

P3. How can I determine the degree to which a single value accurately represents the data?

Measures of variability are the class of statistics that were created to solve this problem. The term 'variability' is synonymous with differences or dispersion, and can be equated with error. A measure of variability serves a dual purpose. First, it quantifies the accuracy of a single value at representing a distribution. Second, it quantifies the degree to which scores in a distribution differ from one another. These two goals are highly related. Differences from a single representative value can be used to quantify the degree to which scores differ from one another *and* can be used to quantify the accuracy of that value at representing a distribution. For instance, the difference of each score from the mean, not only measures how well the mean represents the distribution, but also whether the scores are clustered around a single value or are spread apart. Not all measures of variability accomplish this dual purpose. Some only focus on describing the dispersion of scores (i.e., the degree to which scores differs from one another), whereas others are more ambitious and attempt to describe the dispersion of scores *and* the accuracy of a measure of central tendency. Nonetheless, we should consider the goal of a measure of variability as twofold: 1) describe the degree to which scores differ from another; and 2) describe the degree to which a single value represents a distribution accurately. If we can create a measure of variability that accomplishes this goal, then in conjunction with a measure of central tendency, we will be able to increase our accuracy and efficiency at describing a data set.

4.1 The Range

Given the twofold purpose of a measure of variability, it might be easier to begin our discussion of variability by focusing only on dispersion. Afterwards, we can attempt to describe the accuracy of a measure of central tendency. Three distributions of grades are presented below. For each distribution, try to answer the following question: How can you quantify the degree to which the grades differ from one another? Make a note of the strategy that you are using. Try it!

1. $x_i = \{4,4,4,4,4,4,4,4\}$

2. $x_i = \{4,4,4,4,4,4,4,3,3,3,3,3\}$

3. $x_i = \{4,4,4,4,4,4,4,100,100\}$

The grades in the first distribution do not differ from one another. Thus the variability in the distribution is zero. Conversely, the grades in the second distribution vary between 4 and 3. At a maximum there is a difference of 1 point among the scores, and it is within reason to quantify the degree of dispersion in this distribution as being equal to 1. The grades in the third distribution also differ from one another, but the magnitude of the difference has increased in comparison to the previous distribution. The grades in this distribution vary between 4 and 100, resulting in a 96 point difference among the scores. Accordingly, it is reasonable to quantify the variability in this distribution as being equal to 96.

Our strategy for quantifying dispersion involved in calculating the difference between the maximum and minimum score in the distribution. This difference is called the range. The range is one method for describing the degree in which scores in a distribution differ from one another. As the difference among scores increases, the value of the range also increases. Thus one partial measure of variability is:

S3-1. Calculate the difference between the maximum and minimum score in the distribution. The value of the difference is called the range.

The range is a simple, quick, and easy method for describing dispersion. Its grossness, however, is also its greatest limitation. To see this, use the range to describe the variability in the following distributions. Does the range represent the degree to which scores differ from one another accurately?

1. $x_i = \{1,1,2,3,1,3,2,1,100\}$

2. $x_i = \{0,1,1,1,1,1,1,1,1,1,1,1,1,5,5,8,9,9,10,22\}$

The range for the first distribution is equal to 99. A value of 99 does not appear to be representative of the *typical* difference among scores. While the majority of scores only differ by 1 or 2 points, the presence of an anomalous value (i.e., 100) inflates the range and distorts the variability in the distribution. Similar issues arise for the range in the second distribution. The range for this distribution is equal to 22, but most of the scores in the distribution do not differ from each other by 22 points. Although the *maximum* difference among scores is 22 points, the *typical* difference among scores is much less.

The forgoing distributions reveal that the range has some serious limitations at describing the variability of quantitative data. The most prominent of them is that the range is not a precise measurement of variability. Why does the range lack precision?

The range does not make use of all the scores in the distribution; it only focuses on the difference between the maximum and minimum values. Consequently the range only describes the maximum degree to which the scores differ from one another rather than the typical degree to which scores differ from one another. Furthermore, given that the range is based on only two scores, it cannot provide any indication as to whether the majority of scores are clustered together or spread apart.

Although the range is a crude measure of variability, it does have one significant benefit. With a slight modification, the range can be used to describe the dispersion of nominal data. Rather than defining the range as the difference between the maximum and minimum scores, the range for nominal data is defined as the maximum number of qualitatively distinct categories. For example, if a distribution of nominal data contains five different categories, then at a maximum there are five different values that a score can have.

The range for nominal data is similar to the range for quantitative data in that it describes the maximum degree to which scores can differ from one another. However, both definitions of the range fail to take into account a measure of central tendency. And this is perhaps the most significant limitation of the range. The range cannot be used to assess how well a single value represents a distribution accurately, which is something that needs to be determined.

4.2 Mean Deviations

A full solution to P3, one which simultaneously quantifies the dispersion of scores and the accuracy of a measure of central tendency, is needed. It is time to confront this issue. As a starting point, we need to consider which measure of central tendency will have its accuracy quantified by the measure of variability. Ideally, the measure of variability should take into account all the scores in a distribution and should be sensitive to the magnitude of difference among scores. This will allow a precise measurement of the dispersion in the distribution and the error associated with a representative value. Bearing in mind that the mean is the only measure of central tendency that considers all of the scores and their magnitude, the mean is apt to contribute to a measure of variability.

Two distributions of grades are presented below. If the mean is selected as a representative value, then the question to be answered is: How can I determine the degree to which the mean accurately represents the data? Make a note of the strategy that you are using. Try it!

1. $x_i = \{4,4,4\}$, $\mu = 4$

2. $x_i = \{5,6,1,5,3\}$, $\mu = 4$

One strategy for assessing the accuracy of the mean is to compare it with every score in the distribution. If the mean differs from a score, then the magnitude of the difference can be used to quantify how well the mean represents that score. Moreover, the degree to which scores differ from the mean is equivalent to the degree to which scores cluster around a single value. Recall that the difference between a score and the mean is called a mean deviation. Mean deviations appear to satisfy the dual purpose of a measure of variability: they describe the accuracy of the mean and the spread of scores.

The mean deviations for the first and second distributions are:

1. $(x_i - \mu)=\{4\text{-}4, 4\text{-}4, 4\text{-}4\} = \{0,0,0\}$.

2. $(x_i - \mu)=\{5\text{-}4, 6\text{-}4, 1\text{-}4, 5\text{-}4, 3\text{-}4\} = \{1,2,-3,1,-1\}$.

The total variability in the first distribution is equal to zero, which makes sense once we realize that all of the scores in the distribution do not differ from each other or from the mean. The magnitude and sign of the mean deviations for the second distribution reveal that all the scores differ from the mean. The sign of the deviation reveals whether the score is greater (i.e., +) or less (i.e., –) than the mean and the magnitude of the deviation reveals the degree to which the score differs from the mean. According to these mean deviations, the maximum difference between a score and the mean is 3 points, and the typical difference is 1 point. To make the interpretation somewhat easier, it would be helpful to summarize the deviations into a single value. Indeed, examining the deviations for a distribution of 100 scores would be tedious and very difficult to interpret. How can we summarize the mean deviations for a distribution?

As discussed in section 3.3, a series of values can be summarized by simply adding them together. The sum of mean for the second distribution is equal to:

2. $\Sigma(x_i - \mu) = 1+2+(-3)+1+(-1) = 0$

This result might have been anticipated if you remembered that the mean balances the sum of positive deviations with the sum of negative deviations. Consequently, the sum of positive and negative mean deviations will always equal zero. This result is not confined to this particular distribution; for any distribution, the sum of mean deviations will always result in zero, regardless of the variability in that distribution.

It appears that we are at an impasse. Although mean deviations assess how well the mean represents the distribution and whether the scores are clustered around a single value, their sum will always result in zero. We need to determine a method that will maintain the magnitude of difference between a score and the mean while preventing the sum of differences from always resulting in zero. Can you think of any strategy, no matter how farfetched, that might accomplish this goal? To help you think about a strategy, reflect on the reason as to why the sum of mean deviations always equals zero. Once that reason is identified, think of a method that will eliminate its influence. Try it!

The reason as to why the sum of mean deviations always equals zero is because of the sign of the deviations: the sum of *negative* deviations will always equal the sum of *positive* deviations. If the sign of a deviation can be removed, then the sum of deviations can result in a non-zero value. Two recourses are available. The first is to simply ignore the sign of the deviations and sum their absolute values. This is known as the sum of absolute mean deviations or *SAD* for short. *SAD* is algebraically expressed as:

$$SAD = \sum(|x_i - \mu|).$$

The sum of absolute mean deviations can result in non-zero values because the absolute value function removes the sign. The sum of absolute deviations for the current distribution is equal to:

2. $\Sigma(|x_i - \mu|) = 1+2+3+1+1 = 5.$

The second solution is to square each mean deviation and then sum the resulting squared values. This strategy is called the sum of squared mean deviations or *SS* for short. The algebraic formulation of *SS* is:

$$SS = \sum (x_i - \mu)^2$$

Squaring eliminates negative values while maintaining the magnitude of difference between each score and the mean. For example, the squared deviations for the current distribution are:

2. $(x_i - \mu)^2 = \{1^2,2^2,-3^2,1^2,-1^2\} = \{1,4,9,1,1\}$.

Notice that the greatest mean deviation (i.e., –3) is also the greatest squared deviation (i.e., 9). Squaring maintains the relative difference between a score and the mean, the only information that is removed is the sign of the difference.

SS is often preferred over *SAD* in measures the variability. There are several reasons for the preference, most of which cannot be fully described in the context of this chapter. Interested readers are encouraged to consult Gorard (2005) for a full treatment on the relative advantages of absolute versus squared deviations. Perhaps the most widespread reason, although not necessarily the most intuitive reason, for adopting *SS* over *SAD* is that *SS* is easier to manipulate algebraically. Indeed, the equation for the mean can be derived by minimizing the sum of *squared* deviations (Watier et al. 2011). The ease at which *SS* lends itself to mathematical manipulation partly explains why the majority of higher-level statistical techniques are based on *SS* rather than *SAD*. Accordingly, *SS* will be the approach adopted in this chapter.

Returning to the problem at hand, once the deviations are squared they can be summed to provide a single a value:

2. $SS = \Sigma\,(x_i - \mu)^2 = 1^2+2^2+(-3)^2+1^2+-1^2 = 1+4+9+1+1 = 16$.

SS is a summary of the total degree to which scores differ from the mean. But is *SS* representative of the variability in the distribution? The sum of squared deviations for the second distribution is equal to 16; does a value of 16 reflect the spread of the scores in the distribution or the accuracy of the mean at representing a distribution? It does not appear so.

We have confronted a similar problem before. In section 3.3, we found that the value of a sum is influenced both by the number and magnitude of scores that went into its calculation. A large sum is not necessarily indicative of a distribution that contains large values. Similarly, a large value for *SS* is not necessarily indicative of a distribution that contains large deviations from the mean; it might simply reflect a distribution that contains numerous deviations of a smaller magnitude. How can we work around this problem?

4.3 Variance

The solution is to divide *SS* by the number of scores in the distribution. Doing so is equivalent to interpreting *SS* relative to the number of scores. The resulting quotient is called variance. Variance is representative of the variability in the distribution rather than the number of deviations that went into *SS*. Considering that variance involves dividing a sum by *N*, it has the same structure as a mean. With this in mind, how would you interpret the value of variance? Variance is the average squared difference between a score and the mean. Variance is a measure of variability and thus a tentative solution to P3:

S3-1. Calculate the average squared difference between a score and the mean. The resulting value is called variance.

Variance is a suitable candidate for describing the dispersion of scores and the accuracy of the mean at representing a distribution. It results in a single value that takes into account: 1) all of the scores in the distribution, via *SS*; the number of scores in a distribution, via *N*, and the degree to which the scores differ from the mean, via *SS*. Variance is symbolized by the Greek letter sigma raised to an exponent of 2 (σ^2). Sigma is raised to an exponent because variance measures *squared* differences from the mean. Algebraically, variance is expressed as:

$$\sigma^2 = \frac{\mathbf{SS}}{N} = \frac{\sum (x_i - \mu)^2}{N}$$

The numerator of this equation is measuring the sum of squared differences from the mean. The denominator is measuring the number of scores. By using this equation, the variance for the second distribution is equal to:

$$\sigma^2 = \frac{\mathbf{SS}}{N} = \frac{16}{5} = 3.2\cdot$$

The variance for the second distribution indicates that the average squared difference between a score and the mean is 3.2 units.

Before accepting variance as a final solution to P3, we should explore any limitations it might have. For the following distributions, calculate variance and ask yourself whether the resulting value is reflective of either the spread of scores in the distribution or the accuracy of the mean at representing the distribution. Try it!

1. $x_i = \{2,4,6\}$

2. $x_i = \{5,3,7,3,2,10\}$

The first step in calculating variance is to determine the mean. The mean is obtained by summing all of the scores in a distribution and dividing the sum by the number of scores in a distribution. The means of the first and second distribution are as follows:

1. $$\mu = \frac{\sum x_i}{N} = \frac{2+4+6}{3} = 4$$

2. $$\mu = \frac{\sum x_i}{N} = \frac{5+3+7+3+2+10}{6} = 5$$

Once the mean is known, the sum of squared deviations can be obtained. *SS* is the backbone of variance; it measures the degree to which scores differ from the mean, which is equivalent to measuring the accuracy of the mean and the degree to which scores are spread apart. The calculations for *SS* are as follows:

1. $SS = \Sigma\,(x_i - \mu)^2 = (2\text{-}4)^2+(4\text{-}4)^2+(6\text{-}4)^2 = 8$

2. $SS = \Sigma\,(x_i - \mu)^2 = (5\text{-}5)^2+(3\text{-}5)^2+(7\text{-}5)^2 + (3\text{-}5)^2 + (2\text{-}5)^2 + (10\text{-}5)^2 = 46$

Because a sum is a function of the number of terms that went into its calculation, *SS* does not necessarily describe the variability in the distributions; rather, its value could reflect the number of scores in a distribution. This problem can be resolved by dividing *SS* by the number of scores. Doing so relativizes the sum and prevents the measure of variability from increasing as the number of scores in the distribution increase. The resulting quotient is called variance. The variances for the first and second distributions are as follows:

1. $$\sigma^2 = \frac{\mathbf{SS}}{N} = \frac{8}{3} = 2.67$$

2. $$\sigma^2 = \frac{\mathbf{SS}}{N} = \frac{46}{6} = 7.67$$

The scores, means, and variances for the distributions are presented below. Notice that the scores in the second distribution are more spread out from their mean compared to the scores in the first distribution. Accordingly the variance for the second distribution is larger compared to the variance for the first distribution. As scores differ more and more from the mean, the value of variance increases, which is exactly how a measure of variability should behave.

1. $x_i = \{2,4,6\}$, $\mu = 4$, $\sigma^2 = 2.67$

2. $x_i = \{5,3,7,3,2,10\}$ $\mu = 5$, $\sigma^2 = 7.67$

Take a second glance at the value of variance and the spread of scores in the above distributions. Are the values of variance inflated compared to the actual variability in the distributions? For example, the scores in the first distribution differ from the mean by either 0 or 2 points, yet the variance for this distribution is 2.67. Why is variance greater than the maximum difference in the distribution?

4.4 Standard Deviation

Variance measures the average *squared* difference from the mean. The value of variance is in squared units, which explains why variance is inflated compared with the actual variability in the distribution. Describing variability in squared units makes interpretation less straightforward. How can we transform variance so that it measures variability in original units rather than squared units?

The solution is to take the square root of variance. The resulting value is called the standard deviation. Standard deviation measures the average difference from the mean, which is much easier to interpret than the average *squared* difference from the mean. Thus another solution to P3 is:

S3-3. Calculate the square root of variance. The resulting value is the average difference between a score and the mean. This value is called the standard deviation.

The standard deviation shares all the desirable properties with variance without resulting in an inflated value. It reflects the average difference between a score and the mean in the original units of measurement. Standard deviation is symbolized by sigma (σ) rather than sigma squared (σ^2). Algebraically, standard deviation can be expressed as:

$$\sigma = \sqrt{\sigma^2} = \sqrt{\frac{\mathbf{SS}}{N}} = \sqrt{\frac{\sum (x_i - \mu)^2}{N}}$$

The standard deviations for the above distributions are equal to:

1. $\sigma = \sqrt{\sigma^2} = \sqrt{2.67} = 1.63$

2. $\sigma = \sqrt{\sigma^2} = \sqrt{7.67} = 2.77$

The values of standard deviation are more accurate descriptions of variability compared with variance. For example, consider the variance and standard deviation for the first distribution:

1. $x_i = \{2,4,6\}$, $\mu = 4$, $\sigma^2 = 2.67$, $\sigma = 1.63$

The scores in the first distribution differ from the mean by either 0 or 2 points. Accordingly, the standard deviation is 1.63 points, a value in between the maximum and minimum deviations.

Given that standard deviation is more representative of variability compared with variance, you might be questioning the utility of variance. Variance is not used as a descriptive statistic but rather an inferential statistic. Inferential statistics are the class of statistics that are used to make generalizations on the basis of a limited amount of observations. Generalizations of the variability of a phenomenon tend to be more accurate with variance compared to standard deviation.

4.5 Median Deviations

With the exception of the range, the preceding measures of variability have focused on quantifying the accuracy of the mean. We should also consider a measure of variability that quantifies both the accuracy of the median and the dispersion of ordinal data. Recall that the magnitude of difference among scores is meaningless in ordinal data; the magnitude of a score is used only to determine rank. Nevertheless, the same strategy that was used to quantify the variability around the mean can be adapted to quantify the variability around the median. The important concept to remember is that differences from a single value can be used to assess how accurate that value is at representing a distribution and the degree to which scores cluster around a single value. With this in mind, try to quantify the variability in the following distributions of ordinal data:

1. $x_i = \{1,2,3,4,5,6,100\}$, median = 4.
2. $x_i = \{5,6,7\}$, median = 6.
3. $x_i = \{60,70,80,90\}$, median = 75.

The difference that each score is from the median can be used to describe how well the median represents the distribution. Deviations from the median are called median deviations. The median deviations for the above distribution are:

1. $(x_i - \text{median}) = \{-3,-2,-1, 0, 1, 2, 96\}$.
2. $(x_i - \text{median}) = \{-1,0,1\}$.
3. $(x_i - \text{median}) = \{-15,-5, 5,15\}$

It would be helpful to summarize the deviations into a single value. A single value that describes the accuracy of the median is much easier to communicate than a full list of median deviations.

How can we determine a value that is representative of the typical difference between a score and the median?

Given that the median splits the distribution in half, some of the scores will be considerably greater than the median and others will be considerably less than the median. Summing the median deviations could produce distorted measures of variability because some of the scores are expected to differ greatly from the median. We need to determine the typical difference between a score and the median without summing the scores. How do you think this can be accomplished?

In line with the essence of the median, we could use the value that splits the distribution of median deviations in half. In other words, we could use the median of the median deviations. The median of the median deviations for the above distributions are:

1. median of (x_i – median) = {-3,-2,-1, 0, 1, 2, 96} = 0

2. median of (x_i – median) = {-1,0,1} = 0

3. median of (x_i – median) = {-15,-5, 5,15} = 0

The median of median deviations will always result in a value of zero. This is because the *number* of deviations below the median is equal to the number of deviations above the median. A simple solution to this problem is to ignore the sign of the deviation; that is, use the absolute value of the median deviations. By doing so, the median of the absolute median deviations will no longer always result in a value of zero:

1. median of (| (x_i – median) |) = (0,1,1,2,2,3,96) = 2

2. median of (| (x_i – median) |) = (0,1,1) = 1

3. median of (| (x_i – median) |) = {5,5,15,15} = 10

The median of the absolute median deviations is called the median absolute deviation or *MAD* for short. *MAD* accurately describes the degree in which scores differ from median. For example, the scores, medians, and *MADs* for the three distributions are:

1. x_i = {1,2,3,4,5,6,100}, median = 4, *MAD* = 2

2. x_i = {5,6,7}, median = 6, *MAD* = 1

3. x_i = {60,70,80,90}, median = 75, *MAD* = 10

The *MAD* for the first distribution reflects the fact that most of the scores in the distribution differ from the median by 2 points, even though one score differs from the median by 96 points. The scores in the second distribution that differ from the median do so by 1 point. Accordingly, the *MAD* for this distribution is equal to 1. Similarly, the typical difference among scores in the third distribution is 10 points and the scores differ from the median by 5 or 15 points. Accordingly, the *MAD* for this distribution is equal to 10. Given the usefulness of the *MAD* for describing the dispersion of ordinal data and the accuracy of the median, another solution to P3 is:

S3-4. Calculate the median of the absolute median deviations. The resulting value is called the median absolute deviation.

4.6 Summary

The range, variance, standard deviation, and median absolute deviation are four measures of variability. The purpose of a measure of variability is twofold: to describe the accuracy of a measure of central tendency at representing a distribution and to describe the dispersion of scores. With the exception of the range, each of the aforementioned measures of variability accomplishes this purpose, albeit in various degrees. The range only describes the maximum degree in which scores differ from each other; it does not describe the accuracy of a measure of central tendency. Nonetheless, the range is the only useful measure of variability for nominal data. Variance, and its closely related cousin standard deviation, measure the accuracy of the mean at representing a distribution and describe the spread of scores in a distribution. Variance is the average squared difference between a score and the mean, whereas standard deviation is the average difference between a score and the mean. For data sets where the magnitude of difference among scores is arbitrary (i.e., nominal data), the appropriate measure of variability is the absolute deviation median. Unlike variance or standard deviation, which owing to their use of the sum and the mean are sensitive to outliers, the median absolute deviation is unaffected by the presence of outliers.

4.7 Local-Global Relationships in Measures of Variability

Like the measures of central tendency, the measures of variability act as globalizing functions in that they emerge from local elements in the distribution. With the exception of the range, measures of variability can be considered one level above measures of central tendency in a local-global hierarchy because measures of central tendency are part of a measure of variability. The superordinate status of measures of variability entail that their meaning is dependent on a measure of central tendency. For example, understanding the standard deviation requires an understanding of the mean. The cumulative and hierarchical nature of statistical concepts lends itself well to a local-global framework, and we wish to highlight this point because it may help you organize your thoughts once you are confronted with advanced statistical techniques. If you cannot grasp the meaning of a new statistical concept, revisit its antecedent concepts and reflect on how they are combined to form the new concept.

5. Summary of Descriptive Statistics

Descriptive statistics are required to interpret data. The central problem of descriptive statistics is:

P1. How can I summarize a data set accurately and efficiently?

A tentative solution to this problem is:

S1. Determine a single value that best represents my data and determine the degree to which that single value accurately represents my data.

The tentative solution can be parsed into two separate sub-problems. The first sub-problem is:

P2. How can I determine a single value that best represents my data?

Three solutions to this problem were proposed. They are as follows:

S2-1. Identify the most frequently occurring score in the data set. The value of the most frequently occurring score is called the mode.

S2-2. Identify the score whose value splits the data set in half such that 50% of all scores are greater than that value and 50% of all scores are less than that value. The score that separates the upper and lower halves of the data set is called the median.

S2-3. Add up all of the scores in the data and divide the sum by the number of scores in the data. The resulting value is called the mean.

These solutions are measures of central tendency. Each measure identifies a central value—a value that is representative of the data as a whole. This value can be used to summarize and describe large amounts of observations. Once a central value is identified, the accuracy of that value at representing the data needs to be determined. By doing so, the degree to which observations differ from one another can also be summarized and described. This leads us to the second sub-problem:

P3. How can I determine the degree to which a single value accurately represents the data?

Four solutions to this problem were proposed. They are as follows:

S3-1. Calculate the difference between the maximum and minimum score in the distribution. The value of the difference is called the range.

S3-2. Calculate the average squared difference between a score and the mean. The resulting value is called variance.

S3-3. Calculate the square root of variance. The resulting value is the average difference between a score and the mean. This value is called the standard deviation.

S3-4. Calculate the median of the absolute median deviations. The resulting value is called the median absolute deviation.

These solutions are measures of variability. Each measure, with the exception of the range, describes the accuracy of a measure of central tendency and the degree to which the data is clustered around a single value or spread apart.

Measures of central tendency and of variability describe the centre and spread of a data set, respectively. Together, they summarize a data set accurately and efficiently. The characteristics of the data place constraints on the selection of a descriptive statistic. To give you practice at selecting, calculating, and interpreting descriptive statistics, several problems and data sets are presented below. For each data set, your task is to summarize and describe the data using descriptive statistics. Try it!

5.1 Problems

1. When calculating variance, why is it necessary to use squared deviations from the mean?
2. Why is it important to always include a measure of variability when reporting a measure of central tendency?
3. What is the appropriate measure of central tendency and measure of variability for the following distribution? Why? Assume the data is discrete. {1,3,2,3,2,24,5,2,3}
4. What is the appropriate measure of central tendency and measure of variability for the following distribution? Why? Assume the data is continuous. {1,3,2,3,2,5,2,3,4}
5. What is the appropriate measure of central tendency and measure of variability for the following distribution? Why? Assume the data is continuous. {1,3,2,3,2,5,2,3,6 and above}
6. What aspect of the mean allows for outliers to skew the value of the mean in their direction?

7. For following distribution, determine the mean, variance, and standard deviation: {2,3,1,5,6,7,8,0}. In a sentence or two, describe this distribution (refer to the mean and standard deviation in your description).
8. Data from two experimental conditions are presented below. Assume that the data is continuous. For each condition, try to summarize the data accurately and efficiently. State which condition has the most variability.

Condition 1	Condition 2
17	5
16	10
18	12
12	12
15	4

5.2 Solutions

1. When calculating variance, why is it necessary to use squared deviations from the mean?
   a. Squaring deviations from the mean prevents the sum of deviations from always resulting in zero.
2. Why is it important to always include a measure of variability when reporting a measure of central tendency?
   a. Measures of variability reveal how accurate a measure of central tendency is at representing a distribution of scores
   OR
   b. Measures of variability reveal the spread of the scores in a distribution.
   OR
   c. Measures of central tendency do not reveal how accurate they are at representing the distribution or how the scores are spread out in the distribution.
3. What is the appropriate measure of central tendency for the following distribution? Why? Assume the data is discrete. {1,3,2,3,2,14,5,2,3}
   a. The median and the median absolute deviation.
   b. There are two modes and one outlier (i.e., 14).
4. What is the appropriate measure of central tendency for the following distribution? Why? Assume the data is continuous. {1,3,2,3,2,5,2,3,4}
   c. The mean and standard deviation.
   d. There are two modes, no outliers and data is continuous.

5. What is the appropriate measure of central tendency for the following distribution? Why? Assume the data is continuous. {1,3,2,3,2,5,2,3,6 and above}
   a. The median and the median absolute deviation.
   b. There are undetermined values in the distribution and two modes.
6. What aspect of the mean allows for outliers to skew the value of the mean in their direction?
   a. All the scores in a distribution are summed. The outlier will contribute disproportionately to the sum relative to the other scores. This has the effect of driving the value of the mean closer to the outlier.
7. For following distribution, determine the mean, variance, and standard deviation: {2,3,1,5,6,7,8,0}. In a sentence or two, describe this distribution (refer to the mean and standard deviation in your description).
   a. Mean = (2+3+1+5+6+7+8+0)/8 = 32/8 = 4
   b. $\text{Variance}=((2-4)^2+(3-4)^2+(1-4)^2+(5-4)^2+(6-4)^2+(7-4)^2+(8-4)^2+(0-4)^2)/8$
      $= (4+1+9+1+4+9+16+16)/8$
      $= 60/8$
      $= 7.5$
   c. Standard Deviation = $\sqrt{7.5} = 2.73$
   d. The sample mean for this distribution is 4 and the average difference between a score and the mean is 2.73 units.
8. Data from two experimental conditions are presented below. Assume that the data is continuous. For each condition, try to accurately and efficiency summarize the data. State which condition has the most variability.

Condition 1	Condition 2
17	5
16	10
18	12
12	12
15	4

   a. Condition 1
      i. Mean $= (17+16+18+12+15)/5$
      $= 78/5$
      $= 15.6$

ii. Variance $= ((17-15.6)^2+(16-15.6)^2+(18-15.6)^2+(12-15.6)^2+(15-15.6)^2)/5$
$= (1.96+0.16+5.76+12.96+0.36)/5$
$= 4.24$

iii. Std. Deviation $= \sqrt{4.24} = 2.06$

b. Condition 2

i. Mean $= (5+10+12+12+4)/5$
$= 43/5$
$= 8.6$

ii. Variance $= ((5-8.6)^2+(10-8.6)^2+(12-8.6)^2+(12-8.6)^2+(4-8.6)^2)/5$
$= (12.96+1.96+11.56+11.56+21.16)/5$
$= 11.84$

iii. Std. Deviation $= \sqrt{11.84} = 3.44$

c. Condition 2 has more variability compared with Condition 1

References

Gorard, S. 2005. Revisiting a 90-Year-Old Debate: The advantages of the mean deviation. British Journal of Educational Studies, *53*(4): 417–430.

Watier, N.N., Lamontagne, C. and Chartier, S. 2011. What does the mean mean? Journal of Statistics Education, *19*(2).

CHAPTER 2

Probability

José I. Barragués, Adolfo Morais* and *Jenaro Guisasola*

1. The Problem

"I'm going to be late for work", "it will probably rain in the morning", "the unemployment rate may rise above 17% next year", "the economic situation is expected to improve if there is a change of Government". In our daily life we perceive chance natural events affecting us to a greater or lesser extent, but we have no control over them. This morning I was late for work because I was stuck in a traffic jam caused by an accident due to the rain. I would have avoided the traffic if I had left home five minutes earlier, but I was watching the worrying predictions about the unemployment rate trend on TV, which might improve if there is a change of Government. So, if yesterday there had been a change of Government, perhaps today I would not have arrived late for work.

To foresee events well in advance is a powerful ability to assess constantly the consequences of making decisions, avoiding risks, overcoming obstacles and achieving success in future. Our mind is faced with the difficult task of providing criteria with which to predict what will happen in the future. We suspect that many potential events are related, but in most situations it is impossible to establish this relationship precisely. However, we have a remarkable ability to value (with or without success) the odds in favor and against the occurrence of a given event. We are also able to use the evidence provided by our experience to establish with a degree of confidence how plausible an event is. We have intuitive resources that allow us to judge

Polytechnical College of San Sebastian. University of the Basque Country, Spain.
*Corresponding author

random situations and make decisions. However, many studies show that our intuitions about chance and probability can lead to errors of judgment.[1] There it follows three examples:

Example 1. Linda is a clever, 31-year old single girl. When she was a student she was very concerned about issues of discrimination and social justice. Indicate which of the following two situations (1) or (2) you think is most likely:

1) Linda is currently employed in a bank.
2) Linda is currently employed in a bank and she is also an activist supporting the feminist movement.

Example 2. Let us suppose one picks a word at random (of three or more letters) from a text written in English. Is it more likely that the word starts with R or that R is the third letter?

Example 3. Let us suppose that two coins are flipped and that it is known that at least one came up heads. Which of the following two situations (1) or (2) do you think is more likely?

1) The other coin also came up heads.
2) The other coin came up tails.

With regard to Example 1, it is usual to think that situation (2) is more likely than situation (1). Linda's description seems more apt for a person who is active in social issues such as feminism than for a bank employee. Note, however, that situation (1) includes a single event (to be a bank employee) while situation (2) is more restrictive because it also includes a second event (to be a bank employee and also to be an activist supporting the feminist movement). Thus, situation (1) is more likely than situation (2).

With regard to Example 2, people can more easily think of examples of words that begin with the letter R than words containing the letter R in the third position. Consequently, most people think that it is more likely that the letter R is in the first position. However, in English some consonants such as R and K are actually more frequent in the third position than in the first one.

With regard to Example 3, since the unknown result may be heads (H) or tails (T), it seems that both events have the same probability (50%). However, this is not correct. If it is known that the outcome of one of the coins was H, then the outcome TT is ruled out. Thus, there are just three

[1] Kahneman, Slovic and Tversky (1982) first studied systematically certain common patterns of erroneous reasoning. For an analysis of several of them see Barragués, Guisasola and Morais (2006).

possible outcomes: HH, TH and HT. Each of these outcomes has the same probability (33%). However, if the statement had been that the first coin came up H, then the probability of H in the second coin would be 50%, because in this case the outcomes TH and TT would have been ruled out.

In many practical situations it is necessary to assess accurately the probability of events that might occur. Here are some examples:

Example 4. Let us suppose that 5% of the production of a machine is defective and that parts are packaged in large batches. To check all pieces of a batch can be expensive and slow. Therefore, a quality control test must be used that is capable of removing batches containing faulty parts. A quality control plan operates as follows: a sample batch of 10 units is selected, if no part is faulty, the batch is accepted; if more than one part is faulty, the batch is rejected; if there is exactly one faulty part, then a second sample of 10 units is selected and it is accepted only if the second batch contains no faulty items. Let us suppose that checking a part costs one dollar. Some pertinent questions are: What percentage of batches will be accepted? How much on average will it cost to inspect each batch?

Example 5. Let us suppose we are investigating the reliability of a test for a medical disease. It is known that most tests are prone to failure. That is, it is possible that a person is sick and that the test fails to detect it (false negative) and it is also possible that the test may yield a positive for a healthy individual (false positive). One way to obtain data on the effectiveness of a test is to conduct controlled experiments on subjects for whom it is already known for a fact whether they are sick or not. If the test is conducted on a large number of sick patients, the probability of a false negative p can be estimated. If the test is conducted on a large number of healthy patients, probability of a false positive q can be obtained the. However, in a real situation it is unknown whether the patient is sick or not. If the test shows positive, what is the probability that the patient is really sick? What if the test shows negative? Will it be possible to determine these probabilities from the known values of p and q?

Example 6. Figure 1 shows a distribution network through which a product is transported from point (a) to point (b). The product may be for instance an electric current or a telephone call. Let us suppose that it is a local computer network that connects users placed at (a) and (b). The intermediate nodes 1–7 are computers that receive the data from the previous node and forward it to the next node. There are several routes the data can follow from (a) to (b). For example, one possible path is 1367. It means that if in a given time computers 1, 3, 6 and 7, are operating, then it does not matter if the rest are functioning or not, because communication is assured. Similarly, if computers 1, 4, 5 and 7 are in operation, communication will occur

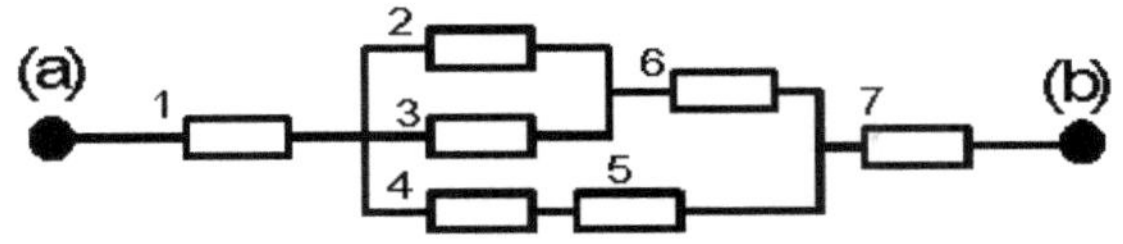

Figure 1. Computer Network.

whether or not the remaining computers are or not. Let us assume that for each node its probability to transmit information is known. What is the probability that the communication is possible from (a) to (b)? We may also consider other questions in addition to the reliability of the network. Let us suppose that one of the computers 2 or 5 is faulty, what is in this case the probability of the network being operational? And if it is guaranteed that any of the computers 4 or 6 is operational at all times, what is in this case the probability of the network being operational?

Example 7. Let us suppose an urn U contains four balls. The balls may be white or black, but we do not know how many balls of each color there are in the urn. Let us suppose we randomly and indefinitely draw a ball, write down the ball's color and return the ball to the urn before the next draw. Can we somehow determine the number of balls of each color? For example, if we take out five balls and get a white ball every time, all we can say for sure is that not all the balls in the urn are black. In this case the possible contents of the urn are U(0n,4b), U(1n,3b), U(2n,2b), U(3n,1b). According to the available information, what is the probability of each of these four possible arrangements? What if we draw ten balls and the ball is always white? Intuition tells us that the possibility that all the balls are white (U(0n,4b)). However, perhaps the urn contains a single white ball (U(3n,1b)) which we randomly extracted again and again. But let's suppose that the eleventh comes up a black ball. Thus, the possible contents of the urn are U(1n,3b), U(2n,2b), U(3n,1b), so now what is the probability for each arrangement?

Example 8. Let us suppose you are the head of a large land-haulage company and that it is your responsibility to procure fuel. The price of fuel is highly variable, and therefore the company will save a lot of money if you purchase it before the price increases. On the other hand, you should buy fuel before a possible price drop. The point is that OPEC is scheduled to meet next week to decide its policy on oil production for the next three months. It is known that when oil production increases the price of gasoline decreases. It seems that OPEC will increase production for at least the first two months and perhaps the third. However, it is known that one member will do everything possible to reduce production, thereby increasing the price. You should decide to buy now or use the company's fuel reserves and postpone the purchase by three months. How should the decision be made?

The above examples show complex situations for which intuition about chance provides little or no valid information for making decisions. It is therefore necessary to develop methods of calculating probabilities that can be applied in various practical situations: gambling, quality-control, risk-assessment, reliability studies, etc. It is also necessary to clarify the meaning of the calculated probability values. For example, if a fair coin is flipped 120 times, we can calculate that the probability of it coming up heads in over 55 of the total number of flips is $p = 0.79$. But what exactly does this probability value mean? Thus, the problem we intend to solve is as follows:

THE PROBLEM

One should develop methods to calculate the probability that can be applied in a variety of practical situations. In addition, the meaning of the calculated probability value should be understood.

2. Model and Reality

Consider the values X, Y, Z, W as defined below. Which of them do you think are random?

X = "Maximum temperature (°C) to be measured in Reno (NV) exactly in 10 years time";

Y = "Maximum temperature (°C) measured in Reno (NV) exactly 10 years ago";

Z = "Outcome (H/T) obtained when flipping a coin";

W = "Speed (m/s) at which a free-falling object released from a height of 100 meters hits the ground."

The reader may conclude that value X is random since it is not possible to predict. In contrast, the reader may think that value Y is not random since it refers to a past phenomenon and it is sufficient to check Reno's weather records for the maximum temperature at the time. OK. But Let us suppose you do not have access to such meteorological records. In such a situation, which involves less uncertainty: a prediction about the value of Y or about the value of X?

Regarding the variable Z, Let us suppose that we show you the following sequence of 10 heads and tails: HHHTTHTTHT. Are you able to predict the value (H or T) of the eleventh outcome? The value of the eleventh position of the sequence is uncertain for you. But this uncertainty disappears if we tell you that this sequence was generated from the first decimal places of the number $\pi \approx 3.14159265358979323846$. If the decimal place is between 0 and 4, we wrote in the sequence H, otherwise, we wrote T.

Finally, the value W can be considered non-random because Physics teaches us how to calculate it from the initial conditions. Are you sure? Have you taken into account factors such as friction? Do we know the exact value of the acceleration of gravity in that geographical location? Do you know the exact height from which the object was dropped? If we measure the final speed, do we expect the value to fit the prediction made by physics' equations?

What does all this mean? It is often said that natural phenomena are classified as «deterministic phenomena» and «random phenomena». It is explained that for deterministic phenomena the final outcome can be predicted based on certain factors and known initial conditions. In contrast, there is no procedure for random phenomena to make this prediction, because they involve factors of a random nature, and so the final result may be different each time you perform the experiment. This classification is illustrated with examples such as «the flip of a coin» (random phenomenon) and «time for an object to hit the ground after its release» (deterministic phenomenon). Thus, it seems that real phenomena are random or deterministic and that using certain equations a natural phenomenon may be fully described. Actually, what we classify as «deterministic» or «random» are not real phenomena, but the models we use to analyze these phenomena.

Let us suppose that we would like to predict the distance travelled in 10 s by a vehicle running with constant acceleration a=2 m/s^2, and initial velocity v_0=25 m/s. The following deterministic model to predict the value of S(t) at instant t may be used:

$$S(t)=v_0 t+\frac{1}{2}at^2 \Rightarrow \begin{pmatrix} v_0=25 \\ a=2 \\ t=10 \end{pmatrix} \Rightarrow S(10)=350 \text{ m} \qquad (1)$$

Figure 2 shows the predicted S(t) obtained by the deterministic model (1) for each $t\in[0,10]$. This prediction may be sufficiently accurate for many applications. The deterministic model (1) could be improved by adding other known parameters such as the wheels' friction with the ground, the aerodynamics of the vehicle, etc.

But now let us suppose that we wish to study the space covered by a large number of vehicles on a road. The vehicles arrive at random and their exact values of acceleration (a) and initial velocity (v_0) are unknown. From our point of view, we can consider that (a) and (v_0) are random. To define this situation in more detail, Let us suppose that $a \in [1,2.5]$ and $v_0 \in [24,32]$. At each instant $t \in [0,10]$ the position S(t) at which the vehicle is located is now a random value that depends on the random values (a) and (v_0). In Fig. 3 we show the graphs of S(t) for the extreme values $a=1$, $v_0=24$ (bottom graph) and $a=2.5$, $v_0= 32$ (upper graph). Any other graph of S(t) for $a \in [1,2.5]$ and $v_0 \in [24,32]$ will be between the border graphs of Fig. 3.

Note that these two graphs define, for each value of $t \in [0,10]$, the interval in which the value of S(t) is located. For example, for t = 5, the random value S(5) is in the interval [132.5,191.25]. Likewise, the final position of the vehicle S(10) is a random value in the interval [290,445]. Note also how the uncertainty about the position of the vehicle S(t) increases as the value of t increases (the interval containing S(t) is also wider).

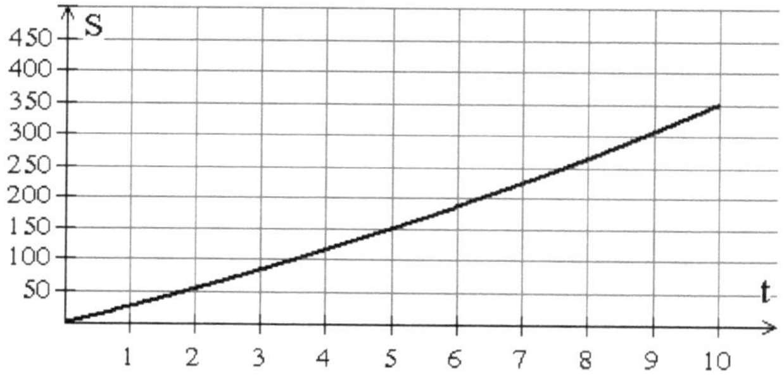

Figure 2. Deterministic model.

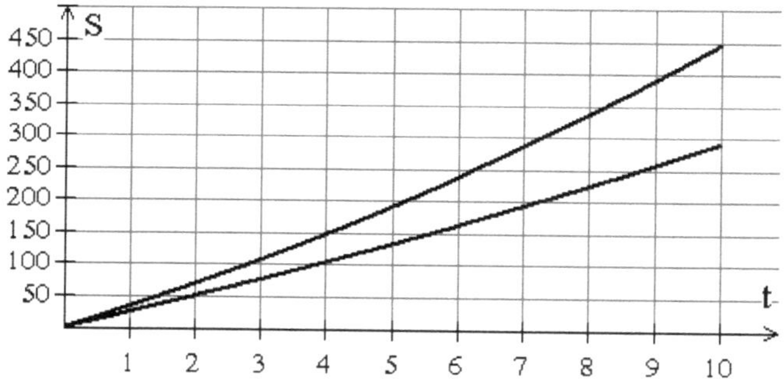

Figure 3. Graphs plotted with extreme values of a and v_0.

Now look at Fig. 4, which shows the presence of chance in the model. At each time $t \in [0,10]$ we simulated[2] a couple of random values of (a) and (v_0) and calculated S(t) in the model (1). Following this method, we generate the graph plotted in Fig. 4, which is located between the two border graphs.

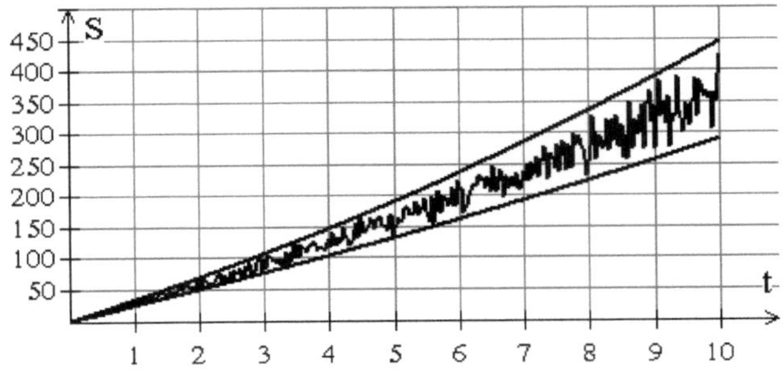

Figure 4. Visualizing uncertainty on S(t) for each $t \in [0,10]$.

3. Event and Probability

Let us continue with the discussion of our example. The next task will be to use this probabilistic model to formulate predictions about the position of the vehicle S at time t=10. We know that $S(10) \in [290,445]$. Instead of trying to predict the exact value of S(10), the idea is to make predictions about the position of S(10) in different sub-intervals within [290,445]. For example, how likely is the occurrence of $290 \leq S(10) < 336$? And the event $S(10) < 401$? To measure this likelihood we will use the concept of relative frequency. The relative frequency of a sub-interval is our *estimate* of the probability of that sub-interval.

We will start decomposing the interval [290,445] in the eight sub-intervals $I_1, I_2, \ldots, I_8$, as shown in the first column of Table 1. Each random value S(10) will be in one and only one of these sub-intervals. Then we simulate 10,000 pairs of random values $a \in [1,2.5]$ and $v_0 \in [24,32]$. For each pair (a,v_0) we calculate the value $S(10) = 10v_0 + 50a$, which is also random. The

[2] Pocket calculators often incorporate a RANDOM function that generates a random value in [0,1] which is distributed uniformly in the interval. For this utility, Excel has the built-in function called RAND. The values generated by this type of generator are often called pseudo-random because they are obtained by deterministic algorithms. However, as it has been discussed in this section, if the calculation algorithm is unknown, the uncertainty of the generated value is exactly the same, so these values can be considered fully random. To generate a random value Y included in the interval [p,q], it is sufficient to generate a random value $X \in [0,1]$ and use the equation $Y = p + (q-p)X$. Therefore, to generate the two random values $a \in [1,2.5]$ and $v_0 \in [24,32]$, it is sufficient to generate the random values $X_1, X_2 \in [0,1]$ and use the equations $a = 1 + 1.5X_1$, $v_0 = 24 + 8X_2$.

second column of Table 1 shows the relative frequency of each sub-interval I_i, where i=1,...,8. The relative frequency of each sub-interval is our *estimate* of the probability that S(10) falls into the sub-interval. Figure 5 shows the histogram of the data-distribution (for simplicity, the histogram sub-intervals are shown with the same length, but in fact they are of different lengths).

Exercise 1. Using Table 1 data, estimate the probability p(A) of each of the following situations:

a) $A = 290 \leq S(10) < 336$
b) $A = S(10) < 401$
c) $A = S(10) < 336$ or $S(10) \geq 387$
d) $A = 290 \leq S(10) \leq 445$

Table 1. Outcomes of the simulation.

Subinterval	Relative frequency
I_1=[290,310)	0.032
I_2= [310,320)	0.0397
I_3= [320,336)	0.1048
I_4= [336,366)	0.3012
I_5= [366,378)	0.145
I_6= [378,387)	0.0954
I_7= [387,401)	0.1162
I_8= [401,445]	0.1657

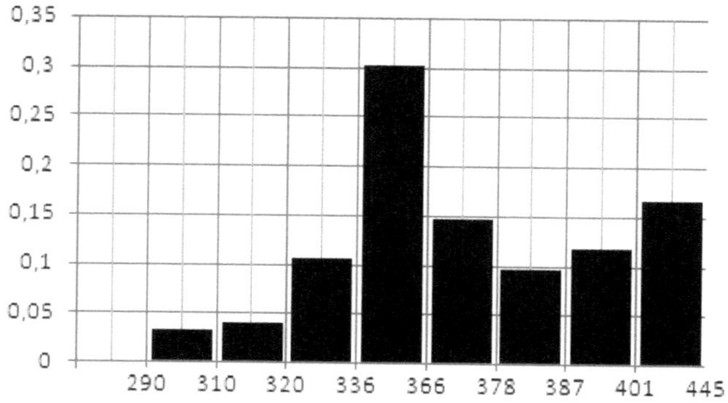

Figure 5. Relative frequency distribution.

Exercise 2. Observe the roulette wheel in Fig. 6. After playing many times, Let us suppose we have proved that even numbers are about twice as frequent as odd ones. Calculate the probability p(A) of the following situations:

a) A = Odd
b) A = Even
c) A = Odd and less than 7
d) A = Even or black
e) A = Even and greater than 5 or odd and black
f) A = Black and multiple of 4 and greater than 7
g) A = Multiple of 5 and odd
h) A = White or even or 11

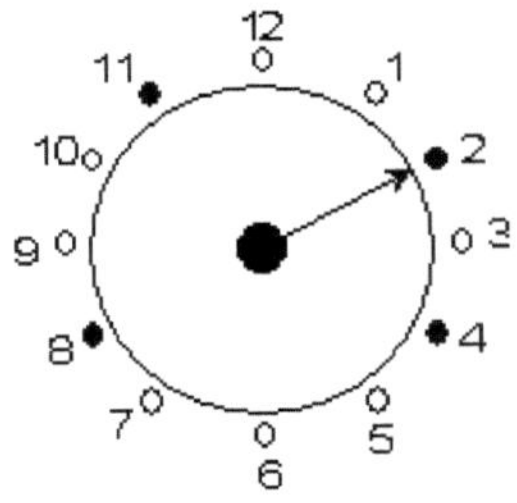

Figure 6. Fudged roulette.

We will introduce the terminology for the various concepts that have emerged. Please, look at Theory-summary Table 1.

Theory-summary Table 1

Random Experiment: An experiment whose outcome cannot be predicted in the established conditions.

Sample Space: The set of possible outcomes that can occur when running a randomized experiment.

Elementary event: Each of the elements of the sample space. Each time you run the random experiment and one and only one elementary event occurs.

Probability of an elementary event: If S is an elementary event of a random experiment, the probability of that event p (S) is the value to which the relative frequency of S gets closer when increasing the sample size.

Event: Any set of elementary events (that is, any subset of the sample space).

Probability of an event: The addition of the probabilities of the elementary events that form the event.

Exercise 3. Use the definitions of Theory-summary Table 1 in Exercises 1 and 2.

Exercise 4. Let us suppose a sample space consisting of n elementary events. How many events can be defined?

Example 9. From an urn containing three white balls and 5 black balls, balls are randomly drawn out and returned to the urn. Figure 7 shows the number of extractions on the horizontal axis and the relative frequency of the event A = "white ball" on the vertical axis. As shown, if the number of extractions is small, there is a great variability in the relative frequency of A. However, the frequency is stabilized for large samples. We can *estimate* the value of p(A) by the relative frequency of A using the full sample (N=2000) to obtain p(A)≈742/2000=0.371. Assuming that each ball has the same chance of being chosen, the exact value of the probability of A is p(A)=3/5=0.375, which almost coincides with the estimated value. However, the relative frequency has a large variability for small samples, as shown in Fig. 7. We can measure this variability. The standard deviation value (of the population) of nine relative frequencies from N=2 to N=150 is σ=0.109, while for the thirteen relative frequencies from N=200 to N=2000 the deviation σ=0.017 (a sixth of the previous deviation). Nevertheless, pay attention to the following point: if we increase the sample size, we cannot ensure that the relative frequency will be closer to p(A) with a smaller sample. In this example for a sample size of N=20 a better estimate of the probability is obtained than for N=50.

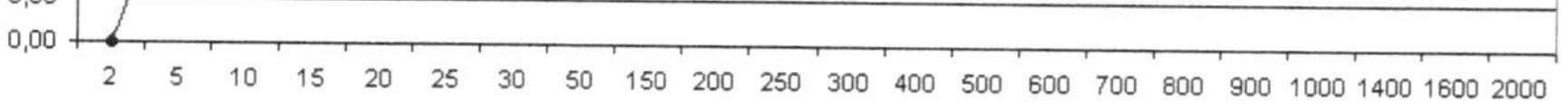

Figure 7. Evolution of the relative frequency of the event A="white ball".

Let us comment on the ideas that have emerged so far:

- In Example 9, the value of the probability of an event has been estimated using relative frequency. However, this empirical method which assigns probabilities to elementary events has some limitations that should be underlined. Firstly, data should be collected through randomized experiments performed under identical conditions.

However, many interesting practical situations are impossible to replicate under identical conditions. Think for example of situations in daily life, whether social, medical, economic or historical. The second objection to this frequency-based method of interpreting probability is that the value of the probability of an event is never known, but is estimated from a sample. If a different sample is used, the estimate will also be different.

- To estimate the probability of an event using relative frequency, it is extremely important to have a sufficiently large sample. To illustrate this, observe the histograms in Fig. 8. The histogram in Fig. 5 is plotted using a simulated sample of 10,000 values. In contrast, for the same model, the three histograms of the upper row in Fig. 8 were calculated by means of respective samples of size N=100. Notice how different the histograms are. The three histograms in the middle row are very similar to each other, and have been calculated using samples of size N=1000. Finally, the histograms of the last row are calculated using samples of size N=10,000 and they are practically identical to those obtained with samples of size N=1000. That means that in this random experiment a sample size N=100 is not sufficient to estimate the probability of different elementary events, but the sample sizes of N=1000 and N=10,000 provide similar probability estimates.
- A common mistake is to interpret the value of the probability p(A) of an event A as a prediction about whether the event A will occur in the next performance of the random experiment. In fact, there is a tendency to interpret a value p(A)>0.5 as predicting "the event A will

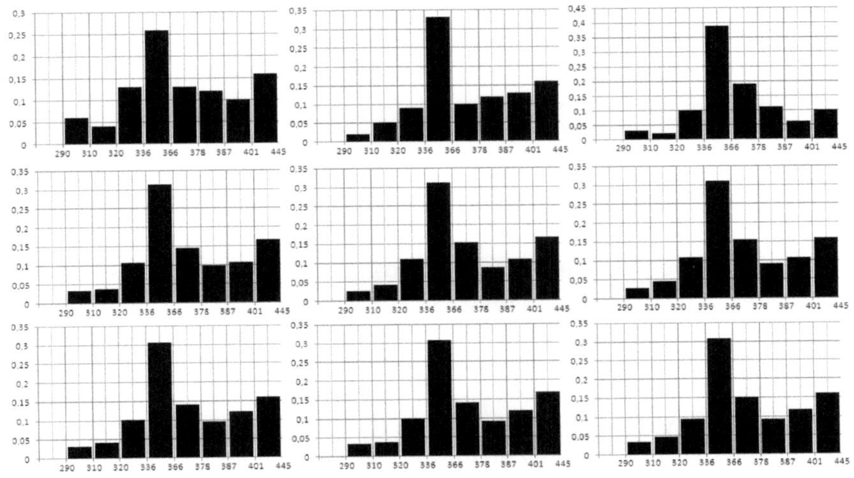

Figure 8. In rows, size distributions of samples N=100, N=1000 and N=10000.

occur" and a value p(A)<0.5 as predicting "the event A will not occur". One perceives uncertainty only if p(A)=0.5. However, predicting an individual result is, most of the time, what really matters. For example there are available data (i.e., frequencies) about divorces, airplane accidents, disease treatments, stock market investments, etc. Yet there are still those who will consult a fortune teller because the teller provides customized "predictions" to questions like: will *my* marriage fail?; will *my* plane crash?; will *I* recover from *my* illness?; will *I* earn money on *my* investment? The answer offered by probability is not easy to understand. Moreover, even if it is understood, it may not be satisfactory.

- Probability values of the elementary events in Exercise 2 have been calculated on the assumption p(even)=2p(odd). Similarly, if we assume that a coin is fair, we may take p(C)=p(X)=1/2. If we assume that a die is not loaded, we could take p(i)=1/6 where i=1,...,6. These assumptions are actually about the relative frequency of elementary events. The probabilistic model thus constructed will be useful to make predictions for a given random experiment as long as the assumptions are valid for that experiment. In practical applications, before calculating, one should think about the assumptions to be made to obtain a solution. Then one should express these hypotheses explicitly.

Exercise 5. A traffic light that regulates the traffic at a cross-road may be in one of the following four states: Red (R), Green (G), Amber (A) or Flashing Amber (FA). What is the probability that at a given moment the position of the traffic light will be R or G?

Exercise 6. Let us suppose that you plan to run a randomized experiment for N times. Let p(A) be the probability of some event A.

a) Explain how to use the value p(A) to predict the number of times that the event A will occur.
b) Use this result for the events in Exercises 1 and 2, assuming N=1250.

4. Unions and Intersections of Events

In Section 3 we presented a method to calculate the probability of events (see Table 1). The starting point was to plot events as sets. From this point of view, an event A is simply a subset of the sample space E. The performance of the experiment produces a random occurrence of one and only one elementary event. But there is also the occurrence of all events that contain this elementary event. For example, in Exercise 2 imagine that we spin the wheel and get the number 11. In this case, the elementary event that occurred was S={11}. But there have been other events such as A="odd"={1,3,5,7,9,11}, B="greater than 7"={8,9,10,11,12} and C="black"={2,4,8,11}. *Any* event that contains the elementary event S={11} has occurred. In order to calculate the probability p(A) of any event, firstly the event A should be expressed by means of all the elementary events that are part of it. The elementary events are the parts from which events are built. The sample space E has all parts that are available to form events. The value of p(A) is calculated by adding up the probabilities of all elementary events that form A.

Exercise 7. Express formally this method for the calculation of the probability.

Let us explore in a little more detail this method of calculating the probability of an event. Let us suppose we are studying the distribution of the number of traffic accidents in the city. In Fig. 9, diagram E represents the eight districts of a city, $D_1,\ldots, D_8$. An accident may occur in any of the eight districts. The diagrams A, B, C and D represent areas of the city districts formed by grouping districts. In probability terminology, E is the sample space and A, B, C and D are events that may occur.

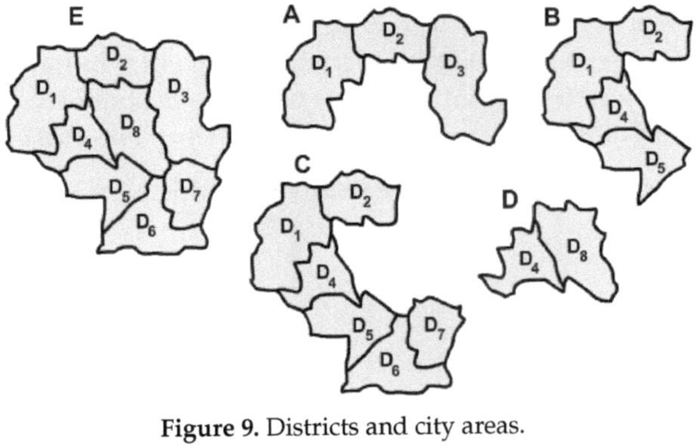

Figure 9. Districts and city areas.

Exercise 8

a) Let us suppose that an accident has occurred in the district D_1: which of the events A, B, C, D have occurred?
b) Calculate p(A), p(B), p(C) and p(D).
c) Consider the event R="have an accident in A or D". Could you think of a way to write p(N) as a function of p(A) and p(D)?
d) Describe formally the event M="have an accident in A and B". Calculate p(M).
e) Describe the event formally N="have an accident in A or B". Calculate p(N). Could you think of a way to write p(N) as a function of p(A), p(B) and p(M)?
f) Describe formally the event H="have an accident outside of A". Could you write p(H) as a function of p(A)?

g) Describe which new findings have been obtained in this exercise.

Table 2 contains a theory-summary of the findings.

Theory-summary Table 2

Probability of an event's opposite event: $p(\overline{A})=1-p(A)$ **Probability of the union event:** If A and B are events, then in general $p(A\cup B)=p(A)+p(B)-p(A\cap B)$. **Compatible and incompatible events:** Two events A and B are incompatible if they cannot occur simultaneously. The condition of incompatibility of the events A and B is $A\cap B=\varnothing$. In this case, $p(A\cap B)=0$ and therefore $p(A\cup B)=p(A)+p(B)$. But if the events A and B may occur simultaneously, they are called compatible events, then $A\cap B\neq\varnothing$.

Note in Fig. 10 the interpretation of the formula $p(A\cup B)=p(A)+p(B)-p(A\cap B)$. The black dots represent the elementary events. Events A and B are formed by grouping elementary events. Event $A\cup B$ consists of the elementary events that are in A or B (and that includes the elementary events that are in both). Event $A\cap B$ consists of elementary events that are in both A and B. Notice how when calculating $p(A)+p(B)$, the probability of the common set $A\cap B$ is calculated twice. Therefore, $p(A\cup B)=p(A)+p(B)-p(A\cap B)$.

Exercise 9. Let A, B and C be three events. Provide a method for calculating $p(A\cup B\cup C)$.

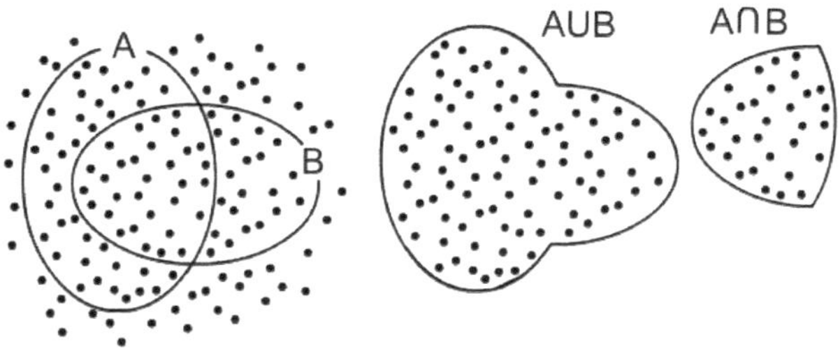

Figure 10. Interpretation of the equation $p(A\cup B)=p(A)+p(B)-p(A\cap B)$.

5. Events that are Difficult to Express in Terms of Elementary Events

In Section 4 we obtained the equation $p(A\cup B)=p(A)+p(B)-p(A\cap B)$, which is very interesting for the practical calculation of the probabilities of events: to compute $p(A\cup B)$ it will not be necessary to express the event $A\cup B$ by means of elementary events (as we have done so far), it will be sufficient to use the values of $p(A)$, $p(B)$ and $p(A \cup B)$. This is especially useful in sample spaces that contain many items. Let us consider for example the computer network in Example 6. Please re-read the details of that example. Let us suppose that the probability of each of the seven individual computers being operational, $p(1), p(2), \ldots, p(7)$ is known. To calculate the probability of the event F="communication exists between (a) and (b)", we can try to express F in terms of the elementary events that form F. To simplify the notation, we will note the events i="the i-th computer is operational", $\overline{i}$ ="the i-th computer is not operational", where $i=1,\ldots,7$. Thus, for example, the elementary event $S=1\overline{2}\overline{3}45\overline{6}7$ means "1457 computers are operational and computers 2, 3, 6 are not operational". How many elementary events are there? Each of the seven computers has two possible states (YES/NO), so that there are $2^7=128$ possible states of the network. Now we should determine which elementary events are favorable to the event F, but this task is very laborious:

$$F=\left\{1234567, 12345\overline{6}7. 1234\overline{5}67, 1\overline{2}\overline{3}45\overline{6}7, 1\overline{2}\overline{3}4567, \ldots\right\}.$$

The difficulties do not end there. What is the probability of each of the elementary events? How could, for example, p(1234567) be calculated? In summary, we have found a strategy that seems useful to calculate the probability of an event A. This strategy consists in writing the event A by means of elementary events and then calculate the value of p(A) as the sum of the probabilities of all of them. However, this strategy may be impractical in sample spaces that contain a large number of elementary events. We need to find another strategy. Note in equation (2) an alternative way to write F:

$$F=1 \cap 7 \cap (((2 \cup 3) \cap 6) \cup (4 \cap 5)) \qquad (2)$$

Let us see how we got to equation (2) from the arrangement in Fig. 1. Imagine that the seven nodes in Fig. 1 represent bridges that may or may not be open to traffic. You are at point (a) and you wish to get to (b) via the network bridge. Clearly, bridges 1 and 7 should be open. Therefore, the event includes the condition $1 \cap 7$. Once you have passed through bridge 1 there are two possible paths: the subnet formed by bridges 2-3-6 or subnet formed by 4-5. Some (or perhaps both) of these two subnets must necessarily be open to traffic. The operation of the subnet 2-3-6 is expressed as $(2 \cup 3) \cap 6$. The operation of the subnet 4-5 is written as $4 \cap 5$. Adding all conditions, we obtain the expression (2). Now, if we use the relation $p(A \cup B)=p(A)+p(B)-p(A \cap B)$, taking $A=1 \cap 7 \cap (((2 \cup 3) \cap 6)$, $B=1 \cap 7 \cap (4 \cap 5)$, then:

$$p(F) = p(1 \cap 7 \cap (((2 \cup 3) \cap 6) \cup (4 \cap 5))) = p(1 \cap 7 \cap (2 \cup 3) \cap 6) \cup 1 \cap 7 \cap (4 \cap 5))) =$$
$$= p(1 \cap 7 \cap (2 \cup 3) \cap 6)+p(1 \cap 7 \cap (4 \cap 5))-p(1 \cap 7 \cap (2 \cup 3) \cap 6 \cap (4 \cap 5)) \qquad (3)$$

However, here we face a new difficulty when using the expression (3): given any two events A and B, we do not know how to calculate $p(A \cap B)$. This is the next issue to be addressed.

6. Calculation of p(A∩B): Conditional Probability

Let us suppose that the social club *New Sunset* has a total of 155 members, who are men and women of various ages. Table 2 shows the distribution by gender and age. As you can see, we have established five age intervals $I_1, \ldots, I_5$ from 14 to 50 years.

Let us suppose that a person is selected at random and they all have the same probability p=1/155 of being selected. This means that p(W)=77/155, p(M)=78/155, $p(R_1)=9/155$, $p(R_2)=35/155$, etc. These probabilities are calculated based on the total number of people (155). Let us suppose that you choose a person. Would it be more likely to be W or M? The value of p(M) is slightly higher than p(W). If we repeated the experiment a number of times, the relative frequencies of each event would be very similar, approximately 77/155=0.497 and 78/155=0.503 respectively. But Let us suppose we have the following additional information about the choice, "the person's age is in the range I_3". The gender is uncertain, so are the odds of the events W and M still the same now? Of course not! Now the probability is 25 out of 32 for event W and 7 out of 32 for event M. If it is known that the age of a person is in the interval I_3, it is more likely that she is a woman. These two probabilities are not calculated on the full sample space consisting of 155 people, but on the subspace I_3, which has only 32 people. The odds in this new situation are written as follows: $p(W/R_3)=25/32=0.78$, $p(M/R_3)=7/32=0.22$.

Table 2. Gender and age distribution.

	I_1 = [14,18)	I_2 = [18,20)	I_3 = [20,25)	I_4 = [25,40)	I_5 = [40,50]	total
W	6	12	25	32	2	77
M	3	23	7	36	9	78
total	9	35	32	68	11	155

Exercise 10. Using the data from Table 2, calculate the probabilities of the following events and interpret their meanings.

a) A="The person's age is in the range I_3".
b) B="The person's age is in the range I_3, if he is known to be a man".
c) C="A person is a man, if he is known to be under 18 years".
d) D="The person's age is under 25 years".

e) E="The person's age is under 25 years, if she is known to be a woman".
f) F="The person's age is under 25 years and she is also a woman".

We will give a name to the new concept p(A/B).

Theory-summary Table 3

Conditional probability: Let E be a sample space and A, B two events. The value p(A/B) will be called conditional probability of event A on condition of event B. This value is calculated as follows: $p(A/B)=p(A\cap B)/p(B)$. The meanings of p(A/B) are: p(A/B) is the probability of event A given that event B has occurred. p(A/B) is also the value of the probability of event A, but recalculated in view of the new information B. p(A/B) is also the probability of event A, NOT calculated on the entire sample space E, but on a reduced sample space; the subspace consisting only of elementary events that form B. Finally, the value 100p(A/B) is approximately the long-term percentage of times that the event A occurs but calculated NOT on the total times that the experiment is repeated, but on the times that the event B occurs. **Probability of the intersection:** $p(A\cap B)=p(B)p(A/B)=p(A)p(B/A)$.

Exercise 11. Let us return to Exercise 2. In the following paragraphs a pair of events A and B are defined. The task is to calculate the values p(A), p(A/B) and to interpret the difference between these values.

a) A="odd", B="black"
b) A="black", B="odd"
c) A="multiple of 5", B="black"
d) A="black", B="multiple of 5"
e) A="greater than 1", B="black"

7. Probabilistic Independence

Remember that in Section 5 we were trying to calculate the probability of a computer network running. At that time we succeeded in expressing the event F= "communication exists between (a) and (b)" by unions and intersections of events 1, 2, 3, 4, 5, 6 and 7, thus:

$$p(F)=p(1\cap 7\cap (2\cup 3)\cap 6)+p(1\cap 7\cap 4\cap 5)-p(1\cap 7\cap (2\cup 3)\cap 6\cap 4\cap 5) \quad (4)$$

However, to continue the calculations in (4), in Section 5 we encounter the difficulty of estimating the probability of the intersection of the two events A and B, that is, to calculate the probability of A and B. Well, in Section 6 we obtained the expression (5):

$$p(A \cap B)=p(A)/p(B/A)=p(B)/p(A/B) \quad (5)$$

Can we now proceed with the calculations in (4)? Consider for example the first term of (4). Note that this is to calculate the probability of the intersection of four events: 1, 7, $2\cup3$ and 6. To use (5), we note for example $A=1\cap7\cap(2\cup3)$, B=6. Then:

$$p(1\cap7\cap(2\cup3)\cap6)=p(1\cap7\cap(2\cup3))p(6/1\cap7\cap(2\cup3)) \quad (6)$$

If we apply the same procedure twice to the first factor, thus:

$$p(1\cap7\cap(2\cup3)\cap6)=p(1\cap7)p(2\cup3/1\cap7)p(6/1\cap7\cap(2\cup3))=$$
$$= p(1)p(7/1)p(2\cup3/1\cap7)p(6/1\cap7\cap(2\cup3)) \quad (7)$$

How could we calculate the different conditional probabilities in (7)?

Exercise 12. Look at the theory-summary Table 3 to review the meaning of p(A/B). How can the three values p(7/1), $p(2\cup3/1\cap7)$ and $p(6/1\cap7\cap(2\cup3))$ of (7) be calculated?

After Exercise 12, assuming the hypothesis that the seven computers in the network operate independently, we can express p(F) in (4) as a function of the individual probabilities p(1) to p(7):

$$p(F)=p(1)p(7)p(2\cup3)p(6)+p(1)p(7)p(4)p(5)-p(1)p(7)p(2\cup3)p(6)p(4\cap5)=$$
$$p(1)p(7)p(6)(p(2)+p(3)-p(2)p(3))+$$
$$+p(1)p(7)p(4)p(5)-p(1)p(7)p(6)p(4)p(5)(p(2)+p(3)-p(2)p(3))$$

Table 3. Gender and course distribution.

Year	Girls	Boys	Total
1st	30	20	50
2nd	24	16	40
Total	54	36	90

Let us explore in a little more detail the meaning of independence between two events A and B. Consider the following examples:

Example 10

a) You have an evenly balanced coin and an urn U(2b,5n). The coin is flipped and a ball is drawn from the bowl. What is the probability of getting heads and a black ball?
b) There is an evenly balanced coin and two urns U1(2b,5n), U2(4b,7n). The coin is flipped. If it comes up heads, then a ball from U1 is extracted,

if tails is obtained, the ball is drawn from U2. What is the probability of obtaining heads and a black ball?

In case (a), clearly the event n="black ball" is independent of the event H="heads", because information about the outcome of the coin toss does not alter the probability of obtaining a black ball. In this case, p(H∩n)=p(H) p(n)=(1/2)(5/7)=5/14. However, in situation (b) the probability of the event n= "black ball" *depends on* the outcome of the flip. Moreover, p(n/H)=5/7, p(n/T)=7/11, p(H∩n)=p(H)p(n/H)=(1/2)(5/7)=5/14. In paragraph (b), it is easily accepted that the probability of event H depends on the event n="black ball". However, it is more difficult to accept that the probability of event H also depends on the event n="black ball". Discuss this issue after solving the following exercise.

Exercise 13. On a table we have four cards: two aces and two kings. We place them face down and mix them. Obviously, if now we draw a random card, the probability of getting an ACE is identical to the probability of getting a KING (i.e., 0.5). Well, what we do is to draw a random card and replace it without looking at it to see what it is. Then, from the remaining three cards we draw another one, which happens to be an ACE. According to this second output, is the probability that the first card was an ACE now equal to greater or less than 0.5?

We have a test to study the probabilistic independence of two events A and B. The test compares the values p(A/B) and p(A). If p(A/B)=p(A), then the event occurrence is independent of B. If p(A/B)>p(A), it means that the occurrence of event B increases the expectation of the occurrence of event A. If p(A/B)<p(A), it means that the occurrence of event B reduces the expected occurrence of event A. In Example 10, there is a clear dependence or independence of the events. However, Let us observe the following example:

Example 11. Let us suppose that a group of 90 students go hiking. Table 3 shows the gender and course distribution. Randomly we select one of the schools. Is the event "to be a girl" independent of the event "to be a first-year student"? Is the event "to be a first-year student" independent of the event "to be a girl"? Will the event "to be a second-year student" independent of the event "to be a boy"? Is the event "to be a boy" independent of the event "to be a second-year student"?

Note G="to be a girl" and 1= "to be a first-year student". In this case, p(G)=54/90=3/5, p(G/1)=30/50. Thus the event "to be a girl" is independent of the event "to be a fist-year student". Moreover, p(1)=50/90, p(1/G)=30/54=50/90. Thus, the event "to be a first-year student" is independent of the event "to be a girl". Confirm that in this case there is independence between any possible combination of gender and course.

This example shows a bizarre result that should be analyzed from a general perspective. Given two events A and B, if A is independent of B, is B then also independent of A? Are opposite events independent? Note that if these results were true in general, then we would not say "A is independent of B" or "B is independent of A", but we would rather say "A and B are independent".

Exercise 14. Let us suppose A is independent of B. Demonstrate that as a consequence B is independent of A. Analyze the independence of the events $\overline{A}$ and B, and $\overline{A}$ and $\overline{B}$.

To conclude our exploration of the meaning of probabilistic independence, look at the situation in the following example.

Example 12. Consider the sample space E and the two events in Fig. 11. Assume the hypothesis of the equiprobability of the 24 elementary events in E. Are A and B independent?

In this case, p(A)=8/24=1/3, p(A/B)=4/12=1/3. Thus, the two events are independent. Note the graphic interpretation of independence: the weight (probabilistically speaking) of the event A in the sample space is identical to the weight that the part of A and B has in the subspace B.

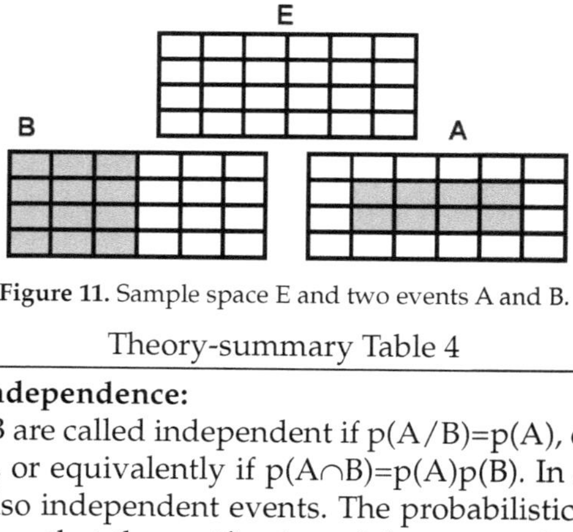

Figure 11. Sample space E and two events A and B.

Theory-summary Table 4

Probabilistic independence:
Two events A, B are called independent if p(A/B)=p(A), or equivalently if p(B/A)=p(B), or equivalently if p(A∩B)=p(A)p(B). In addition, $\overline{A}$, $\overline{B}$ and $\overline{A}$, B are also independent events. The probabilistic independence of A and B means that the verification of the events does not alter the probability that the other event will be verified.

Probabilistic dependence:
The events A and B are called dependent if they are not independent, that is, if p(A/B)≠p(A), or equivalently if p(B/A)≠p(B), or equivalently if p(A∩B)≠p(A)p(B). The probabilistic dependence of A and B means that if one of the events has been verified, it modifies the probability of verifying the other event.

Exercise 15. Let us consider an urn U(3b,5n). The experiment consists of drawing a ball and afterwards drawing a second ball without returning the first ball to the urn. We repeated the experiment 1048 times, obtaining the results shown in Table 4. The aim is to estimate, calculate and interpret the probabilities of the following events: b1, n1, b2, n2, b1 and b2, b1 and n2, n1 and b2, n1 and n2, b2/b1, b2/n1, n2/b1, n2/n1, n1/n2, b1/b2 b1/n2, b1 and b2, b1 and n2, n1 and b2, n1 or n2.

Exercise 16. Let two events be A and B. Prove that using the values of p(A), p(B) and p(A/B) it is possible to obtain the following values: $p(\overline{A})$, $p(\overline{B})$, $p(A \cap B)$, $p(A \cup B)$, $p(B/A)$, $p(A/\overline{B})$, $p(B/\overline{A})$, $p(\overline{B}/A)$, $p(\overline{A}/B)$, $p(\overline{B}/\overline{A})$, $p(\overline{A \cap B})$, $p(\overline{A \cup B})$.

Exercise 17. Formally analyze Example 1.

Table 4. Distribution of outcomes.

Event	Frequency
b1 and b2	104
b1 and n2	294
n1 and b2	251
n1 and n2	399
TOTAL	1048

8. Total Probability Theorem. Bayes' Theorem

Example 13. Let us suppose your work involves repairing computers and you know from experience that the most common types of errors affect the screen (SC) in 15% of cases, the video card (VC) in 10%, the motherboard (MC) in 15%, are caused by viruses (VR) in 35% of the cases, or by problems with the software (SW) which occur in the remaining 25% of the malfunctions. Let us suppose that a computer is delivered to your workshop with the following symptoms: the computer turns on but the screen is blank. Where should you start looking for the fault?

A first approach to the solution of Example 13 may consist in starting looking for the fault from the most common failure (VR). In the long run, this strategy will be successful in 35% of the cases, but you will be looking in the wrong place in the remaining 65% of the cases. This strategy may be appropriate if you don't have any information about the fault. However, in this case there is a symptom (BS="blank screen") that can guide you to the most likely source of the fault. Your technical experience indicates that the BS symptom probabilities of each of the possible sources of failure are not now p(SC)=0.15, p(VC)=0.1, p(CM)=0.15, p(VR)=0.35 and p(SW)=0.25. The updated probabilities will be calculated in the view of the new (BS) data. That is, the new probabilities are p(SC/BS), p(VC/BS), p(MC/BS), p(VR/BS) and p(SW/BS). Figure 12(a) shows a plot of this situation. The certain event E= "the computer is broken" is divided into five mutually exclusive events E=SC∪VC∪MC∪VR∪SW. Figure 12(b) shows the event BS, which may overlap with each of the five events. This allows to break BS down into five incompatible events (in pairs) and finally to calculate p(BS):

$$BS=(BS\cap SC)\cup(BS\cap VC)\cup(BS\cap MC)\cup(BS\cap VR)\cup(BS\cap SW)$$
$$p(BS)=p(BS\cap SC)+p(BS\cap VC)+p(BS\cap MC)+p(BS\cap VR)+p(BS\cap SW)= \quad (8)$$
$$=p(SC)p(BS/SC)+p(VC)p(BS/VC)+p(MC)p(BS/MC)+p(VR)p(BS/VR)+$$
$$+p(SW)p(BS/SW)$$

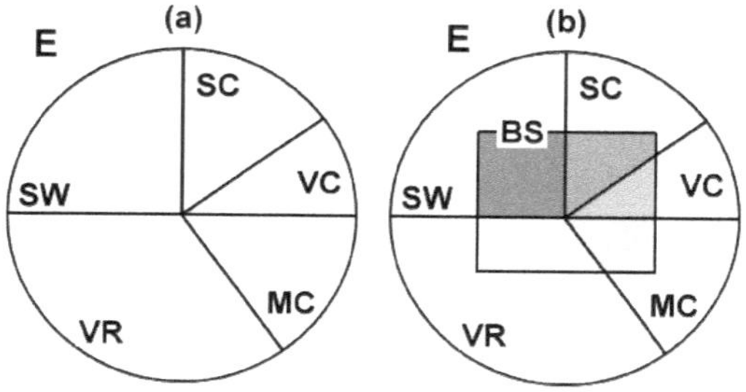

Figure 12. Breakdown of BS.

Let us suppose now that, in your experience as a technician, you know the BS symptom appears:

- In 45% of the cases in which the screen is faulty (p(BS/CS)=0.45).
- In 50% of the cases in which the video card is faulty (p(BS/BV)=0.5).
- In 10% of the cases in which the motherboard is faulty (p(BS/MC)=0.1).
- In 5% of the cases in which there is a virus problem (p(BS/VR)=0.05).
- In 15% of the cases in which there is a problem with the software (p(BS/SW)=0.15).

Using (8), p(BS)=(0.15)(0.45)+(0.1)(0.5)+(0.15)(0.1)+(0.35)(0.05)+(0.25)(0.15)=0.18. In other words, on average, 18% of the computers present the BS symptom, whatever the source of their fault.

Note, however, that what really interests us are the probabilities of the events SC, VC, MC, VR and SW, calculated in view of the new information BS. That is, we want to compute the values p(SC/BS), p(VC/BS), p(MC/BS), p(VR/BS) and p(SW/BS).

Exercise 18. Obtain an expression for p(SC/BS), p(VC/BS), p(MC/BS), p(VR/BS) and p(SW/BS). Calculate these values and decide where to start looking for the failure.

The equation (8) can be easily generalized, obtaining the *Total probability theorem*. The expression for obtaining the probabilities of the various possible reasons (SC, VC, MC, VR and SW in this case) derived from a known effect (in this case BS) is called Bayes' theorem. Let us formally enunciate these two useful results.

Theory-summary Table 5

Let $F_1, F_2, \ldots, F_n$ be events which form a partition of the space E, i.e.:

$$E=F_1 \cup F_2 \cup \ldots \cup F_n,\ F_i \cap F_j = \varnothing\ \forall i \neq j$$

Let F be an event. Then:

Total probability theorem:

$$p(F)=p(F_1)p(F/F_1)+\ldots+p(F_n)p(F/F_n) \tag{9}$$

Bayes' theorem:

$$p(F_i/F)=\frac{p(F_i \cap F)}{p(F)}=\frac{p(F_i)p(F/F_i)}{p(F_1)p(F/F_1)+\ldots+p(F_n)p(F/F_n)} \quad i=1,\ldots,n \tag{10}$$

Let us consider Example 13 in more detail and the results (9) and (10) which we have obtained from it. Given a faulty computer, the source of the malfunction is uncertain. There are five possible faults: SC, VC, MC, VR and SW. The hypothesis is that all faulty computers present one and only one of the failures. Thus, the certain event E="the computer is broken" is broken down into a set of five mutually exclusive events: $E=SC \cup VC \cup MC \cup VR \cup SW$. How we need to calculate the probability of the event BS= "black screen". Your experience as a computer repair technician tells you how likely it is to encounter a BS symptom. That is, you know the values p(BS/CS), p(BS/VC), p(BS/MC), p(BS/VR) and p(BS/SW). With this data, how can p(BS) be calculated? The idea is to write BS as a function of the events SC, VC, MC, VR, SW and then use the Total Probability Theorem (9) for calculating p(BS). Next use Bayes' theorem (10) to calculate the probabilities of the possible causes SC, VC, MC, VR and SW based on the known symptom BS. We performed these operations in (8).

We used this procedure previously. Consider Exercise 13 and its solution. From the standpoint of the results (9) and (10), in this case the partition of the certain event is $E=ACE_1 \cup KING_1$. Now the known symptom ACE_2 and the possible causes are the events $KING_1$ or ACE_1. The aim is to calculate the probability of case ACE_1 based on the known symptom ACE_2. That is, our goal is to calculate $p(ACE_1/ACE_2)$. Following an identical procedure to that used in Example 13, firstly we write ACE_2 in terms of ACE_1 and $KING_1$, next using (9) we calculate $p(ACE_2)$ and finally we use (10) to evaluate $p(ACE_1/ACE_2)$:

$$ACE_2=(ACE_2 \cap ACE_1) \cup (ACE_2 \cap KING_1)$$

$$p(ACE_2)=p(ACE_1)p(ACE_2/ACE_1)+p(KING_1)p(ACE_2/KING_1)=$$
$$=(1/2)(1/3)+(1/2)(2/3)=1/2$$

$$p(ACE_1/ACE_2)=p(ACE_1)p(ACE_2/ACE_1)/p(ACE_2)=(1/2)(1/3)/(1/2)=1/3$$

Exercise 19. Let us suppose there are two urns U1(9b,1n) and U2(1b,9n). An urn is chosen at random and a ball is drawn, which happens to be white. To which urn did the ball most likely belong?

9. Laplace's Rule: Combinatorial

Throughout Sections 2 and 3 we developed a simple procedure to calculate the probability of an event A. Firstly, we determined the sample space associated with the random experiment $E=\{S_1, S_2, ..., S_n\}$. After that we expressed the event A by the elementary events that form it. Let us suppose for example that the event A consists of the first k elementary events ($k \leq n$), $A=\{S_1, S_2, ..., S_k\}$ ($p(S_1)+\ldots+p(S_n)=1$). The value p(A) is calculated by adding the probabilities of the elementary events of A, i.e., $p(A)=p(S_1)+\ldots+p(S_k)$. However, we found that this procedure may be difficult or impossible to implement in many practical situations. This method of probability calculation requires us to determine all elementary events of E, all elementary events that form it and also to know the probability of each of them. This procedure can be viable for sample spaces which have few elementary events, such as the fudged roulette analyzed in Exercise 2. Remember the computer network in Example 6 which we discussed in Section 5: it was simply not possible to use the very laborious method of calculating probability.

However, Let us suppose that all elementary events have the same probability, i.e., Let us suppose there is equiprobability. Based on this hypothesis, how can you calculate the probability of an event A?

Exercise 20. Let us suppose that the n elementary events $E = \{S_1, S_2, ..., S_n\}$ are equally likely. Let A be $= \{S_1, S_2, ..., S_k\}$. Calculate p(A).

Theory-summary Table 6

Laplace's Rule:
Let us suppose that the n elementary events of the space E have the same probability (p). In this case p=1/n. Under this condition of equiprobability, it does not matter *what* elementary events form the event A, but *how many*. Assume that an event A comprises k elementary events. The value of k is called "number of cases favorable to the event A". The total number of elements in E is called "number of possible cases." Then, the value of p(A) is calculated as follows:
$$p(A)=\frac{k}{n}=\frac{\text{number of cases favorable to A}}{\text{number of possible cases}} \qquad (11)$$

According to Laplace's rule (11), also known as the classical conception of probability, the probability of an event A is equal to the ratio of the number of possibilities favorable to A and the total number of possibilities. Note that this rule applies only when each of the possibilities (elementary events) have the same probability, i.e., under the condition of equiprobability. A common use of this rule is for non-fraudulent gambling.

In some practical situations it is simple to account favorable and possible cases. For example, Let us suppose you flip a coin three times. What is the probability of the event A= "come up one heads at least once"? In this case, there are n=8 are possible cases, E={HHH, HHT, HTH, THH, HTT, THT, TTH, TTT}, of which k=7 are favorable to event A. Therefore p(A)=7/8. However, in many practical situations it would be very laborious to specify each of the favorable and possible cases. For example, a common lottery game in many countries rewards a sequence of six numbers selected in any order from 49. What is the probability of winning in this game? In this situation it would be very laborious to specify each of the possible ways to choose from 49 numbers 6 of them. Fortunately, to account for the possible cases we may use the so-called *combinatorial calculation* rules. To apply combinatorial calculation rules we start from a set C consisting of n different elements. The aim is to calculate the number of possible ways to choose m elements from the set of n elements. In practical applications of combinatory rules we must answer two questions: is the order in which the elements are chosen relevant? Can the same item be chosen more than once?

Exercise 21. In these situations carry out the following tasks: (1) determine the set C and the values of m and n; (2) determine whether the choosing order is relevant; (3) determine whether the elements can be chosen several times.

a) Five-digit numbers can be formed by rolling a dice five times.
b) Numbers that can be formed with 16 bits (0/1).
c) A jury will be selected with three people from a group of 20. The jury will consist of a president, a secretary and a member. Find the number of possible ways to choose the jury.
d) A restaurant offers a three-course menu to be chosen freely from a list of 10 choices. Find the number of possible three-course combinations.
e) A committee will choose three people from a group of 20. Find the number of possible 3-candidate combinations.
f) 10 runners take part in a race. Find the number of ways to cross the finishing line (assuming two runners never reach the tape simultaneously).

Below you can find a glossary of the terminology used in combinatorial counting rules. Consider the set C of n objects, from which you wish to choose m elements.

- *A variation with repetition* is a selection of items in which the order is relevant and the elements can be different or repeated. Two variations with repetition are equal if they are formed by the same elements and the elements are arranged in the same order. In Exercise 21(a), the numbers 32416, 23441 and 23414 are examples of different variations with repetition. The number of variations with repetition that can be formed is noted by $VR_{n,m}$ and it is calculated as follows:

  $$VR_{n,m}=n^m$$

- *A variation* is a selection of items in which the order is relevant and items cannot be chosen more than once. Two variations without repetition are equal if they are formed by the same elements and if the elements are arranged in the same order. In Exercise 21(c), the arrangements 13-6-17 and 13-17-6 are examples of different variations without repetition. The number of variations that can be formed without repetition is denoted $V_{n,m}$ and it is calculated as follows:

  $$V_{n,m}=n(n-1)(n-2)...(n-m+1)$$

- *A combination with repetition* is a selection of items in which the order is not relevant and an element can be chosen more than once. Two combinations with repetition are equal if they consist of the same elements, but are placed in a different order. In Exercise 21(d), the arrangements 8-8-2 and 3-7-1 are examples of different combinations with repetition. The number of combinations that can be formed with repetition is denoted $CR_{n,m}$ and it is calculated as follows:

$$CR_{n,m} = \binom{n+m-1}{m}$$

- *A combination* is a selection of items in which the order is not relevant and an element cannot be chosen more than once. Two combinations are equal if they are formed with the same elements and they are placed in a different order. In Exercise 21(e), 8-7-2 and 3-7-1 elections are examples of different combinations. The number of combinations that can be formed without repetition is denoted by $C_{n,m}$ and it is calculated as follows:

$$C_{n,m} = \binom{n}{m}$$

- *A permutation* is an arrangement of all elements of C. In Exercise 21(f) there are two examples of permutations 10-6-5-7-9-8-4-3-1 and 2-4-1-5-3-10-8-6-7-9 -2. Note that a permutation without repetition is a variation in which all the elements of the set C (i.e., m=n) are used. The number of permutations that can be formed is denoted P_n and it is calculated as follows:

$$P_n = V_{n,n} = n(n-1)(n-2)...(n-n+1) = n!$$

There is an additional situation in which the above combinatorial expressions are not applicable. Imagine having six green pieces cloth, four red, three blue, one white and one black. The task is to produce a horizontal tape joining the 15 multicolored cloths. In how many ways can it be done? It is assumed that the pieces of cloth of the same color are indistinguishable. So, that is to arrange the 15 items, knowing that six elements are indistinguishable (green pieces of cloth), four elements are indistinguishable (red pieces of cloth), three elements are indistinguishable (blue pieces of cloth), and finally there are two different additional elements (black and white pieces of cloth). Naturally, if two pieces of cloth of the same color are exchanged, the permutation is identical. These arrangements of elements are called *permutations with repetition*. Let us write it in more detail:

- Consider the set C comprising n elements, where n_1 elements are indistinguishable from each other, n_2 elements are indistinguishable, and so on where n_k are indistinguishable. Naturally, $n_1+n_2+\ldots+n_k=n$. A *permutation with repetition* is an arrangement of all elements of C. The number of permutations with repetition that can be formed is denoted $PR_n^{n_1,n_2,\ldots,n_k}$. In order to calculate this number, simply divide the total number of permutations (P_n) and the number of sorts that are obtained by permuting equal elements. That is:

$$PR_n^{n_1,n_2,\ldots,n_k} = \frac{P_n}{P_{n_1} P_{n_2} \ldots P_{n_k}}$$

Exercise 22. Using combinatorial calculation rules, calculate the number of possibilities for each of the scenarios for Exercise 21 and for the example of colored pieces of cloth.

Exercise 23. Explain why $V_{n,m}=C_{n,m}P_m$.

Exercise 24

a) We roll a dice five times. Calculate the probability of:
   a1) Obtaining 4, 2, 5, 6, 1.
   a2) Obtaining 4, 4, 5, 5, 5.

b) Complete a random test of 14 questions (YES/NO). Calculate the probability that an individual taking test gives as many YES as NO responses.

c) Letters a, b, c, d, f, g, h, i are randomly arranged. Calculate the probability that:
   c1) letters a, b, c, d are together and in that order.
   c2) letters a, b, c, d are together but in any order.

d) A restaurant offers a three course menu to be put together by choosing from a list of 10 choices. Let us suppose we choose a random menu. Calculate the probability that the three course choices are different.

e) If a lottery rewards a sequence of 6 numbers chosen at random from a list of 49, what is the probability of winning?

f) A die is rolled three times and scores are added. The 9 and 10 sums can be obtained in six different ways. For 9 the possibilities are (621), (531), (522), (441), (432) and (333); for 10 they are (631), (622), (541), (532), (442) and (433). Thus, the probability of obtaining 9 is equal to the probability of obtaining 10. Do you agree with this statement?

Exercise 25. A common lottery game in many countries rewards a sequence of six numbers selected in any order from numbers 1 to 49. We know that all sequences are equally likely. But if one examines the historical data, probably one will find that the sequence 1-2-3-4-5-6 has never come up. How do you explain it?

Exercise 26. Let us suppose we have a drum containing ten numbered balls, from 0 to 9. We shake the drum, draw a ball, write down its number and place it back in the drum. The same procedure is repeated five times. Look at these three possible outcomes: 22211, 12345 and 83056. Which one of them seems more likely? Which one seems less likely? Does the situation change at all if the drum contains 100,000 numbers and three are drawn?

Exercise 27

a) We have five cards numbered from 1 to 5. They are arranged randomly. Which of the following arrangements is most likely: 1-2-3-4-5 or 3-5-4-1-2?
b) We have five cards printed with the following symbols 🗁,🖹,🗏,🗐,🗄. They are arranged randomly. Which of the following arrangements is most likely: 🗁-🖹-🗏-🗐-🗄 or 🗏-🗄-🗐-🗁-🖹?

10. Axiomatic of Probability

The French mathematician Pierre Simon de Laplace (1749–1827) carried out the first rigorous attempt to define probability (equation (11)), although the idea of measuring the probability of an event as the ratio of favorable cases to possible ones is older. However, this classical conception of probability leads to significant problems. If the probability of an event that can happen only in a finite number of modes is the ratio of the number of favorable cases to the number of possible cases, then the scope of probability is quite narrow. It is tantamount to gambling. Moreover, Laplace knew that for using this rule, all results should be equally likely. Nowadays we call this condition the *equiprobability hypothesis*. The problem is that the concept of probability is used in Laplace' definition of probability. It is what is called a *circular definition*, which is invalid because a concept is defined using the concept that one aims to define.

On the other hand, the frequency conception of probability also causes different problems. Firstly, the value of probability cannot be calculated using relative frequency, but just an approximation of it. As soon as we perform another sequence of experiments, the value of the relative frequency changes, so what is the value of the probability? In addition, we cannot be certain that by increasing the sample size the relative frequency will be closer to probability (see Example 9). For example, if one flips an equally balanced coin many times, it seems that one can expect the proportion of heads to approach 0.5. However, it is possible with a large sample that one might obtain a frequency far from 0.5. We cannot be certain that the relative frequency will be a convergent sequence to the probability value. What kind of capricious convergence is that?

What does probability mean? For some, the probability of an event A is shorthand for the percentage of times that the event A occurs. Others have suggested that probability is simply a matter of subjective belief, an expression of a personal opinion. In this latter view, probability is interpreted as the degree of belief or conviction about whether or not an event will occur. This would be a personal judgment of an unpredictable phenomenon. Let us consider for example a technician who repairs computers. Given certain symptoms of failure, the technician can be guided by his/her intuition about what is the cause of the failure. However, this subjective conception of probability also poses difficulties because two people can assess probabilities differently. In summary, there are different ways of understanding the meaning of probability, but they all pose challenges.

What is the position of mathematicians on this subject? They have developed an abstract theory for the calculus of probability. This means that probability is defined, managed and calculated without giving a particular interpretation. To understand what this theory is about, you will find a conversation below, that you might have had with the Russian mathematician Andrey Nikolayevich Kolmogorov, who in 1933 proposed this abstract formulation of probability theory.

Conversation with AN Kolmogorov

Kolmogorov—I will try to explain the details of our abstract theory of probability. Abstract means that it is not associated with any particular interpretation of probability. That means you can use this theory to calculate the probability of an event and then interpret the value in the way you find most convenient. Let us start with three definitions:

Definition 1: We call the sample space of possible outcomes of a random experiment' set E.

Definition 2: If E is the sample space, we will call the event to any subset A of E.

You—I've handled these concepts and I know what they mean. I see nothing new.

Kolmogorov—The novelty is that E may now be an infinite set. Until now, you have only handled finite sample spaces. Complex, but in real-life situations, the sample space may be continuous, such as the three-dimensional position in space. Imagine for example that we are analyzing the height or the weight of a population. The set of possible values of these variables is not finite. Let us suppose, for example, that the stature of the people of the population is between 1.3 m and 1.85 m. Any value between 1.3 and 1.85 is a possible outcome for the height of a randomly selected person.

You—I have a question. You mean that any value in the interval [1.3,1.85] is a possible outcome that can be obtained by measuring the height of a person of that population. But the instrument we use to measure heights has limited accuracy (inches or millimeters). The measuring instrument only provides a finite set of possible measurements. So, we could choose a finite sample space and apply what we have studied so far. Is not that so?

Kolmogorov—I see that you are very quick. Admittedly, the set of possible outcomes that the instrument can give is always finite. The variable H="stature of a person" runs continuously in this case the interval [1.3,1.85], but in practice we can only measure a finite number of values within this range. However, only in very few practical situations will we be interested in calculating the probability that the variable reaches an exactly certain value. For example, we would seldom be interested in calculating the probability of events such as "H=1.458" or "H=1.743". It is more useful to calculate the probability that the height H of a person falls in a range, for example between 1.45 m and 1.75 m ($p(1.45 \leq H \leq 1.75)$). In order to analyze a variable which takes its values in an interval, what we need is a theoretical model to handle the continuity.

You—Okay. What else?

Kolmogorov—Once you have defined the sample space E, an event A is simply a subset of E. This idea of managing events as subsets of the sample space is not new. After that you must find a way to calculate a probability p value for each event A. This will be called probability value of A and will be noted as p(A). Now...

You—[Kolmogorov Interrupting] Wait, wait. Are you telling me that I must be the one who finds a way to calculate the probability p(A) of each event A?

Kolmogorov—Right. I leave you in charge of the task of designing a way which associates each event A with a value p(A), which we will call the probability of A. Do not panic, because you have already done it before. For example, when using Laplace' rule to calculate the probability of each event A you divided the number of favorable cases and the number of possible cases. Note that this rule is simply a way to associate a value p(A) to each event A. What happens is that Laplace' rule is not valid in many situations. For example, it can only be used under equiprobability of all elementary events. And, of course, it cannot be used when the sample space is infinite. I mean that your job is to find in each problem how to calculate p(A) for each event A, because there is no way to do that using a valid procedure for all problems. For example, consider the variable R="stature of a person". This is not the same as considering the variable N="Sum of side scores when rolling two dice".

You—I do understand. I should find a way to associate each event A to its p(A) value. But is this not precisely the most difficult issue? In what way does this probability theory ease my job?

Kolmogorov—It may be that to find a way to measure the probability of each event A is the most important hurdle that you must overcome. Among other things it depends on the interpretation given to the probability. This theory of probability in general does not define how to calculate the probability of each event A. However, it makes it easy to perform the calculation of p(A) for complex events. For example, if you have already decided which values to assign to p(A), p(B) and $p(A \cap B)$, then this theory will tell you that you can directly calculate the value of $p(A \cup B)$ as follows:

$$p(A \cup B) = p(A) + p(B) - p(A \cap B)$$

You—But I already knew this formula.

Kolmogorov—Yes, but you only knew the frequency interpretation of probability. And it could only be used in sample spaces that have a finite number of elements. From now on you can always use it regardless of the sample space and of the interpretation that you make of the probability.

You—It is hard to understand that any of us can come up with a different way of assessing the probability. Thus, there is no "probability" but "possible probabilities".

Kolmogorov—You are totally right my friend.

You—And what if a colleague and I do not agree on how to calculate the probability value?

Kolmogorov—What happens if your colleague and you do not speak the same language? If you wish to talk, both of you will learn a common language. You will have to agree on the meaning that both will give to probability. Anyway, if you and your colleague construct various probabilistic models to study the same real phenomenon, you may compare both models experimentally to see which one can make more accurate predictions.

But do not worry; there is a range of widely used procedures that evaluate probability. You and your colleague can use these procedures in most situations. One such method is, of course, Laplace' rule. which is valid in many situations, although in many others it is not.

You—But there are so many different ways to measure probability...

Kolmogorov—This is not about organizing a competition to find the most bizarre way of measuring probability. The point is that in each application, in each practical case, we need to choose the most appropriate way to measure probability, because it allows us to make decisions. In short, a method that allows us to make predictions successfully. Or does it seem to you that reality is so simple that one mathematical model will be sufficient to handle uncertainty for all situations? Believe me, there are many different interesting situations quite unlike those you have handled. [Kolmogorov takes a paper and with rapid strokes draws Fig. 13]

The entire area E may be a metal plate on which rust is deposited at random, or an engine part subjected to fatigue, in which a crack may appear; or a geographical area also contaminated by a toxic substance which is dispersed at random. In all these cases we are interested in estimating the probability of any sub-region A.

If we study the rusted plate, p(A) may be the likelihood that on area A an excessive amount of oxide is deposited. If we studied the risk of fissures appearing on a piece, p(A) may be the probability that the area A may show a fissure during the first 1000 hours of operation of the part. If it was a contaminated geographical area, p(A) may be the probability that the nucleus of population A reaches a dangerous contamination level in 24 hours; the areas of highest probability A will have to be evacuated urgently.

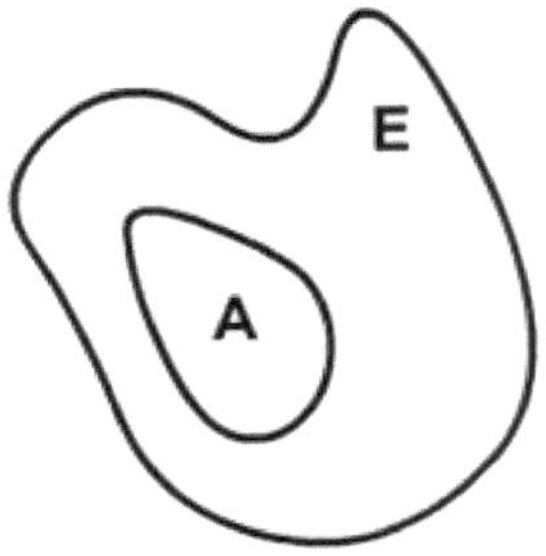

Figure 13. Example of a sample space and an event.

You should understand that in each of these examples it will be necessary to study separately how the probability of each region A can be assessed. And you should also understand that no Supreme Intelligence tells us what criteria should be used. It will be your job to use the available data to build a predictive model that allows you to make decisions.

In addition, the frequency interpretation of probability can be very useful in many cases. In the example of the polluted area, you should understand that it would not be very popular to deliberately contaminate region E a large number of times to find out in which areas A will be more dangerous, in order to figure out how to act in future pollution-related accidents.

What I would like you to understand is that once we have decided how to calculate probability, we can use a theory of probability that is able to encompass many ways of interpreting probability: frequency, counting possible and favorable cases, and also a subjective interpretation of probability. Let me ask you a tricky question. Would you buy the lottery number 44444?

You—Well...

Kolmogorov—I do understand. You have assigned a tiny value to p(44444). You have not evaluated this probability numerically, naturally, you do not have an exact value for p(44444). However, you believe this value is much lower than p(83578), for example. You believe that the number 44444 is less common than the number 83578.

You—But although sometimes I get carried away by hunches, for aversions to certain combinations of numbers, or by fortune-tellers' advices, I know that if five random numbers are drawn many times, the combination 44444 appears with the same frequency as any other combination, e.g., 83578. Each combination appears on an average once every 100,000 times. Therefore, I know that every combination is equally likely.

Kolmogorov—Indeed. When you buy a lottery ticket either being advised by any of the mentioned ways, or formally evaluating the probability $p(44444)=1/VR_{10.5}=1/100000=0.00001$, I would like to show you that the models you are using to account for award prospects are two different ones: the subjective model and the frequency model.

You—I insist that I knew that if you run the experiment to draw five numbers a large number of times, the relative frequency of each number will be about the same. This is experimentally testable.

Kolmogorov—Right. So the second model, the frequency one is appropriate for experiments that can be repeated indefinitely. But I'm sure you will continue to reject lottery numbers such as 44444 or 12121 despite knowing that the relative frequency of any number will be about the same after running the experiment many times.

You—Well, Mr. Kolmogorov, you were explaining to me the calculus theory of probability that mathematicians have defined and which is not limited to any particular way of interpreting probability.

Kolmogorov—I said that you should take care to specify the method of calculating p(A) for each event A. And in return, our theory will provide the means to simplify the calculation, plus many useful results. Actually, the machinery of probability starts once it has been specified how to calculate p(A) for each event A.

You—I must figure out how to calculate p(A) for every A. Well, in the examples I have worked on so far this is not very difficult. For example, if I am flipping an evenly balanced coin twice, I proceed as follows:

1) I set the hypothesis of equiprobability.
2) E={CC, CX, XC, XX}
3) I set the probability values p(CC)=p(CX)=p(XC)=p(XX)=1/4

I know this is a good model. Using it I can predict the result of flipping the coin a large number of times with great accuracy. But I wonder if any way is valid to define the probability function.

Kolmogorov—Not all ways are valid. We require a minimum for the probability function. The requirements are only three, which in Mathematics we usually call axioms. Every profession has its jargon. Here are the three axioms A1, A2 and A3 on which probability theory[3] is based:

A1: $p(A) \geq 0$ for any event A

A2: $p(E) = 1$

A3: $p(A \cup B) = p(A) + p(B)$ if $A \cup B = \varnothing$

---

[3] Up to now we have considered that an event A is any subset of the sample space E (see Theory-summary Table 1). This idea has worked well for finite sample spaces. If the sample space E is finite, the number of possible subsets of E is also finite (see Exercise 4) and to define a probability on E we just need to specify how to calculate the value of p(A) for each of them. However, if the sample space is infinite, it might be difficult or impossible to specify how to calculate p(A) for each possible subset A of E. The solution is to define the probability p(A) only for subsets of E at your discretion. We call events just the chosen ones and define the probability for them only. We are free to set the collection of events Σ for which we will define the probability, but this Σ collection should meet minimum requirements. In the language of set theory, the requirements are that Σ should form a σ-algebra of E. The three conditions which Σ must meet to be a σ-algebra of E are:
1) $\varnothing \in \Sigma$
2) If $A \in \Sigma$, then $\overline{A} \in \Sigma$
3) If $A_i \in \Sigma$ for i=1,2,3..., then $A_1 \cup A_2 \cup A_3 \cup \ldots \in \Sigma$
The structure (E, Σ, P) is called probabilistic space. It is easy to find the justification for these three minimum conditions which Σ must meet. This is so that the set operations between events result in sets that are also events. For example, if Σ is a σ-algebra of E, given A, $B \in \Sigma$, then also $A \cap B \in \Sigma$. Demonstrate that this statement is true.
Finally, note an important additional detail. The condition 3) requires that the union of any countable infinite collection of events is also an event. Instead, in order to apply the axiom A3 it would be sufficient to ensure that if A, $B \in \Sigma$ then also $A \cap B \in \Sigma$. Why is condition 3) more demanding than necessary? Indeed, the axiom A3 may be written in a more general version, which is:
A3: $p(A_1 \cup A_2 \cup A_3 \ldots) = p(A_1) + p(A_2) + p(A_3) + \ldots \quad A_i \cap A_j = \varnothing \; \forall i \neq j$

You—I am disappointed. I also knew these three properties.

Kolmogorov—I may have disappointed you because these are very basic axioms. But the objective is precisely that, to require a minimum number of conditions. In addition, just remember that they are not properties, but axioms. The axioms in any mathematical theory are agreed minimum conditions required to make a formal statement. You call them properties because you demonstrated that they were true. But before that you interpreted probability as a frequency.

You—I think I understand what you mean. Working with the frequency interpretation of probability, I derived an idea of the meaning of probability and afterwards I demonstrated A1, A2 and A3. I also demonstrated other properties such as $p(A \cup B)=p(A)+p(B)-p(A \cap B)$.

Kolmogorov—The difference is that A1, A2 and A3 are now not demonstrable. These are conditions that we agree to impose on the probability function. The rest are properties that will be demonstrable from A1, A2 and A3, all are derived from the axioms. For example, from A1, A2 and A3 it can be proved that $p(A \cup B)=p(A)+p(B)-p(A \cap B)$ and also that $p(\overline{A})=1-p(A)$.

You—But, why these three axioms? Why not others? For example, Let us suppose that I choose B1, B2 and B3 as alternative axioms where:

B1: $p(\overline{A}) = 1 - p(A)$

B2: $p(\varnothing) = 0$

B3: $p(E) = 1$

Kolmogorov—This system is redundant, i.e., some of the conditions can be deduced from the others. Note that we can deduce B2 from B1 and B3. In addition, from B1 and B2 we can deduce B3.

Exercise 28. Prove that Kolmogorov is right.

My friend, what you have chosen is not an axiomatic system. It is not easy to build a good axiomatic system. An axiom is a requirement. Therefore, the number of axioms should be as small as possible. In addition, an axiom cannot be inferred from the others (as in your choice) because in that case it would be a property, not an axiom. In your proposed axiomatic system B1 and B3 are axioms and B2 a property. Or B1 and B2 are axioms and B3 a property.

You—Then the axiomatic system A1, A2 and A3 is really good. It requires few conditions and provides many properties. I will list the ones I know. Let us suppose that A and B are events. Thus:

P1: $p(\varnothing)=0$

P2: $0 \leq p(A) \leq 1$

P3: $p(\bar{A})=1-p(A)$

P4: $p(A \cup B)=p(A)+p(B)-p(A \cap B)$

P5: $p(A/B)=p(A \cap B)/p(B)$

P6: Let us suppose that the sample space E is a finite set and that its n elements have the same probability p. Then:

P6a: $p=1/n$

P6b: p(A) is the quotient between the number of elements that form A (favorable cases) and the number k of elements of E (possible cases).

Kolmogorov—Not only can you list the properties P1 to P6 but you can also demonstrate them using only A1, A2 and A3. If you do not wish to take on this job or you do not consider yourself capable of doing it, you may study the demonstration made by another colleague. However, it is an interesting exercise to try to find the proof of a theorem by yourself, even if you do not succeed. Can you prove P1 to P4 now?

Exercise 29. Demonstrate P1 to P4. Remember: your tools to demonstrate these properties are just A1, A2 and A3.

You—But I feel somehow disappointed. The demands of A1, A2 and A3 are minimal, so that anyone can invent probability calculation functions that fulfill these axioms. Therefore, the discussions can last forever until we come to an agreement about what is the best way to evaluate probability. But what is less rewarding is that it seems that the frequency interpretation of probability has faded, has lost prominence. This axiomatic interpretation of the term «relative frequency» has vanished.

Kolmogorov—In the axiomatic the term relative frequency is missing, and either this or any other method is suggested to assess probability, simply because there is no exclusive way. That is the idea, to build a theory that includes many possible interpretations of probability.

But I have good news for you: there is a very important property the "law of large numbers" that is deduced from our axiomatic system and presents a formidable argument supporting (under some conditions) the frequency interpretation of probability that you seem to like.

You—If I am right, this is an additional property, demonstrable in the axiomatic system A1, A2, A3. I am eager to know about this law.

Kolmogorov—Well, here goes. Please pay attention because it is not easy to understand. Let us suppose that we are able to repeat a random experiment indefinitely, under the same conditions. This is the requirement. If this assumption is not met, the law of large numbers is not valid.

You—I will often be able to repeat the random experiment practically under the same conditions, for example in gambling, and also if I take a random sample of items.

Kolmogorov—Well, suppose that you perform any of the experiments roll a dice or inspect an article. Additionally, suppose that A is a possible event. In the examples, may be A="Get the number 3" (rolling a dice) and A="The part is faulty" (inspecting an article). We are only interested in whether the event A occurs in each repetition of the experiment.

Let us denote $f_n(A)$ the relative frequency of the event A when the experiment is performed n times. You are convinced that the relative frequency $f_n(A)$ gets closer to p(A) as n increases. Well, the Law of Large Numbers states that it is "almost" true. I say "almost" because certainty exists only for the certain event or for the empty event.

We repeat the random experiment n times and calculate $f_n(A)$. Surely you do not need reminding, but $f_n(A)$ is a random value, while p(A) is a constant value. The question is, how close will $f_n(A)$ be to p(A)? Since $f_n(A)$ is a random number, the distance from $f_n(A)$ to p(A) is random, it depends on the sample. We should not expect that the value $f_n(A)$ is exactly the same as the value p(A). However, this may occur in a specific sample [Kolmogorov draws Fig. 14 and shows it to you].

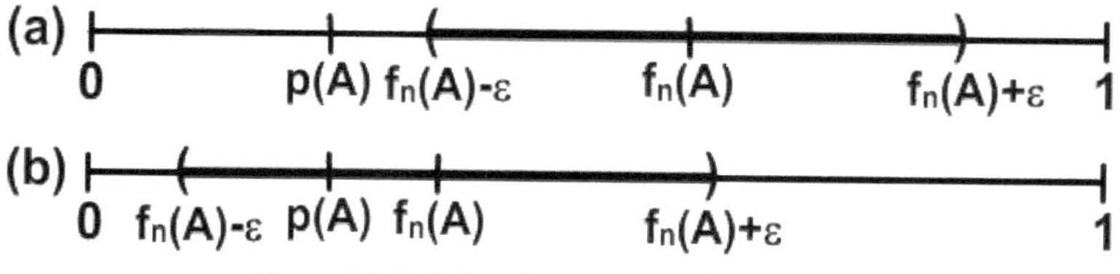

Figure 14. Relative frequency and probability.

Please observe Figs. 14(a) and 14(b). We have chosen an interval centered at $f_n(A)$ and with radius $\varepsilon > 0$. In Fig. 14(a) the interval $(f_n(A) - \varepsilon, f_n(A) + \varepsilon)$ does not contain the value p(A). However, if another sample of the same size is chosen, p(A) may be included, as in Fig. 14(b). For example, Let us suppose we flip a coin 300 times. Let us make the hypothesis that p(C)=0.5. Consider the random interval $(f_{300}(C) - 0.04, p_{300}(C) + 0.04)$, i.e., we chose the radius $\varepsilon = 0.04$. We carried out two series of 300 flips and suppose the outcomes were for the first series 177 heads and for the second 141 heads. The relative frequencies are 177/300=0.59 and 141/300=0.47. The intervals are (0.59–0.04,0.59+0.04)=(0.55,0.63) and (0.47–0.04,0.47+0.04)= (0.43, 0.51).

The first interval contains the value p(C)=0.5 but the second does not. Of course, if we had chosen a different value of ε, the result would be different. For example, if ε=0.1, both intervals include the value p(C) = 0.5. However, if ε=0.0001, none of them include p(C).

We cannot predict whether the interval $(f_n(A) - \varepsilon, f_n(A) + \varepsilon)$ will include the value p(A) or not. This will occur with a specific probability. Nor do we know the value of that probability. That is, we do not know the probability of the event $p(A) \in (f_n(A) - \varepsilon, f_n(A) + \varepsilon)$. Well, listen to what the law of large numbers says:

No matter how small the value of $\varepsilon > 0$ is, as we increase the sample size n, the probability of success $p(A) \in (f_n(A) - \varepsilon, f_n(A) + \varepsilon)$ gets closer to 1.

Let me explain it in another way:

No matter how small (ε=0.1, ε=0.01, ε=0.001, ...) is the radius $\varepsilon > 0$ of the interval $(f_n(A) - \varepsilon, f_n(A) + \varepsilon)$; even if the value of q is as close to 1 as we wish (q=0.8, q=0.9, q=0.99, q=0.999, ...), it is possible to choose a sample size n large enough such that the probability of the event $p(A) \in (f_n(A) - \varepsilon, f_n(A) + \varepsilon)$ is even greater than q.

You—I mean that if I wanted, for example, in the long-term, that more than 80% of the intervals $(f_n(A) - \varepsilon, f_n(A) + \varepsilon)$ contained the value p(A)...

Kolmogorov—... you just need a sample large enough. It is not certain that all p(A) are in the interval $(f_n(A)-\varepsilon, f_n(A)+\varepsilon)$, but you can ensure with a probability greater than 0.8 that this will be the case. In the long term, at least 80% of the intervals $(f_n(A)-\varepsilon, f_n(A)+\varepsilon)$ will include the value p(A). You can ensure "with any degree of certainty" that the relative frequency $f_n(A)$ is "close" to the value p(A). No matter how demanding you are in the definitions of "certainty" and "proximity". All that is needed is to repeat the experiment a large enough number of times and you will achieve this "certainty" and that "proximity".

You—I see. The "degree of certainty" is the value q, the "proximity" is the value ε. I think I have grasped it, but I need an example. Could you give me one?

Kolmogorov—Sure. We return to the example of the coin. If the coin is fair, the probability of getting heads is p(HEADS) = 0.5. But in general, the value of p(HEADS) is an unknown value that can be estimated by the relative frequency f_n(HEADS). For example, if you flip the coin 150 times you may obtain a relative frequency f_{150}(HEADS) = 0.48, and therefore the estimate is p(HEADS)$\approx$0.48. In general, the distance between the relative frequency f_n(HEADS) and the number p(HEADS) cannot be predicted, it varies with each series of flips. Now, we look at a small interval centered at f_n(HEADS), for example (f_n(HEADS) –0.04, f_n(HEADS) +0.04). If we flip the coin n times, this interval may contain the value p(HEADS) or it may not. We cannot predict whether this will happen, but we can get an idea of how likely it is. What is the probability that the interval, (f_n(HEADS) –0.04, f_n(HEADS) +0.04), "captures" the real value of p(HEADS)? Well, as the value of n increases, this probability also increases. We can make this probability as large as necessary, simply by increasing n. Thus, there exists a value of n for which this probability will be greater than 0.7. And there will be another value of n such that this probability will be greater than 0.85. No matter how close to one the q value is, it is always possible to find a sample size n such that the probability that the interval, (f_n(HEADS) –0.04,f_n(HEADS) +0.04), includes the value p(HEADS) is still greater than q.

You—Let me see if I understand you. Firstly, I choose a sample size n large enough so that the probability of success, p(A)∈(f_n(HEADS)-0.04,f_n(HEADS)+0.04), is greater than 0.85. Then I flip the coin n times and Let us say that for example I get f_n(HEADS)=0.48. So the interval is, (f_n(HEADS)-0.04,f_n(HEADS)+0.04)=(0.48-0.04,0.48+0.04)=(0.44,0.52). I do not know the exact value of p(HEADS), but the range (0.44,0.52) has a probability of at least 0.85 of containing the true value of p(HEADS).

Kolmogorov—Wrong! If you refer particularly to the interval (0.44,0.52), we cannot talk about the frequency interpretation of probability. It is as if you drew a ball from a drum and tried to calculate the probability of that specific ball to be white. This method is about analyzing a regular outcome which happens when a random experiment is planned to be run many times under the same conditions.

You—Sorry! I meant that in the long run at least 85% of intervals which follow the rule (f_n(HEADS)–0.04,f_n(HEADS)+0.04), will include the unknown value of p(HEADS). I cannot tell if the interval (0.44,0.52) is within the set of intervals that do contain the value of p(HEADS).

Kolmogorov—That is better. We do not know the value of p(HEADS), but the best estimate you have of p(HEADS) is the calculated relative frequency, p(HEADS)≈0.48. The interval (0.44,0.52) is called *the confidence interval* and the value of q=0.85 *the confidence level.* You are confident that p(HEADS)∈(0.44,0.52), and that confidence relies on the fact that more than 85% of samples of size n confirm that, p(HEADS)∈(f_n(HEADS) –0.04,f_n(HEADS)+0.04).

I have shown the relevance of the sample size to ensure that the relative frequency f_n(A) of an event A is close to the probability p(A) with a degree of confidence q and an accuracy of ε. However, in real life, people do not carry out hundreds or thousands of tests, and generally we do not have such a broad experimental reference for decision making. Many times we assume that a sample or a series of tests is representative of a certain situation, when it really is too small to be reliable.[4]

You—Okay, but how does one determine the sample size n? That is, supposing you have chosen a radius $\varepsilon > 0$ for the confidence interval, and a confidence level q so that $0<q<1$. Now I need to know how many times I should run the experiment to be sure that the true value of p(A) is within more than 100q% of the confidence intervals $(f_n(A)-\varepsilon, f_n(A)+\varepsilon)$. Is there a way to calculate n?

Kolmogorov—Absolutely. But to understand how to determine the sample size you will need to know an important probability distribution, called normal or Gaussian distribution. Be patient and keep working.[5]

You—Okay. I believe the Law of Large Numbers is a strong support for the concept of frequency probability. I think I begin to understand the meaning of probability.

Kolmogorov—The Law of Large Numbers was discovered by the Swiss mathematician Jakob Bernoulli (1654–1705) after twenty years of constant efforts. It was revealed in a posthumous work published in 1713. Jakob and I often have long conversations on the subject and we have not managed to agree on the real meaning of probability. But we keep going.[6]

Exercise 30. Write formally the Law of Large Numbers (this formal statement fits in a row).

11. Computer Activities

RANDBETWEEN( ) and RAND( ) functions are built into EXCEL to assist in generating random numbers, which can design complex random experiments. The function RANDBETWEEN(Bottom, Top) generates with equiprobability a random integer in the range from Bottom to Top. The RAND( ) function generates a random real uniformly distributed in the

---

[4] In the computer activity 2 you may check that the relative frequency of small samples has a large variability. Relative frequency stability is achieved only for sufficiently large samples.

[5] In computer activity 3 you may experiment with ε and q values, and obtain the value n.

[6] Andrey Nikolayevich Kolmogorov died in Moscow in 1987 at the age of 84. He made contributions to many fields of mathematics, including topology, approximation theory, functional analysis and the history and methodology of mathematics. Please take this imaginary conversation as a modest tribute to his work.

interval [0,1]. A uniform distribution The term «distribution uniform» means that the generated random numbers using RAND( ) do not have any tendency to accumulate on any area of [0,1]. For example, if you use RAND( ) to generate a large number of random values and divide the interval [0,1] into five sub-intervals of length 1/5, you could check that about 20% of the numbers fall in each of the intervals, thus the associated histogram is "flat". If you just want to generate a random value in an arbitrary interval [a,b], it is sufficient to evaluate the expression RAND( )*(b-a)+a.

The RAND( ) function can be used to simulate the occurrence of an event A with known probability p(A). The procedure is as follows. If $0 \le \text{RAND}() < p(A)$, we say that the event A has occurred, and otherwise we say that the event never occurred. Following the same logic, we can simulate the outcome of a finite-sample-space random experiment. Let us suppose that $E=\{S_1,S_2,S_3\}$ and that $p_1=p(S_1)$, $p_2=p(S_2)$, $p_3=1-p_1-p_2$ values are known. Then, generating a value U=RAND( ), the result of the experiment is: S_1 if $0 \le U < p(S_1)$; S_2 if $p(S_1) \le U < p(S_1)+p(S_2)$; otherwise, the result is S_3.

Task 1. Who built the Egyptian Pyramids?

The spreadsheet titled **random numbers** includes the following random sequence generators:

(0-1) digits: Generates a page of zeros and ones.

(1-49) digits: Generates a page of integers between 1 and 49.

(0-9) digits: Generates a page of integers between 0 and 9.

(A-Z) letters: Generates a page of characters, from A to Z, enabling one to establish the probability that the generated character is a vowel.

The proposed activity is as follows. Pressing <F9> generate pages and pages of random sequences and find any part that deserves to be highlighted, that seems bizarre for its arrangement, shape, symmetry, subjective meaning, etc. For example, you may find your own name on a page of random letters or the shape of the *Enterprise* in the binary matrix of (1–0) digits. This backs up the idea that a random sequence can display symmetries, patterns and family structures, and that this does not indicate that the "bizarre" sequence has been originated deliberately by someone nor that it is a mysterious message written by chance.

You may accept the idea that chance could generate fragments in which you recognize familiar structures. We recognize the shape of an animal in a cloud, a face in a rock formation or the silhouette of a stranger in a shade. Our mind is very efficient at recognizing something familiar in disorder, if only vaguely.[7] However, it is more difficult to accept that a previously elected «bizarre» sequence is more likely than a specific «messy» sequence. For example, you may think that by flipping a coin six times, the HHHHH sequence is less common than the HHTHTT. The activity that we propose below is to compare these wrong intuitions against experimental data. The spreadsheet called **coin** simulates 20,000 sequences of six coin flips and then counts the number of times that a set sequence has been obtained. Investigate whether a "bizarre" sequence really appears at random less frequently than other "messy" sequences. Using the probability value $(1/2^6)$, predict the number of times you will get such sequences. Check how well the prediction worked.

Task 2. Random experiment

The spreadsheet **random experiment** contains a simulator that reproduces a randomized experiment with four elementary events S_1, S_2, S_3 and S_4, whose probabilities are known. The simulator repeats the random experiment a set number of times and calculates the number of occurrences of each elementary event, showing the percentages obtained in a pie chart.

a) Consider a random experiment with four elementary events (cards, dice, roulette, polls,...). Calculate the probability of each of them. Using the simulator, estimate the probability of each elementary event and verify that you get the expected values. Start by taking a small number of replications of the experiment and gradually increasing it. Check that if the sample size is small, the estimated probability values are highly variable and are not useful. Repeatedly press <F9> to generate new simulations of the same size and check on the pie chart that the percentages show large swings. However, if the sample size is large enough, you can check that the number of occurrences of each elementary event can be accurately predicted.
b) Calculate the probability of the event S_1 if it is known that the event S_3 is not verified. Estimate this probability value using the simulator. HINT: Note that the hidden concept is conditional probability. To estimate by simulation the value of $p(S_1/\{S_1, S_2, S_4\})$, first count the number of times that any of the elementary events S_1, S_2 or S_4 have occurred. Next, within these occurrences, count the number of times that S_1 has happened.

[7] The renowned Martin Gardner (1914–2010), an American science writer and philosopher, devoted much effort to exposing scientific fraud and pseudo-scientific mumbo jumbo. In The Great Stone Face (Gardner 1988) he wrote how a photograph of the surface of Mars taken by the Viking satellite in 1976, which seemed to show a human face, had inspired considerable literature and pseudo-scientific and fiction films. Subsequent high-resolution photographs taken by NASA's Mars Global Surveyor, showed that the Great Stone Face was just a rock formation photographed in low resolution by chance from a certain angle, which led to such fanciful interpretation. Gardner wrote similar stories in the same book.

Task 3. Law of Large Numbers

The **LLN** spreadsheet allows you to explore the meaning of the Law of Large Numbers in greater depth. As explained in Section 10, this law states that the probability of success $p(A) \in (f_n(A)-\varepsilon, f_n(A)+\varepsilon)$ can be made as close to unity as desired, merely by increasing the size of n.

a) Choose two separate values for the radius of the confidence interval ($\varepsilon > 0$) and the confidence level ($q \in (0,1)$). Experimentally estimate the sample size n that must be chosen to ensure that the event, $p(A) \in (f_n(A)-\varepsilon, f_n(A)+\varepsilon)$, occurs with a probability greater than q. For example, if you choose n=500 and $\varepsilon=0.02$, you will find that the value of p(A) is more than 60% of the confidence intervals of the form $(f_n(A)-\varepsilon, f_n(A)+\varepsilon)$. In addition, depending on the value of p(A), the reliability can be much higher than 60%.
b) Experiment with different values of ε and q and analyze what effect a change in these parameters has on the value of n.
c) Analyze experimentally the following statement: "If the radius ε of the confidence interval is halved, then the associated sample size n is multiplied by four."

Task 3. Bayesian Magic

To perform this task you need an audience, as well as four black and four white balls. You should ask a partner to stand with his back to you, to take four of the eight balls and to put them in a bag. You just know that the bag contains four balls, but do not know how many there are of each color. Note that there are five possible compositions of the bag, from 0-4 to 4-0.

With a solemn gesture, announce to the audience that you plan to guess how many balls of each color there are in the bag.

You will need to ask your partner to repeat the following operation at times: randomly draw a ball, show the color and replace the ball in the bag. Using Bayes' theorem (10), recalculate the probability of each of the five possible compositions of the bag according to new data. You can start with equiprobability in the composition of the drum (p(0-4) = p(1-3) = p(2-2) = p(3-1) = p(4-0) = 1/5) or you may want to use your intuition and surmise that the partner will choose two balls of each color, in which case the values might be p(0-4) = 0.05, p(1-3) = 0.05, p(2-2) = 0.8, p(3-1) = 0.05, p(4-0) = 0.05. No matter the initial estimate of the probabilities, you end up guessing right. As you receive data from the extractions, you have more and more certainty about the composition of the bag.

Decide some criteria to consider that you have enough information to guess the composition. Conduct a study on the number of withdrawals that will be needed following the set criteria and the initial estimate of the probabilities. For the operations use the **Bayesian Magic** worksheet. Use these ideas to solve the situation in Example 7. By the way, if you want to have a future as a wizard, ensure the employee is honest: the employee must remove the balls at random.

Task 4. Computer Network

In this task we will simulate the operation of the computer network from Example 6 in order to discover its properties. Remember that in Sections 5 and 7 we derived the calculation of the probability of that network performance in terms of the probabilities of each of the seven computers, p(1) to p(7). However, if the network is very complicated, this calculation can be very laborious. Computer simulation can provide a much simpler solution. In fact, the simulation of randomized experiments is a method used by scientists and engineers to solve problems.

In order to simulate the operation of the network, firstly the operation of each of the seven computers is simulated individually. The procedure described at the beginning of this section is used. These simulations produce seven values 1/0 (1=The computer is operational, 0=The computer is not operational). Let us denote these seven values C_1 to C_7. To combine these seven values into a single value that indicates whether the network works or not, the idea is to perform the following arithmetic operation:

$$C=C_1\left((C_2+C_3)C_6+C_4C_5\right)C_7 \tag{12}$$

The network works if the C value obtained (12) is not 0. Use the **net computer** spreadsheet to perform the following activities:

a) Set the values of p(1) to p(7). Simulate the operation of the network and confirm that the estimated probability coincides very accurately with the theoretical value.
b) Suppose that one of the computers numbered 3 or 4 is operational. What is the probability of the network performance? Analyze it experimentally and theoretically.
c) Suppose that the network is operational. What is the probability that either of the computers numbered 3 or 4 is operational? Analyze this experimentally and theoretically.
d) Suppose that one of the computers numbered 2, 4 or 6 is operational. What is the probability of the network performance? Analyze it experimentally and theoretically.
e) What is the probability of operation of the subnet consisting of 2, 3 and 4 computers? Analyze it experimentally and theoretically.

Task 5. Let's make a Deal

This is a simulation of the TV game show that we will discuss in the solved problem number 12. The spreadsheet Let's make a Deal allows us to select the door behind which is the prize and the number of simulations that are performed. Analyze how the strategy for the door switch has been implemented in the simulation. Confirm that this strategy is a winner on approximately 66% of occasions.

12. Solved Problems

Problem 1. For the following statement, the task is to: (1) interpret its meaning from the point of view of the frequency conception of probability; (2) calculate the required probability values; (3) interpret the obtained probability values.

An urn contains two white balls and five black ones. We draw a ball randomly. Without looking at the color and without replacing the ball in the box, we take a second ball.

a) What is the probability that the second ball was white, knowing that the first ball was white?
b) What is the probability that the first ball was white, knowing that the second ball was white?

Solution

For paragraph (a) it is easy to understand that the color of the second ball depends on the color of the first one, and that it should be calculated $p(b_2/b_1)$. But difficulties arise when interpreting the meaning of paragraph (b). Some people consider it absurd to try to predict the color of the second ball before knowing the color of the first ball. It is not easy to accept that the event "the first ball is white" depends on the event "the second ball is white". Note that the meaning of probabilistic independence is not the same as causal independence. The difficulty disappears if we interpret the statement as the meaning of frequency probability. Let us consider this interpretation:

A bowl contains two white and two black balls. Let us think of a random experiment using this bowl and Let us predict approximately the results we would get if we were running the experiment. The randomized experiment is as follows. Draw a ball at random and write down its color. Without returning the ball to the bowl, we take a second ball and we note its color. We return both balls to the bowl and repeat the experiment several times. Sometimes the two balls are white, and sometimes the two balls are black, at times the first ball is black and the second is white, and at times the first ball is white and the second is black. Finally, we have obtained the following four values A, B, C and D:

A = number of occurrences of b1 and b2

B = number of occurrences of b1 and n2

C = number of occurrences of n1 and b2

D = number of occurrences of n1 and n2

A, B, C and D are values which we would know if we performed the experiment. Naturally, the value N = A + B + C + D is the total number of times that we have performed the experiment. Now Let us make our prediction.

a) We consider only those outcomes for which the first ball was white. Among them, we aim at trying to estimate the relative frequency of the event "the second ball is white". That is, we aim at estimating the value A/(A+B) which we would obtain if we ran the experiment.
b) Now we take only those outcomes for which the second ball was white. Among them, we aim at trying to estimate the relative frequency of the event "the first ball is white." That is, we aim at estimating the value A/(A+C) which we would obtain if we ran the experiment.

Next, we calculate the requested probability values:

a) $p(b_2/b_1)=\frac{1}{6}\simeq 0.17$

b) $b_2=(b_2\cap b_1)\cup(b_2\cap n_1)\Rightarrow p(b_2)=p(b_2\cap b_1)+p(b_2\cap b_1)=$

$$=p(b_1)p(b_2/b_1)+p(n_1)p(b_2/n_1)=\frac{2}{7}\frac{1}{6}+\frac{5}{7}\frac{4}{6}=\frac{11}{21}$$

$$p(b_1/b_2)=\frac{p(b_2\cap b_1)}{p(b_2)}=\frac{p(b_1)p(b_2/b_1)}{p(b_2)}=\frac{\frac{2}{7}\frac{1}{6}}{\frac{11}{21}}=\frac{1}{11}\simeq 0.09$$

The interpretation of these results is that 17% of the times when the first ball is white, the second ball is also white. 9% of the times when the second ball is white, the first ball is also white.

Problem 2. Solve Example 4.

Solution

Let D = "The part is faulty", p(D) = 0.05, X="number of faulty parts in the first sample of ten units", Y="number of faulty parts in the second sample of ten units". A_1="the batch is accepted on the first inspection", A_2= "the batch is accepted on the second inspection", A="the batch is accepted". Note that A_1 and A_2 are mutually exclusive events and the variables X and Y are independent. Then:

$$p(A)=p(A_1 \cup A_2)=p(A_1)+p(A_2)$$
$$p(A_1)=p(X=0)$$
$$p(A_2)=p((X=1)\cap(Y=0))=p(X=1)p(Y=0)$$

So that X=1 there must be exactly one faulty part in the sample of ten. However, this faulty part may appear in any of the ten positions. Thus:

$$p(X=1)=10p(\overline{D_1}\cap\overline{D_2}\cap...\cap\overline{D_9}\cap D_{10})=$$
$$=10p(\overline{D})^9 p(D)=10(0.95)^9(0.05)\simeq 0.31$$
$$p(X=0)=p(Y=0)=p(\overline{D_1}\cap\overline{D_2}\cap...\cap\overline{D_9}\cap\overline{D_{10}})=$$
$$=p(\overline{D})^{10}=(0.95)^{10}\simeq 0.6$$
$$p(A_1)\simeq 0.6,\ p(A_2)\simeq 0.19$$
$$p(A)\simeq 0.19+0.6=0.79$$

Let Z be "price of inspecting a batch". Z is a random variable and its possible values are Z=10 and Z=20. In Chapter 3 we will study random variables and their values in detail. For now, it is sufficient to say that the average value μ_z of a random variable Z is approximately the arithmetic mean of a large sample of values of Z. How is μ_Z calculated? Each possible value of Z is multiplied by the probability of that value, i.e., the probability that Z takes that value. Afterwards all the obtained numbers are added. Thus:

$$\mu_Z=10p(Z=10)+20p(Z=20)=10p(X\neq 1)+20p(X=1)=$$
$$=10(1-p(X=1))+20p(X=1)=10+10p(X=1)=13.1\,\$$$

Let us suppose that p(D) is an unknown value. The solution can be expressed in terms of the parameter p=p(D):

$$p(A)=(1-p)^{10}+10p(1-p)^{19},\ \mu_Z=10+100p(1-p)^9$$

Notice in Fig. 15 the graphs of p(A) and μ_Z in terms of p = p(D). The value μ_Z=13.87 is the maximum possible average cost and it is achieved for p = 0.1, with p(A) = 0.48. Analyze and interpret the properties of both functions.

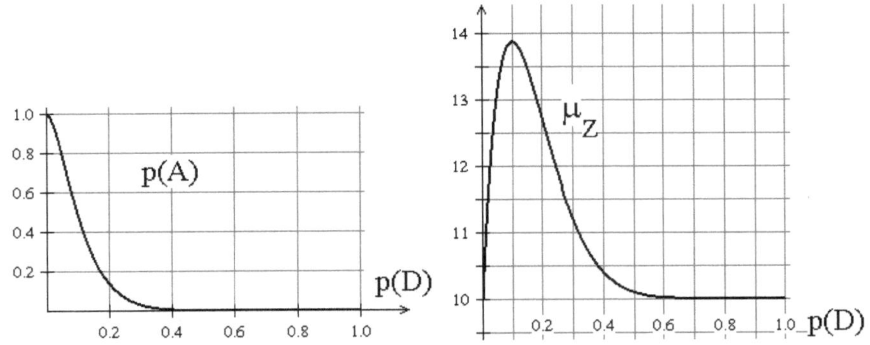

Figure 15. Graph of p(A) and μ_Z.

Problem 3. Solve Example 5.

Solution

Note (+) = "positive test" (-) = "negative test", S = "the testee is sick", H = "the testee is not sick". To solve the problem we use the Total probability theorem (9) and Bayes' theorem (10):

$$p(+)=p(S)p(+/S)+p(H)p(+/H); \quad p(S/+)=\frac{p(S)p(+/S)}{p(+)} \tag{13}$$

Let us suppose that this is a disease that affects only one in 2,000 people (p(S)=0.0005). In addition, we will establish the following test reliability parameters. The test is positive in 99% of the sick cases and in 4% of the healthy ones. So, we are assuming that most sick people are detected by the test and only 4% are false positive. With these data, we can expect that a positive test indicates a high probability that the subject is sick, that is, we can expect a high value of p(S/+), is it so? Let us use (13) to perform the calculation:

$$p(+)=(0.0005)(0.99)+(0.9995)(0.04) \simeq 0.0405$$

$$p(S/+)=\frac{(0.0005)(0.99)}{0.0405} \simeq 0.0122$$

The obtained value p(S/+) is the probability of occurrence of the event S in view of the new information +. Thus, only 1.2% of people who test positive are really sick. Does it surprise you? The explanation is that the proportion of sick people in the entire population is very small. See Fig. 16. We have represented the graphs of p(S/+) and p(+) depending on the parameter p(S). As can be observed:

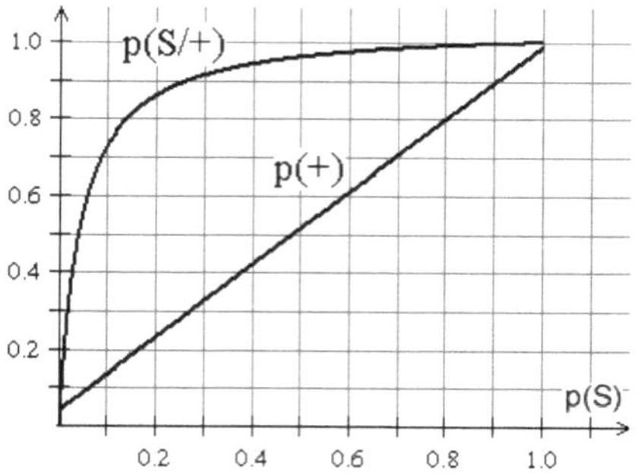

Figure 16. Graph of p(S/+) as a function of p(S).

- A small value of p(S) (a small proportion of the population affected by the disease), also implies a small value of p(S/+). This would change if the incidence of the disease in the population were much higher. Imagine for example that it is a flu epidemic that affects 25% of the population (p(S) = 0.25). Using (8) we find that p(S/+) = 0.89.
- Note how p(S/+) increases its value when increasing p(S), from p(S/+)=0 for p(S) = 0 to p(S/+)=1 for p(S)=1.
- The value of p(+) is the probability that the result is positive, having no information on whether or not the person is sick. Notice also how the value of p(+) increases when increasing p(S). For p(S)=0 (no sick patients), we have p(+)=0.04 because the test will be positive just for 4% of false positives. For p(S)=1 (the whole population is sick), we have p(+)=0.99 because the test will be positive for 99% of testees.

Let us suppose we use the same medical test for a person for whom the test was positive. What now is the probability that this person is sick? Now the person has not been chosen at random from the entire population. This is a person chosen from the population of people for whom the test was positive. This population comprises sick people (S) and healthy people (H). The new probabilities are now p(S) = 0.0122 and p(H) = 0.9878. If we use again (13):

$$p(+)=(0.0122)(0.99)+(0.9878)(0.04) \simeq 0.0516$$

$$p(S/+)=\frac{(0.0122)(0.99)}{0.0516} \simeq 0.2341$$

And if we use the test a third time for the same patient and the patient tests positive:

$$p(+)=(0.2341)(0.99)+(0.7659)(0.04) \simeq 0.2624$$

$$p(S/+)=\frac{(0.2341)(0.99)}{0.2624} \simeq 0.8832$$

Note how after the third positive test, we are highly confident that the person is really sick (p(S/+)=0.88). This recursive calculation process of p(S/+) can be shown graphically. Observe Fig. 17. We have plotted the graph of p(S/+) and the bisector of the first quadrant y = x. Starting from a value p(S)=p_1 we find that p_2=p(S/+) on the OY axis. Taking this p_2 value to the bisector, we place p_2 on the axis OX. Then using p_2 we find that p_3=p(S/+) and the process is repeated. Observe how as tests are positive, we obtain a sequence of probabilities p_1, p_2, p_3,... converging to unity: it is increasingly likely that the person is really sick.

This procedure is based on Bayes' theorem (10). Let us review the value of p(S) in view of new available data. In this case, these revisions have been made on the basis of objective data (the test results). However, in a real situation, doctors may also use their beliefs (subjective probability) based on their own professional experience and personal interpretation of the patient's history, habits, family history, the results of other tests, etc. This example shows how, in general, you may consider personal beliefs in the calculation of the probability of an event. The subjectively assessed probabilities are combined with the application of Bayes' theorem to re-calculate the probability of an event in the view of new available information and personal interpretations of such information. This is a general procedure that can be used to assess risk in various areas: investments, business, weather, politics, insurance policies, etc. For example, insurance companies advertize "good driver" discounts in their policies: every year that passes without the driver having had an accident increases the probability that the driver belongs to the group of drivers who have a low risk of having an accident. Computer task 3 allows you to experiment with the subjective conception of probability.

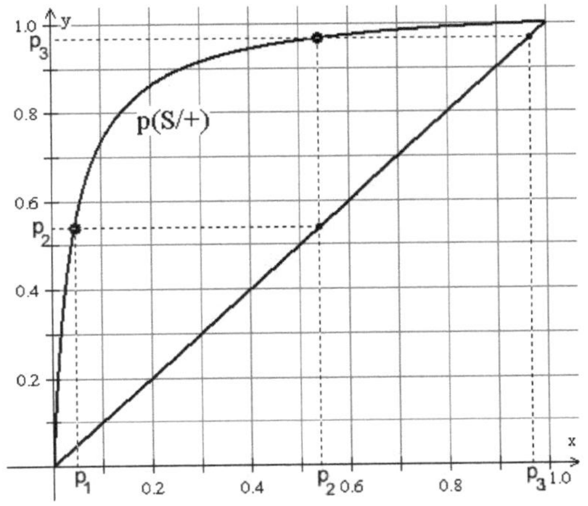

Figure 17. Successive revisions of p(S).

Problem 4. An urn contains five white balls and three black balls. The first ball is extracted and without returning it to the urn a second ball is drawn. Would you play any of the following games?

a) You win if you get a white ball on the second draw.
b) Repeat the game until you get a white ball on the second draw. You win if the first ball was black.
c) Repeat the game until you get a white ball in the first draw. You win if the second ball is black.
d) You win if both balls are white.
e) You win if any of the balls are white.

Solution

a) $p(b_2)=p(b_2 \cap b_1)+p(b_2 \cap n_1)=$
$=p(b_1)p(b_2/b_1)+p(n_1)p(b_2/n1)=$
$=\frac{5}{8}\frac{4}{7}+\frac{3}{8}\frac{5}{7}=\frac{5}{8}\simeq 0.62$

b) $p(n_1/b_2)=\frac{p(n_1 \cap b_2)}{p(b_2)}=\frac{p(n_1)p(b_2/n_1)}{p(b_2)}=\frac{\frac{3}{8}\frac{5}{7}}{\frac{5}{8}}=\frac{3}{7}\simeq 0.36$

c) $p(n_2/b_1)=\frac{3}{7}\simeq 0.43$

d) $p(b_1 \cap b_2)=p(b_1)p(b_2\,/\,b_1)=\frac{5}{14}\simeq 0.36$

e) $p(b_1 \cup b_2)=p(b_1)+p(b_2)\text{-}p(b_1 \cap b_2)=\frac{5}{8}+\frac{5}{8}-\frac{5}{14}=\frac{25}{28}\simeq 0.89$

Problem 5. Solve Example 8.

Solution

Let S_i="The price grows the i-th month", where i=1,2,3. The sample space is:

$$E=\left\{S_1S_2S_3, \overline{S_1}S_2S_3, S_1\overline{S_2}S_3, S_1S_2\overline{S_3}, \overline{S_1S_2}S_3, \overline{S_1}S_2\overline{S_3}, S_1\overline{S_2S_3}, \overline{S_1S_2S_3}\right\}$$

You must choose a criterion to make the decision. A possible criterion is to buy the fuel if the probability of the event A= "the oil price increases in more than a month" is greater than 0.8. Therefore, the fuel will be purchased if:

$$p(A)>0.8 \Leftrightarrow p(S_1S_2S_3 \cup \overline{S_1}S_2S_3 \cup S_1\overline{S_2}S_3 \cup S_1S_2\overline{S_3})>0.8 \Leftrightarrow$$

$$\Leftrightarrow p(S_1S_2S_3)+p(\overline{S_1}S_2S_3)+p(S_1\overline{S_2}S_3)+p(S_1S_2\overline{S_3})>0.8$$

At this point a hypothesis can be established that simplifies the solution of the problem. If we assume the hypothesis that the events S_i are independent in pairs and further that $p(S_i)=p$, where i=1,2,3 then:

$$p(S_1S_2S_3)=p^3, p(\overline{S_1}S_2S_3)=p(S_1\overline{S_2}S_3)=p(S_1S_2\overline{S_3})=(1-p)p^2$$

$$p(A)>0.8 \Leftrightarrow p^3+3(1-p)p^2>0.8$$

Observe Fig. 18 where the graph of p(A) is plotted. The graph reaches the value p(A)=0.8 if p(S)≈0.71. Thus, on the basis of these assumptions we will have to buy fuel when the expected p(S)> 0.71.

However, these two hypotheses may be unrealistic. We will propose a solution based on the subjective interpretation of probability. Let us suppose you are an expert in the energy sector. The current international political

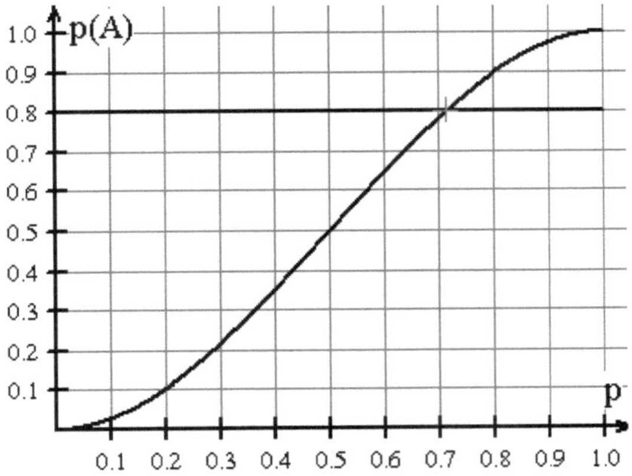

Figure 18. Graph of p (A) as a function of p=p(S).

and economic situation, the opinion of your colleagues and your sharp nose for business lead you to make the following estimates:

$$p(S_1S_2S_3) = 0.4,\ p(\overline{S_1}S_2S_3) = 0.2,\ p(S_1\overline{S_2}S_3) = 0.1,\ p(S_1S_2\overline{S_3}) = 0.24$$

$$p(\overline{S_1S_2}S_3) = 0.01,\ p(\overline{S_1}S_2\overline{S_3}) = 0.025,\ p(S_1\overline{S_2S_3}) = 0.1,\ p(\overline{S_1S_2S_3}) = 0.001$$

$$(0.4 + 0.2 + 0.1 + 0.24 + 0.01 + 0.025 + 0.024 + 0.001 = 1)$$

$$p(A) = 0.4 + 0.2 + 0.1 + 0.24 = 0.94 > 0.8$$

As a consequence, the decision is to buy fuel at this time.

Problem 6. Figure 19 shows a vertical structure and a ball which is released from any of the points A, B or C, then the ball follows a certain path and ends up in basket 1 or 2.

a) Calculate the probabilities of falling into each of the baskets.
b) If the ball has fallen into basket 2, what is the most likely point of departure?

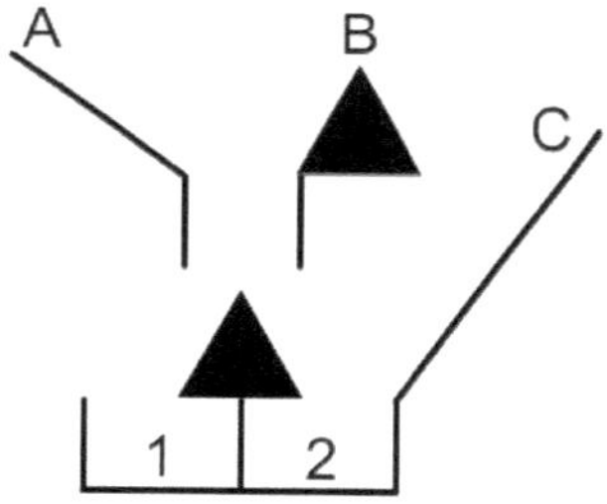

Figure 19. Random path of a ball.

Solution

We will set up two hypotheses. Firstly, Let us suppose that the points A, B and C from which the ball departs are chosen with equiprobability (p(A)=p(B)=p(C)=1/3). Then note that the ball takes right-left in the two top corners of the triangles. Well, Let us suppose also equiprobability in both directions (R= "right", L= "left", p(R)=p(L)=1/2).

a) $1 = (1 \cap A) \cup (1 \cap B) \cup (1 \cap C)$

$$p(1) = p(A)p(1/A) + p(B)p(1/B) + p(C)p(1/C) =$$

$$= \frac{1}{3}\left(\frac{1}{2} + \frac{1}{4} + 0\right) = \frac{1}{4}$$

Thus p(2) = 1-p(1) = 3/4. Or:

$$p(2) = p(A)p(2/A)+p(B)p(2/B)+p(C)p(2/C)=$$

$$=\frac{1}{3}\left(\frac{1}{2}+\frac{1}{4}+\frac{1}{2}+1\right)=\frac{3}{4}$$

b) $$p(A/2)=\frac{p(A\cap 2)}{p(2)}=\frac{p(A)p(2/A)}{p(2)}=\frac{\frac{1}{3}\frac{1}{2}}{\frac{3}{4}}=\frac{2}{9}$$

$$p(B/2)=\frac{p(B\cap 2)}{p(2)}=\frac{p(B)p(2/B)}{p(2)}=\frac{\frac{1}{3}\left(\frac{1}{4}+\frac{1}{2}\right)}{\frac{3}{4}}=\frac{3}{9}$$

Thus p(C/2) = 1-p(A/2)-p(B/2) = 4/9. Or:

$$p(C/2)=\frac{p(C\cap 2)}{p(2)}=\frac{p(C)p(2/C)}{p(2)}=\frac{\frac{1}{3}}{\frac{3}{4}}=\frac{4}{9}$$ (C is the most likely origin)

Problem 7. Let us suppose that you have three boxes. One box contains a prize and the other two are empty. Three players try to choose the box containing the prize. The first player chooses one of the boxes. If the box contains the prize, the player wins. If the first player misses the prize, the second player chooses one of the remaining boxes and if the box contains the prize the second player wins. If the second player misses the prize, the prize will be for the third player. The question is: if you played in the contest, in which position would you rather play 1st, 2nd or 3rd?

Solution

Let us suppose that each of the three boxes has a probability of 1/3 of containing the prize. Let G_1="first player wins", G_2="second player wins", G_3="the third player wins." So it seems that $p(G_1)=1/3$, $p(G_2)=1/2$, $p(G_3)=1$. Is that so? The sum of these three probabilities is greater than 1. What's wrong with this calculation? Note that for the second player to win, not only must the player choose the correct box from the two remaining options, but in addition, the first player must miss the prize. Similarly, for the third player to win players 1 and 2 must first miss the prize. Player 1 has the advantage that no other player could choose the prize before him. But the first player's disadvantage is having three boxes to choose from. Player 2 has the advantage of having to choose from just two boxes. But the second player's disadvantage is that the first player must first miss the prize. Player 3 has the advantage of winning for sure if he/she ends the game. But the disadvantage is that players 1 and 2 must first miss the prize. Let us now make the calculations correctly. Let A_1= "Player 1 wins", A_2= "Player 2 wins", A_3= "Player 3 wins", then:

$$p(G_1) = p(A_1) = \frac{1}{3}$$

$$p(G_2) = p(\overline{G_1} \cap A_2) = p(\overline{G_1})p(A_2 / \overline{G_1}) = \frac{2}{3}\frac{1}{2} = \frac{1}{3}$$

$$p(G_3) = 1 - \frac{2}{3} = \frac{1}{3}$$

Thus the probability of winning is independent of the order in which the game is played.

Problem 8. Your friend Peter is very fond of physical experiments. He has called you to let you know about the latest one he has in mind. He has put some nails in a vertical panel, as shown in Fig. 20.

When the ball is dropped from the top part, it bounces between the nails and ends up in one of the holes in the bottom part. Peter's aim is to predict in which hole the ball will finish. I have carefully measured the position.

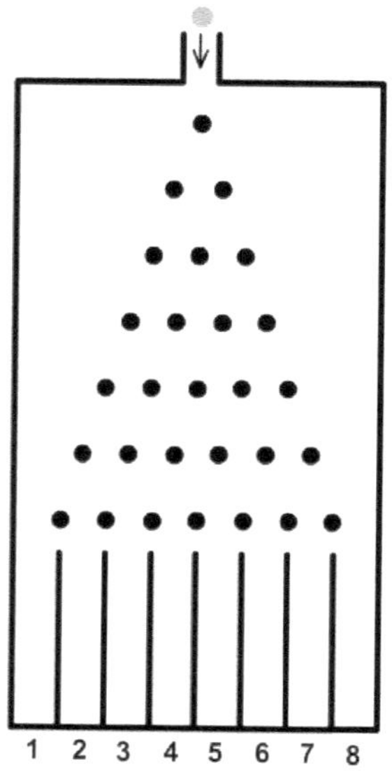

Figure 20. Random path of a ball.

He has carefully measured the position of the nails, the weight and the diameter of the ball and the characteristics of the materials used to make it. His aim is to use all this information and the principles and laws of Physics (for impacts and movement of objects) to calculate the expected trajectory followed by the ball in its fall, and thus to be able to predict which hole it will fall into. Nevertheless, Peter does not see clearly how to start, which physical laws to use and how to do it. For that reason, he has requested for your help. You should advise him on the following matters as follows:

a) He explains his approach to the problem. Please let him know your opinion.
b) Propose a solution to the problem.

Solution

a) In his approach, Peter sets out to trace the exact path that the ball will follow and thereby predict the hole it will drop into. However, there are many factors which influence the trajectory of the ball and it would be impossible to take into account small changes in the initial position of the ball, inaccuracies in the placement of the nails, friction, ball imperfections and materials, etc. As a consequence, you must advise Peter to abandon this method to find the solution.
b) Let us find a probabilistic solution. The probabilistic solution will not allow you to predict the ball's path for the next time you throw it. The probabilistic approach allows you to make a prediction about what will happen if you throw the ball a large number of times. Thus you will be able to calculate the percentage of balls which end up in each of the eight holes with great accuracy.
Note by p(i), i=1,...,8 the probabilities of each of the holes. A probabilistic solution consists in throwing a large number of balls (N), counting the number of balls in each hole (n_i) and estimating p(i) $\approx n_i/N$ where i=1,...,8.
However, if we make the hypothesis that the ball goes right or left with equiprobability on each nail, we can build a theoretical probabilistic model. For every nail, note R="Right", L="Left". All the possible trajectories have the same probability. So what makes the holes have different probabilities? The number of favorable paths is not the same for each hole. To apply Laplace's rule, let us calculate the number of possible paths and the number of favorable paths for each hole.
Number of possible paths: the ball follows one of the two directions {R, L} seven times. Therefore, we just need to choose a direction R or L seven times in order. The number of possible paths will be $VR_2^7 = 2^7 = 128$.

Number of favorable paths for holes 1 and 8: there is only one favorable path for each one (RRRRRRR and LLLLLLL respectively); thus p(1) = 1/128.

Number of favorable paths for the holes 2 and 7: to end up in hole number 2, the ball must go six times in the direction L and once R; to end up in hole number 7, the ball must go six times in the direction R direction and once L. Thus, there are seven possibilities for each case, p(2) = p(7) = 7/128.

Number of favorable paths for holes 3 and 6: to finish in hole number 3, the ball must go five times in the direction L and twice R; to finish in hole 6, the ball must go five times R and twice L. Thus $p(3)=p(6)=C_{7,2}/128=21/128$.

Number of favorable paths for holes 4 and 5: to end up in hole number 4, the ball must go four times L and three times R ; to end up in hole number 5, the ball must go four times R and three times L one. Thus $p(4)=p(5)=C_{7,3}/128=35/128$.

Figure 21 shows how to calculate the number of possible paths favorable to each hole. This representation of combinatorial numbers is called Pascal's triangle, in honor of the mathematician and philosopher Blaise Pascal (1623–1662). Each triangle number is obtained by adding the upper row's numbers on its right and left (by definition 0!=1). The device shown in Fig. 20 is known as Galton's machine in honor of the British scientist Sir Francis Galton (1822–1911).

Problem 9. Let us suppose that a player has two coins. One of them has two heads and the other heads and tails. The player chooses one of the coins and flips it five times, getting five heads.

a) What is the probability that the player has used the two-sided coin?

b) Suppose the player flips the coin N times and gets heads in every flip. What can be concluded?

1
1 1
1 2 1
1 3 3 1
1 4 6 4 1
1 5 10 10 5 1
1 6 15 20 15 6 1
1 7 21 35 35 21 7 1

$\binom{0}{0}$
$\binom{1}{0}\ \binom{1}{1}$
$\binom{2}{0}\ \binom{2}{1}\ \binom{2}{2}$
$\binom{3}{0}\ \binom{3}{1}\ \binom{3}{2}\ \binom{3}{3}$
$\binom{4}{0}\ \binom{4}{1}\ \binom{4}{2}\ \binom{4}{3}\ \binom{4}{4}$
$\binom{5}{0}\ \binom{5}{1}\ \binom{5}{2}\ \binom{5}{3}\ \binom{5}{4}\ \binom{5}{5}$
$\binom{6}{0}\ \binom{6}{1}\ \binom{6}{2}\ \binom{6}{3}\ \binom{6}{4}\ \binom{6}{5}\ \binom{6}{6}$
$\binom{7}{0}\ \binom{7}{1}\ \binom{7}{2}\ \binom{7}{3}\ \binom{7}{4}\ \binom{7}{5}\ \binom{7}{6}\ \binom{7}{7}$

Figure 21. Pascal's Triangle.

c) Suppose we want to be sure to at least 0.9 probability that the player is using the two-sided coin. How many times must the player flip the coin?

Solution

a) The probability that the player will choose a particular coin is unknown. So we denote HH="The player uses the two-headed coin", HT = "The player uses the coin with heads and tails", p(HH) = p, 5H = "five heads are obtained by flipping the coin five times". Suppose that the coin with heads and tails is evenly balanced (p(H/HT) = p(T/HT) = 0.5). This illustrates an important application of Bayes' theorem: the estimation of the probability of an event (p(HH)) from an observation (5H).

$$p(HH/5H) = \frac{p(HH \cap 5H)}{p(5H)} = \frac{p(HH)p(5H/HH)}{p(5H)} =$$

$$= \frac{p(HH)p(5H/HH)}{p(HH)p(5H/HH) + p(HT)p(5H/HT)} =$$

$$= \frac{p}{p + (1-p)(0.5)^5}$$

Note in Fig. 22a the behavior of the value p(HH/5H) according to p. For example, if the player uses the coin HH in 5% of cases, then p(HH) = 0.05 and p(HH/5H) = 0.627. On the other hand, if p(HH) = 0.8, then p(HH/5H) = 0992. As the value of p(HH) increases, the likelihood that the player is using the two-headed coin also increases. But notice

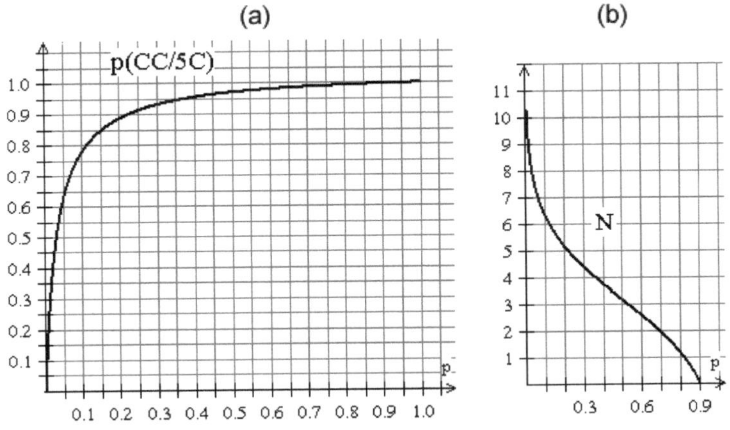

Figure 22. Graphic of p(HH/5H).

how this increased likelihood is non-linear. For example, check that going from p=0.1 to p=0.2, the increase in the value of p(HH/5H) is 0.109. In contrast, going from p=0.6 to p=0.7, the increase in the value of p(HH/5H) is smaller: it is 0.07.

b) Repeating the operations for N releases:

$$p(HH/NH) = \frac{p}{p + (1-p)\left(0.5\right)^N} \tag{14}$$

If $p \neq 0$, when $N \to \infty$ then $p(HH/NH) \to 1$. That is, regardless of the percentage of times that the player uses the two-headed coin, as heads are obtained, the likelihood that the coin used is the two-headed one approaches unity. However, observe that we never reach certainty about which coin is being used.

c) In equation (14) we take p(HH/NH)=0.9 and solve it for N:

$$N(p) = \frac{-1}{\ln 2} \ln\left(\frac{8p}{9(1\text{-}p)}\right)$$

Figure 22b shows the graph of N(p). Notice how the number of flips required (N) decreases when the value of p increases. For example, if p = 0.05 then N = 7418. That is, if the player uses the two-headed coin 5% of the times and the coin is flipped seven times always obtaining heads, we are confident with a degree of probability greater than 0.9 that the player is using the two-headed coin. On the other hand, if p = 0.7, then N=1948, i.e., it will be sufficient to flip the coin twice.

Problem 10. A jury consists of three people, two of whom have probability p of being right as to the verdict while for the third person the probability is 1/2. The jury's decision is made by majority vote. A second jury consists of a single person, that has probability p of being right. Which of the two juries is more likely to be right?

Solution

Let 1= "person 1 is right", 2= "person 2 is right", 3= "person 3 is right", 4= "person 4 is right", p(1) = p(2) = p, p(3) = 1/2, p(4) = p, A_1= "jury 1 is right", A_2= "jury 2 is right", $p(A_2)$= p. Let us suppose that members of the first jury make their decision independently. The following calculations show that both jurors are equally likely to be right in their verdict:

$$p(A_1) = p(1 \cap 2 \cap 3) + p(\bar{1} \cap 2 \cap 3) + p(1 \cap \bar{2} \cap 3) + p(1 \cap 2 \cap \bar{3}) =$$

$$= p(1)p(2)p(3) + p(\bar{1})p(2)p(3) + p(1)p(\bar{2})p(3) + p(1)p(2)p(\bar{3}) =$$

$$= \frac{1}{2}p^2 + \frac{1}{2}(1-p)p + \frac{1}{2}(1-p)p + \frac{1}{2}p^2 = p$$

Problem 11. Do you find the following coincidences surprising?

a) You attend a meeting of N people, and discover that at least two persons share the same birthday.
b) You attend a meeting of N people, and discover that at least one other person's birthday is the same as yours.
c) Suppose that you play a lottery that rewards a sequence of six numbers selected in any order from the numbers 1 to 49. Looking through the history of the results of the 5,000 draws that have been held, you discover that in two draws the same combination of numbers won.

Solution

a) To calculate the probability of the event A= "At least two people share the same birthday", it is easier to calculate the probability of the opposite event $\bar{A}$ = "No pair of persons share their birthday". Assuming the hypothesis that the 366 possible birthdays are equally likely:

$$p(A)=1-p(\bar{A})=1-\frac{V_{366,N}}{VR_{366,N}}$$

This value can be surprisingly high. For example, if N = 23, p(A) = 0.506. So, at a meeting of 23 people, there is a probability greater than 0.5 that at least two people share the same birthday. This probability exceeds the value 0.97 if N=50. The key to this result is that there are many possible pairs of candidates to share birthdays. The situation changes if you are looking for a person whose birthday is the same day as yours, which is precisely what will be analyzed in question b).

b) Let A= "At least someone's birthday is the same day as yours".

$$p(A)=1-p(\bar{A})=1-\frac{VR_{365,N}}{VR_{366,N}}=1-\left(\frac{365}{366}\right)^N$$

For practice, make some calculations. If N=23, then p(A)=0.061; if N=254, then p(A)=0.5009; if N=845, then p(A)=0.900093.

c) As we have seen in Exercise 25, the number of possible outcomes when drawing six numbers from 1 to 49 is 13,983,816. Now, you should realize that this is actually the same problem as in question a), but using N=5000 people and 13,983,816 possible birthdays. If A= "Extract at least twice the same combination", then:

$$p(A)=1-p(\bar{A})=1-\frac{V_{13983816,5000}}{VR_{13983816,5000}}=0.59091$$

Thus, this coincidence is not surprising; there is a probability close to 0.6 that it happens. And if we assume, as in Exercise 25, that the draw took place 18,250 times, we find that this coincidence is virtually certain:

$$p(A)=1-p(\overline{A})=1-\frac{V_{13983816,18250}}{VR_{13983816,18250}}=0.99999$$

However, if you have chosen a random combination and you expect that it has won at least once in the history of the competition, proceeding as in question b), confirm for yourself that you will require a history of 10 million sweepstakes to achieve a probability slightly above 0.5!

Problem 12. *Let's make a Deal* is the name of a famous quiz program that ran from 1963 until 1990 on American television. In one of the contests, the participant had to choose one door out of three. While one of the doors hid a great prize, behind the other two doors there was no prize. Monty Hall, the friendly presenter of the contest, knew where the prize was hidden. Once the player chose a door, Monty opened another door which had no prize from the remaining two and asked the contestant the following question: "Would you like to change your choice of door?". Your task is to decide whether it was advantageous for the participant to switch doors.

Solution

Once you open a door without a prize, it seems clear that the chances of winning are 50%, whether you switch the doors or not. Could there be anything more obvious? One Sunday in September 1990, a reader made a query in the famous newspaper column '*Ask Marilyn*'. This column, which began its publication in 1986, appears in 350 U.S. newspapers with a total circulation of almost 36 million copies. Marilyn vos Savant, already famous among other things for having a very high IQ, gave this reader a surprising response that started a heated debate.[8] Basically, Marilyn said that it was more advantageous for the participant to switch doors, going against the general opinion and what seemed to be obvious.

'*Ask Marilyn*' readers seemed disappointed and reacted with a flood of protest letters. Marilyn had been able to deal with a variety of topics in the column great success. Therefore, how could Marilyn be wrong about such a simple question? Experts in probability, mathematics teachers, mathematicians, Army doctors and universities complained of "the nation's innumeracy" and asked Marilyn to rectify. However, Marilyn maintained her position.

Do not worry if you also think that Marilyn was wrong. The Hungarian Paul Erdös, one of the leading mathematicians of the twentieth century,

[8] Please read the complete story in Mlodinow (2008).

was furious and said "this is impossible". Martin Gardner noted that "in no other branch of mathematics is it so easy for experts to blunder as in probability theory".[9]

Apparently this is a calculation of conditional probability p (choose the door which hides the prize/a door has been opened without a prize). The Monty Hall problem is difficult to understand because the role of the presenter is not noticeable. The presenter opens a door without a prize, but this is not in fact a random choice, because the presenter knows which door hides the prize.

Suppose the participant adopts the strategy of switching the first choice of door. This strategy results in a failure if the participant chooses the door which hides the prize, which occurs with a probability of 1 in 3. But this strategy is successful if the first choice did not hide the prize, which occurs with a probability of 2 in 3. Therefore, the door-switching strategy results in a probability of winning of 2 in 3.

Skeptical? Data from 1963 to 1990 obtained from the 4,500 programs were analyzed and it was found that those who switched doors won twice more frequently than those who stuck to their first choice.

Still skeptical? Let us make the formal calculations:

Let W= "To win the prize", SW= "To choose the door hiding the prize at the first attempt", then:

$$W=(W \cap SW) \cup (W \cap \overline{SW})$$

$$p(W)=p(W \cap SW)+p(W \cap \overline{SW})=$$

$$=p(SW)p(W/SW)+p(\overline{SW})p(W/\overline{SW})=$$

$$= \frac{1}{3} \cdot 0 + \frac{2}{3} \cdot 1 = \frac{2}{3}$$

Are you still skeptical? When Paul Erdös saw the formal proof of the correct answer, he still could not believe it and became more furious. We only have one way to convince you, the same way that finally convinced Erdös: simulation. Perform computer task 5. This is a simulation of the switching door strategy. You will find that approximately 66% of the simulations are successful.

Problem 13. A teacher examines a student as follows the student starts with a mark of N=10. The teacher asks a first question and if the student gives a correct answer, the test ends and the student gets marks of N=10. However,

[9]The Monty Hall problem appeared in the October 1959 issue of Scientific American, in the Mathematical Games section for which Gardner was responsible.

if the student does not provide a correct answer, the marks are reduced by one point (N=9) and the teacher asks a new question. The process continues until the student provides a correct answer to the teacher's question.

a) How many questions will the teacher ask the student?
b) What marks will the student get?

Solution

Observe that the student will get a negative mark N if the number of questions that the teacher asks X is greater than 11. In addition, the number of questions X that the teacher can ask is any integer in the range $[1,\infty)$. Certainly it is unreasonable to think that the teacher and the student will spend the rest of their lives in the exam. However, we will build a probabilistic model in which any integer k contained in the interval $[1,\infty)$ is a possible value of the variable X="number of questions asked by the teacher". Finally, please note that nothing is known about the probability that the student answers each question correctly. This probability may not be constant and may depend on the question itself, on the student's fatigue, etc. However, to model the situation we will make the hypothesis that the probability that the student answers each question correctly is always constant and equal to p $(0<p<1)$. The sample space is E={X=1, X=2, X=3, ...}. The events are any subsets of E, for example A= "$X\leq 6$", B= "$X>10$", C= "$3\leq X\leq 12$", B= "$X\neq 20$", F= "X=-4"=$\varnothing$. Let us see how the probabilistic model verifies Kolmogorov's axiom. Let i C_k= "Answer the k-th question correctly", where k=1,2,3,...

Axiom A1

Based on assumptions of probabilistic independence:

$$p(X=k)=p(\overline{C_1}\cap\overline{C_2}\cap\ldots\cap\overline{C_{k-1}}\cap C_k)=(1-p)^{k-1}p,\ \ k=1,2,3,\ldots$$

Since $0<p<1$, then $(1-p)^{k-1}p>0$, where k=1, 2, 3,... and therefore $p(A)\geq 0$ for any event A.

Axiom A2

$$p(E)=\sum_{k=1}^{\infty}p(X=k)=\sum_{k=1}^{\infty}(1-p)^{k-1}p=p\sum_{k=1}^{\infty}(1-p)^{k-1} \tag{15}$$

Recall that:

$$\sum_{k=0}^{\infty}r^k=\frac{1}{1-r},\quad 0<|r|<1 \tag{16}$$

Using (16) in (15) and taking r = 1-p:

$$p(E)=p\sum_{k=1}^{\infty}(1\text{-}p)^{k\text{-}1}=p\frac{1}{1-(1-p)}=1$$

Axiom A3

Let A={X=x_1, X=x_2, ...}, B={X=y_1, X=y_2, ...}. If $A\cap B=\varnothing$, then $x_i \neq y_j$ for all i, j.

$$p(A\cup B)=p(\{X=x_1,X=x_2,...\}\cup\{X=y_1,X=y_2,...\})=$$
$$=\left(p(X=x_1)+p(X=x_2)+...\right)+\left(p(X=y_1)+p(X=y_2)+...\right)=p(A)+p(B)$$

a) The number X of questions that the teacher will ask the student is an unpredictable value. However, we can estimate this value by the average value of the random variable X. The mean value concept appeared in the solution to problem # 2. The meaning of μ_X is as follows: if you run the experiment a large number of times, the average of the observed values of X will be approximately equal to μ_X. Remember that the average value μ_X of the random variable X is obtained by multiplying each possible value of X by the probability that X reaches this value, and adding all the obtained products.

$$\mu_X=\sum_{k=1}^{\infty}k(1\text{-}p)^{k\text{-}1}p=p\sum_{k=1}^{\infty}k(1\text{-}p)^{k\text{-}1} \qquad (17)$$

Bear in mind that $\sum_{k=1}^{\infty} kr^{k\text{-}1}=\frac{1}{(1-r)^2}, \quad 0<|r|<1$ (18)

Equation (18) is obtained by calculating the derivative of (16) with respect to r. Using (18) in (17) and taking r = 1-p:

$$\mu_X=p\sum_{k=1}^{\infty}k(1\text{-}p)^{k\text{-}1}=\frac{1}{p} \qquad (19)$$

Equation (19) is simple to interpret. If the student has a good chance of answering each question correctly (p close to 1), then the average number of questions μ_X to be put to the student will be close to 1. For example, if p = 0.8, the expected number of questions is μ_X=10/8=1.25. Instead, the value of μ_X will tend towards ∞ as p approaches 0. For example, if p = 0.02 the expected number of questions is μ_X=100/2=50. Note that the mean μ_X of a random variable X does not need to be a possible value of X, as happens in this case.

b) Thus N=11-X. N is a random variable whose mean value μ_N is calculated in the same way as μ_X:

$$\mu_N = \sum_{k=1}^{\infty} (11\text{-}k)(1\text{-}p)^{k-1} p = p\left(11 \sum_{k=1}^{\infty} (1\text{-}p)^{k-1} - \sum_{k=1}^{\infty} k(1\text{-}p)^{k-1} \right) =$$
$$= 11p \sum_{k=1}^{\infty} (1\text{-}p)^{k-1} - p \sum_{k=1}^{\infty} k(1\text{-}p)^{k-1} = 11 - \mu_X = 11 - \frac{1}{p} \qquad (20)$$

For example, if p=0.8, the student's expected mark is μ_N = 11-1.25 = 9.75. But if p=0.02, the expected mark is μ_N = 11-50 = -39. For your own learning, interpret the meaning of equation (20).

13. Unsolved Problems

Problem 1. Let us suppose that a person is lost in the mountains, in a square-shaped region C as shown in Fig. 23a. The rescue team is assessing the probability p(A) that the person is in a specific area $A \subset C$.

a) Define a probability function that is compatible with Kolmogorov's axioms. That is, you must write how to evaluate p(A) for each zone $A \subset C$, while ensuring that it fulfills the Kolmogorov's three axioms.
b) Calculate p(A), p(B) p $(A \cup B)$ for regions A and B of Figs. 23b and 23c.
c) Supposing the person is in B, what is the probability that the person is in A?
d) Calculate p(U), where U is the largest circle that can be drawn in C.
e) Describe a procedure for estimating the value of p.

Problem 2. Someone has drawn the curved shape shown in Fig. 24 on the ground. Describe a procedure for estimating the area enclosed by the curved line, if you only have a kilo of lentils to do it with. What if you also had weighing scales?

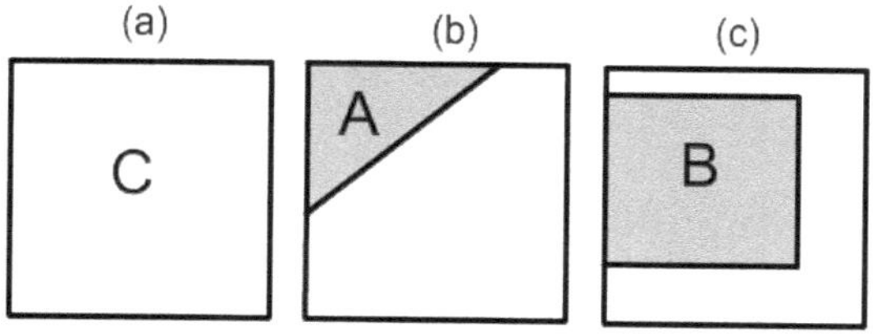

Figure 23. Defining a chance.

Figure 24. What is the area of the region?

Problem 3. The night before their final chemistry exam, three students are partying in another city. When they return to their college, the examination has already finished. They apologize to the teacher and say they had a puncture, and ask if they can re-sit the examination. These students had very good grades throughout the semester, and so the teacher decides to give them another chance. The teacher writes a new exam and places the students in separate rooms. The exam consists of two questions. The first question is about Chemistry and it is worth 5% of the grade. As the students are very good, there is a probability of 0.98 that they will solve it correctly. The second question is worth the remaining 95% of the grade, and is as follows: which wheel was punctured? The teacher considers the answer correct if all students agree on it. What rating do you expect students to get on the exam? What if the number of students was N? HINT: Use the same procedure as in solved problem 2.

Problem 4. A company uses the following quality control system to check the manufactured items. It has a team of four experts who are able to recognize a defective item in 75%, 80%, 70% and 95% of cases, respectively. The first three experts inspect the item and if all agree the item is considered to be OK, then the article is accepted. Otherwise, the fourth expert inspects the item and makes the final decision. What is the probability that a defective item is rejected?

Problem 5. Figure 25 shows a water supply network from the reservoir located to the left, to the city located on the right. The six numbered dots represent water pumping stations. At a set time each of the pumping stations may or may not be faulty.

a) Assume that the city is receiving the water supply. What is the probability that station number 2 is operational?
b) Suppose that the city is not getting water. What is the probability that station number 5 is faulty?
c) Suppose that the city is not getting water. What is the probability that the subnet formed by stations 4-5-6 is not operational?
d) Using the ideas that appeared in computer task 4, build a simulator for this network and use it to estimate the probabilities of questions a)-c).

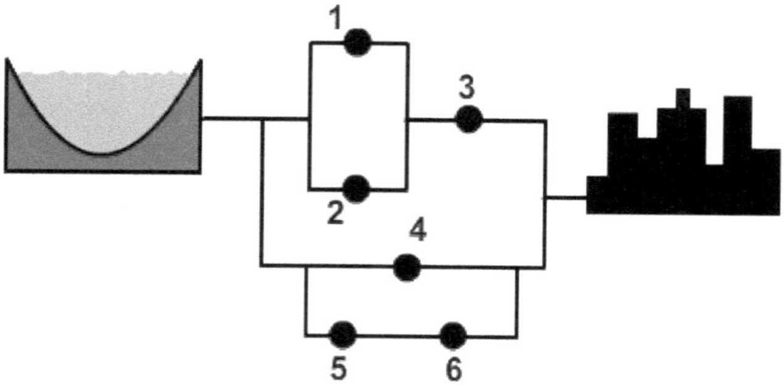

Figure 25. Water supply network.

Problem 6. A continuous random variable takes its values in an interval of real numbers[10] such as [a,b], [a,∞), (−∞,a] or (-∞,∞). Examples of continuous random variables are: the height and weight of a person, the concentration of contaminant in a water sample, the amount of fat in yoghurt, the annual rate of inflation, the price of oil and the waiting time in a line at a bank.

In practice, how are the probabilities of continuous random variables calculated? For example, if you are a chemist who studies a river's pollution, you can take various water samples and define the continuous random variable X= "trichlorethylene concentration in mg/l". Then you will want to calculate probability values such as $p(X\geq 3.56)$, $p(X<8.13)$ and $p(3.5<X<4.7)$.

However, the probability of a random variable is described by the probability density function. See Fig. 26. The graph of the function y(x) represents the probability distribution of a certain variable X defined on the interval [a,b]. The probability that the variable X takes a value in the subinterval $[p,\theta]\subset[a,b]$ is the area limited by the graph of y(x) and the x-axis, for $x\in[p,q]$:

$$p(p\leq X\leq q)=\int_p^q y(x)dx$$

In the example shown in Fig. 26, the areas of the shaded regions correspond to the probabilities of the subintervals $[x_1,x_2]$ and $[y_1,y_2]$. Note

[10] A random variable is called discrete if the set of its possible values is finite or countable infinite. If two dice are rolled, the discrete random variable X = "sum of scores" can take the finite set of values 2, 3, 4, 5, 6, 7, 8, 9, 10, 11, 12. The solved problem 13 discusses a discrete random variable that can take any value of the countable infinite set 1,2,3,4, ...
In order to analyze a discrete random variable, one needs to know the set of possible values $E=\{x_1, x_2, x_3, ...\}$ and the probability of each possible value, $p(X=x_1)$, $p(X=x_2)$, $p(X=x_3)$, ...

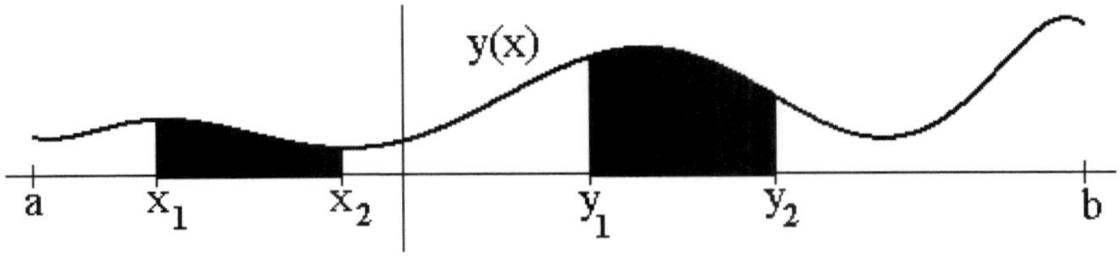

Figure 26. Probability density function.

that both have the same length, yet $p(X \in [x_1,x_2]) < p(X \in [y_1,y_2])$. The graph of the probability density function provides a picture of the distribution of the random variable.

a) Analyze what conditions y(x) must meet so that it represents a probability density function compatible with Kolmogorov's axioms.
b) Show that the following are probability density functions and calculate the requested probabilities.

b1) $y(x)=\frac{12}{223}(x+1)(x-2)(x-3)+\frac{408}{1115}$, $x \in [-1,4]$, $p(-1 \leq X \leq 2)$, $p(X>3)$, $p(X=2.5)$

b2) $y(x)=\frac{1}{\pi(1+x^2)}$, $x \in (-\infty,\infty)$, $p(X \neq 2.5)$, $p(-\infty < X \leq 2)$, $p(4 < X<\infty)$, $p(X>1/X<3)$

b3) Uniform distribution (see footnote 1 and the introduction in paragraph 11).

$y(x)=\frac{1}{b-a}$, $x \in [a,b]$, $p(0 < X<0.2(b-a))$, $p(0.2(b-a) < X<0.4(b-a))$, $p(0.8(b-a) < X<b-a)$

c) Consider a large group of 6-year old children. Graph the possible probability density function y(x) of the random variable X= "Height of a pupil".
Now suppose that the group will add a second large group of 13-year old children. How does it affect the graph of the probability density function?

14. Solutions to Unsolved Exercises

Exercise 1

a) $p(A)=p(I_1)+p(I_2)+p(I_3)=$ 0.032+0.0397+0,1048=0.477
b) $p(A)=p(I_1)+p(I_2)+p(I_3)+p(I_4)+p(I_5)+p(I_6)+p(I_7)=0.7181$
c) $p(A)=p(I_1)+p(I_2)+p(I_3)+p(I_6)+p(I_7)+p(I_8)=1-p(I_4)-p(I_5)=0.5538$
d) $p(A)=p(I_1)+\ldots+p(I_8)=1$ (this is a certain event)

Exercise 2

a) p(Even)=2p(Odd); p(Even)+p(Odd)=1; 3p(Even)=1; p(Odd)=1/3.

b) p(Even)=2/3

c) p(1)=p(3)=p(5)=p(7)=p(9)=p(11)=1/18; p(2)=p(4)=p(6)=p(8)=p(10)==p(12)=1/9

A={1,3,5}; p(A)=p(1)+p(3)+p(5)=1/6

d) A={2,4,6,8,10,11,12}; p(A)=p(2)+p(4)+p(6)+p(8)+p(10)+p(11)+p(12)=13/18

e) A={6,8,10,11,12}; p(A)=p(6)+p(8)+p(10)+p(11)+p(12)=4/9+1/18=1/2

f) A={8}; p(A)=1/9

g) A={5}; p(A)=1/18

h) A={1,2,3,4,5,6,7,8,9,10,11,12}; p(A)=1 (this is a certain event)

Exercise 3

For Exercise 1:

- Random Experiment: it consists of calculating the value S(10) from the random values a and v_0 and determining in which of the subintervals I_i (where i=1, ..., 8) is the value of S(10).
- Sample space E={I_1, I_2, I_3, I_4, I_5, I_6, I_7, I_8}
- Elementary events: I_1, .., I_8.
- Probability of each elementary event: p(I_1)=0.032, p(I_2)=0.0397, p(I_3)=0.1048, p(I_4)=0.3012, p(I_5)=0.145, p(I_6)=0.0954, p(I_7)=0.1162, p(I_8)=0.1657.
- Event: e.g., A={I_4,I_6,I_7}, B={I_2,I_4,I_5,I_8}, C={I_7}, D={I_4,I_5}

Probability of the events: p(A)=p(I_4)+p(I_6)+p(I_7)=0.5128; p(B)=p(I_2)+p(I_4)+p(I_5)++p(I_8)=0.6516; p(C)= p(I_7)=0.1162; p(D)= p(I_4)+p(I_5)=0.4462.

For Exercise 2:

- Random Experiment: Spin the wheel and get a number.
- Sample space E = {1,2,3,4,5,6,7,8,9,10,11,12}
- Elementary events: 1,2,3,4,5,6,7,8,9,10,11,12
- Probability of each elementary event:

p(1)=p(3)=p(5)=p(7)=p(9)=p(11)=1/18; p(2)=p(4)=p(6)=p(8)=p(10)=p(12)==1/9.

- Event: e.g., A = {1,2,3}, B = "black" = {2,4,8,11}, C = "greater than 10"== {11.12}

Probability of the events: p(A)=p(1)+p(2)+p(3)=2/9; p(B)=p(2)+p(4)+p(8)+p(11)=7/18; p(C)=p(11)+p(12)=1/6.

Exercise 4

An event is any subset of E. The number of subsets that can be formed with the n elements of a finite set is 2^n. For example, if you flip a coin, the sample space is E={H, T}, the elementary events are H and T but the number of events that can be formed is $2^2=4$. Surprising? Do not forget that the empty set and the whole set (∅, E) are also possible subsets of E. Thus, the four possible events are: H, T, E, ∅. The event E is the event indeed, and represents any situation that occurs whenever you run the random experiment (e.g., E= "heads or tails"). The event ∅ is the impossible event, and represents a situation that will never happen (for example, ∅= "no heads nor tails"). Obviously, p(E)=1 and p(∅)=0.

Exercise 5

In this case E={R, G, A, FA}. One approach is to assume the equiprobability hypothesis of the four states of the traffic light, i.e., p(R)=p(G)=p(A)=p(FA)=1/4. According to this hypothesis, the desired probability is p({R,G})=p(R)+p(G)=1/2. However, note that in this situation it is unreasonable to assume the equiprobability of all states of the traffic light. In the absence of experimental data, the best approach is to propose a parametric solution: p(R)=a, p(G)=b, p(A)=c, p(FA)=1-abc. p({R,G})=p(R)+p(G)=a+b.

Exercise 6

The estimated value is the absolute frequency of the event A, noted by N_A. If the sample size N is large enough, then $p(A) \approx N_A/N$, and hence we may estimate $N \approx Np(A)$. For example in Exercise 1, if $A=290 \leq S(10) < 336$ then $N_A \approx 1250p(0477) \approx 596$. For Exercise 2, if A= "Odd or black" then $N_A \approx 1250p(13/18) \approx 903$.

Exercise 7

$E=\{S_1, S_2, \ldots, S_n\}$; $p(S_1)+\ldots+p(S_n)=1$; $A=\{S_{n1}, S_{n2}, \ldots, S_{nk}\}$; $p(A)=p(S_{n1})+\ldots+p(S_{nk})$

Exercise 8

a) $A=\{D_1,D_2,D_3\}$; $B=\{D_1,D_2,D_4,D_5\}$ $C=\{D_1,D_2,D_4,D_5,D_6,D_7\}$. $D=\{D_4,D_8\}$. The events A, B and C have occurred, since the elementary event D_1 is in A, B and C. The event D has not occurred.

b) $p(A)=p(D_1)+p(D_2)+p(D_3)$; $p(B)=p(D_1)+p(D_2)+p(D_4)+p(D_5)$; $p(C)=p(D_1)+p(D_2)+p(D_4)+p(D_5)+p(D_6)+p(D_7)$; $p(D)=p(D_4)+p(D_8)$.

c) The event R occurs if any of the elementary events that form A or any of the elementary events that form D occur. Let us use the concept of set union: $R=A \cup D=\{D_1, D_2, D_3, D_4, D_8\}$, p(R)=p(A)+p(D).

d) The event R occurs if an elementary event that is part of A and also part of B occurs. Let us use the concept of the intersection of sets: $C=A\cap B=\{D_1, D_2\}$, $p(M)=p(D_1) + p(D_2)$.

e) $N=A\cup B=\{D_1, D_2, D_3, D_4, D_5\}$, $p(N)=p(A) + p(B) - p(A\cap B)$

f) H is the event consisting of all elementary events that are not contained in A. In the language of set theory, it is the complementary set of A, noted as $\bar{A}$. Notice in Fig. 27 a plot of $\bar{A}$ where the dots represent the sample space of elementary events. In this case $p(\bar{A})=\{D_4, D_5, D_6, D_7, D_8\}$. Overall, $E=A\cup\bar{A}$ and $p(A)+p(\bar{A})=1$, thus $p(\bar{A})=1-p(A)$.

g) If two events A and B contain no common elementary event, then $p(A\cup B)=p(A)+p(B)$, in other words, if $A\cap B=\varnothing$, then $p(A\cup B)=p(A)+p(B)$, yet another way to express this situation is: if events A and B are *incompatible,* that is, if they cannot both occur at once, then the probability of the union event is calculated by adding the probabilities of A and B.

The situation is different if the events A and B contain a common elementary event. In this case, the event $A\cap B$ consists of elementary events that are in A and B. In this situation, then $p(A\cup B)=p(A)+p(B)-p(A\cap B)$. In other words: if events A and B are *compatible,* that is, if both events can occur simultaneously, then the probability of the union event is the sum of their probabilities minus the probability of the part common to A and B.

Exercise 9

$p(A\cup B\cup C)=p((A\cup B)\cup C)=p(A\cup B)+p(C)-p((A\cup B)\cap C)=$

$=p(A)+p(B)-p(A\cap B)+p(C)-p((A\cap C)\cup(B\cap C))=$

$=p(A)+p(B)+p(C)-p(A\cap B)-p(A\cap C)-p(B\cap C))+p(A\cap B\cap C)$. Interpret this result by drawing a graph similar to that of Fig. 10.

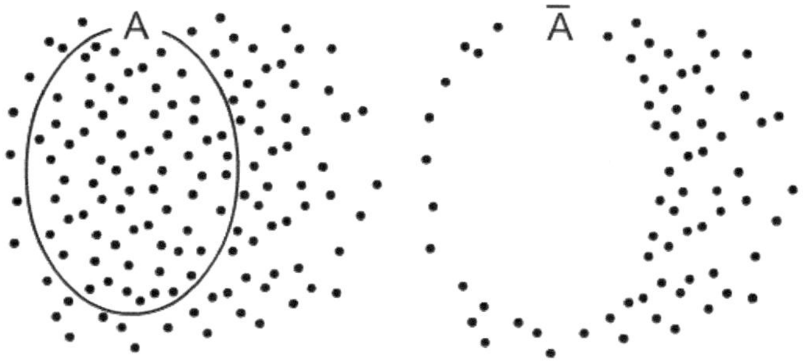

Figure 27. The event $\bar{A}$.

Exercise 10

a) $p(A)=p(I_3)=32/155\approx0.21$. Probability calculated without gender information.
b) $p(B)=p(I_3/M)=7/78\approx0.01$. That is, if one knows that he is a man, the likelihood that his age is in the range I_3 is much lower than if there were no such information.
c) $p(C)=p(M/I_1)=3/9\approx0.33$. Instead, $p(M)=78/155\approx0.5$. Thus, the expectation that it is a man increases if his age is in the interval $I_{1.}$
d) $p(D)=p(I_1\cup I_2\cup I_3)=p(I_1)+p(I_2)+p(I_3)=9/155+35/155+32/155\approx0.49$.
e) $p(E)=p((I_1\cup I_2\cup I_3)/W)=(6+12+25)/77\approx0.56$. Thus, knowing that she is a woman slightly increases the probability that she is less than 25 years old.
f) Note the difference between the event $(I_1\cup I_2\cup I_3)/W$ and the event $(I_1\cup I_2\cup I_3)\cap W$. In the first one, there is no uncertainty about the gender of the person, since it is known that she is a woman. In this case what needs to be calculated is the probability that the age is in the interval $I_1\cup I_2\cup I_3$ knowing that she is a woman. However, in the event $(I_1\cup I_2\cup I_3)\cap W$ the uncertainty is related to gender as well as to age, because this is about calculating the probability that the age is in the interval $I_1\cup I_2\cup I_3$ and also whether or not she is a woman. Given two events A and B, it is important to understand the difference between the events A/B and $A\cap B$. What is the relationship between p(A/B) and $p(A\cap B)$? See Fig. 28. The event $A\cap B$ is contained in the event B. Therefore, $p(A\cap B)$ is the probability of $A\cap B$ in the entire space. But $p(A\cap B)/p(B)$ is the probability in the subspace B of the part of A that is in B. Thus, $p(A\cap B)/p(B)=p(A/B)$. Or equally $p(A\cap B)=p(B)p(A/B)=p(A)p(B/A)$. Using these equations, then:
$p(F)=p((I_1\cup I_2\cup I_3)\cap W)=(6+12+25)/155=43/155$

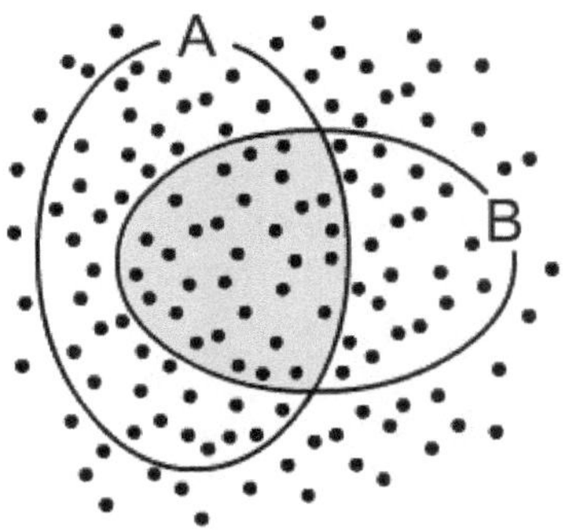

Figure 28. The intersection event.

Likewise:

$p(F)=p(I_1 \cup I_2 \cup I_3)p(W/(I_1 \cup I_2 \cup I_3))=$
$=((9+35+32)/155)((6+12+25)/(9+35+32))=43/155$

Likewise:

$p(F)=p(W)p((I_1 \cup I_2 \cup I_3)/W)=(77/155)((6+12+25)/77)=43/155$

Exercise 11

a) p(odd/black) = p(odd∩black)/p(black) = p(11)/p({2,4,8,11})= = (1/18)/(7/18)=1/7, p(odd)=1/3. Thus, knowing that the result was a black number, it reduces the probability of it being an odd number by 42%.
b) p(black/odd)=p(odd∩black)/p(odd)=(1/18)/(1/3)=1/6, p(black)=7/18. Thus, knowing that the result is odd, it reduces the odds that it is a black number by 42%. The results obtained in questions a) and b) allow us to speculate that in general the ratio p(A/B)p(A) is identical to the ratio p(B/A)p(B) (the ratio is $3/7 \approx 0.42$ in this case). Can you prove that this is true in general?
$p(A/B)/p(A)=p(A \cap B)/p(B)p(A)=p(A)p(B/A)/p(B)p(A)=p(B/A)/p(B)$

Furthermore:

$p(A/B)>p(A) \Rightarrow p(A \cap B)/p(B)>p(A) \Rightarrow p(A)p(B/A)/p(B)>p(A) \Rightarrow p(B/A)>p(B)$

$p(A/B)<p(A) \Rightarrow p(A \cap B)/p(B)<p(A) \Rightarrow p(A)p(B/A)/p(B)<p(A) \Rightarrow p(B/A)<p(B)$

So, knowing that event B has occurred, it increases the expectation that event A will occur. In addition, knowing that event A has occurred, it increases the expectation that event B will occur. Similarly, if knowing that event B has occurred reduces the expectation that event A will occur, then knowing that event A has occurred reduces the expectation that event B will occur.

c) p(multiple of 5/black)=p(multiple of 5∩black)/p(black)= =p(∅)/p(black)=0, p(divisible by 5)=p(5)+p(10)=1/6. Thus, unsurprisingly, knowing that the result is black reduces to 0 the probability that it is a multiple of 5.
d) p(black/multiple of 5)=p(multiple of 5∩black)/p(multiple of 5)=0, p(black)=7/18. Thus, unsurprisingly, knowing that the result is a multiple of 5 reduces to 0 the probability that it is a black number.

e) $p(>1/black)=p(>1 \cap black)/p(black)=p(black)/p(black)=1$, $p(>1)= =1-p(1)=17/18$. Thus, unsurprisingly, knowing that the result is black makes the event "greater than 1" certain.

Exercise 12

Theory-summary Table 3 reads: *p(A/B) is the probability of the event A occurring knowing that the event B occurred. And p(A/B) is also the value of the probability of the event A but recalculated in the view of the new information B.* Thus the value of p(7/1) indicates the probability that computer 7 is in working order knowing that computer 1 is in working order. However, assuming that both computers work independently, each is available or unavailable regardless of the rest of the computers. It is easier to understand this statement if we return to the interpretation of the computer network in Fig. 1 as a road with seven bridges joining the points (a) and (b). Each of the bridges may or may not be closed to traffic, regardless of whether or not the other bridges are shut. In general, if the occurrence of the event B does not affect the probability of the event A, then $p(A/B)=p(A)$ and therefore $p(A \cap B)=p(A)p(B)$.

In our case, $p(7/1)=p(7)$, $p(2 \cup 3/1 \cap 7)=p(2 \cup 3)$ and $p(6/1 \cap 7 \cap (2 \cup 3))=p(6)$.

Exercise 13

It may seem that the chance that the first card is an ACE is not affected by a result that occurs later in time, and therefore the probability that the first card is an ACE is 0.5, regardless of the outcome of the previous card picks. Let us think a little more about this statement. In the first pick, the expectation for the occurrence of the ACE event is p(ACE)=0.5. Imagine that we take out two more cards that prove to be two ACEs. Has something changed in the expectation for the first card? Of course it has. Now, the expectation for an ACE in the first extraction is zero, surely the first extraction was a KING. The difficulty in understanding this situation relies on the confusion of probabilistic independence and causal independence. Obtaining an ACE in the second extraction does not change the card obtained in the first extraction. The result of the first extraction is causally independent of the result of the second extraction. However, the meaning of probabilistic independency is a different one. To understand the situation in probabilistic terms, we should stop thinking about what happens in a single run of the experiment. The question of the statement should be interpreted as follows: "Let us suppose we run the experiment a large number of times. Of these, we will count only those runs when ACE was obtained in the second extraction. In what percentage was the first extraction also an ACE?" The calculation of the probability of an event (ACE in the first extraction) calculated in the view of new information (ACE in the second extraction) is precisely the meaning of conditional probability:

$p(ACE_1/ACE_2)=p(ACE_1 \cap ACE_2)/p(ACE_2)=$
$=p(ACE_1)p(ACE_2/ACE_1)/p(ACE_2)$

$p(ACE_1)=0.5$

$p(ACE_2/ACE_1)=1/3$

$p(ACE_2)=p((ACE_2 \cap ACE_1) \cup (ACE_2 \cap KING_1))=$
$=p(ACE_2 \cap ACE_1)+p(ACE_2 \cap KING_1)= p(ACE_1)p(ACE_2/ACE_1)+p(KING_1)=$
$=p(ACE_2/KING_1)=(1/2)(1/3)+(1/2)(2/3)=1/2$

Therefore: $p(ACE_1/ACE_2)=1/3$, it is expected that approximately one third of the times in which an ACE is obtained in the second extraction, in the first extraction an ACE was also obtained. With respect to paragraph (b) of Example 10, Let us see if the event "get H" depends on the event "black ball". If you do not have any information about the color of the ball, then $p(H)=0.5$. Let us see if the expectation of the outcome of the coin is modified knowing that the ball is black:

$p(H/n)=p(H \cap n)/p(n)=p(H)p(n/H)/p(n)$

To calculate p(n), first "break" the event "black ball" in two incompatible events, "black ball and head" and "black ball and tail":

$n=(H \cap n) \cup (T \cap n)$, $p(n)=p((H \cap n) \cup (T \cap n))=p(H \cap n)+p(T \cap n)=$
$=p(H)p(n/H)+p(T)p(n/T)=$
$=(1/2)(5/7)+(1/2)(7/11)=52/77$.

Therefore, $p(H/n)=p(H)p(n/H)/p(n)=(5/14)/(52/77)=55/104 \approx 0.53$. Thus, if one knows that the ball was black, this fact slightly increases the expectation that the coin was heads. Surprised again? To analyze probabilistic independence, stop thinking in terms of cause-effect or in terms of before-after. You should think that this is a regularity that occurs after running the random experiment a number of times. The result $p(H/n) \approx 0.53$ means that" *if we repeat the random experiment a number of times, approximately 53% of the times in which the ball is black, the coin outcome was heads*".

Exercise 14

It is known that $p(A/B)=p(A)$, the task is to demonstrate that $p(B/A)=p(B)$;

$p(B/A)=p(B \cap A)/p(A)=p(B)p(A/B)/p(A)=p(B)p(A)/p(A)=p(B)$.

$p(\overline{A}/B)=1-p(A/B)=1-p(A)=p(\overline{A})$.

$p(\overline{A}/\overline{B})=1-p(A/\overline{B})=1-p(A\cap\overline{B})/p(\overline{B})=1-p(A)p(\overline{B}/A)/p(\overline{B})=$
$=1-p(A)(1-p(B/A))/p(\overline{B})=$

$=1-p(A)(1-p(B))/p(\overline{B})=1-p(A)p(\overline{B})/p(\overline{B})=1-p(A)=p(\overline{A})$;

Exercise 15

$p(b2)=p(b2\cap b1)+p(b2\cap n1)=p(b1)p(b2/b1)+p(n1)p(b2/n1)=$
$=(3/8)(2/7)+(5/8)(3/7)=3/8$

$p(n2)=p(n2\cap b1)+p(n2\cap n1)=p(b1)p(n2/b1)+p(n1)p(n2/n1)=$
$=(3/8)(5/7)+(5/8)(4/7)=5/8$

$p(b2\cap n1)=p(n1)p(b2/n1)=(5/8)(3/7)=15/56$

$p(n2\cap n1)=p(n1)p(n2/n1)=(5/8)(4/7)=5/14$

$p(n1/b2)=p(n1\cap b2)/p(b2)=(5/8)(3/7)/(3/8)=5/7$

$p(n1/n2)=p(n1\cap n2)/p(n2)=(5/8)(4/7)/(5/8)=4/7$

$p(b1/b2)=1-p(n1/b2)=1-5/7=2/7$

$p(b1/n2)=p(b1\cap n2)/p(n2)=p(b1)p(n2/b1)/p(n2)=(3/8)(5/7)/(5/8)=3/7$

$p(b1\cup b2)=p(b1)+p(b2)-p(b1\cap b2)=3/8+3/8-(3/8)(2/7)=3/4$

$p(b1\cup n2)=p(b1)+p(n2)-p(b1\cap n2)=3/8+5/8-(3/8)(5/7)=41/56$

$p(n1\cup b2)=p(n1)+p(b2)-p(n1\cap b2)=5/8+3/8-(5/8)(3/7)=41/56$

$p(n1\cup n2)=p(n1)+p(n2)-p(n1\cap n2)=5/8+5/8-5/14=50/56$

S	Estimation of p(S)	p(S)	S	Estimation of p(S)	p(S)
b1	398/1048≈0.38	3/8≈0.375	n2/b1	294/398≈0.74	5/7≈0.71
n1	650/1048≈0.62	1-3/8≈0.625	n2/n1	399/650≈0.61	4/7≈0.57
b2	355/1048≈0.34	3/8≈0.375	n1/b2	251/355≈0.71	5/7≈0.71
n2	693/1048≈0.66	5/8≈0.625	n1/n2	399/693≈0.58	4/7≈0.57
b1∩b2	104/1048≈0.099	3/28≈0.11	b1/b2	104/355≈0.29	2/7≈0.29
b1∩n2	294/1048≈0.28	15/56≈0.27	b1/n2	294/693≈0.42	3/7≈0.43
n1∩b2	251/1048≈0.24	15/56≈0.27	b1∪b2	249/1048≈0.62	3/4≈0.75
n1∩n2	399/1048≈0.38	5/14≈0.36	b1∪n2	797/1048≈0.76	41/56≈0.73
b2/b1	104/398≈0.26	2/7≈0.29	n1∪b2	745/1048≈0.72	41/56≈0.73
b2/n1	251/650≈0.39	3/7≈0.43	n1∪n2	745/1048≈0.9	50/56≈0.89

Exercise 16

$p(\overline{A})=1-p(A)$. $p(\overline{B})=1-p(B)$. $p(A\cap B)=p(B)p(A/B)$. $p(A\cup B)=$
$=p(A)+p(B)-p(A\cap B)$. $p(B/A)=p(A\cap B)/p(A)$. $p(A/\overline{B})=$
$=p(A\cap\overline{B})/p(\overline{B})=p(A)p(\overline{B}/A)/p(\overline{B})=p(A)(1-p(B/A))/p(\overline{B})$. $p(B/\overline{A})=$
$=p(B\cap\overline{A})/p(\overline{A})=p(B)p(\overline{A}/B)/p(\overline{A})=p(B)(1-p(A/B))/p(\overline{A})$. $p(\overline{B}/A)=$
$=1-p(B/A)$. $p(\overline{A}/B)=1-p(A/B)$. $p(\overline{B}/\overline{A})=1-p(B/\overline{A})$.

Using De Morgan's law:

$$p(\overline{A\cap B})=p(\overline{A}\cup\overline{B})=p(\overline{A})+p(\overline{B})-p(\overline{A}\cap\overline{B})=p(\overline{A})+p(\overline{B})-p(\overline{A})p(\overline{B}/\overline{A}).$$

$$p(\overline{A\cup B})=p(\overline{A}\cap\overline{B})=p(\overline{A})p(\overline{B}/\overline{A}).$$

Exercise 17

Let B= "to be an employee of a bank", F= "to be an activist of the feminist movement". $p(1)=p(B)$, $p(2)=p(B\cap F)$; knowing that $B\cap F\subset B$ $(B\neq F)$, then $p(B\cap F)<p(B)$.

Exercise 18

$$p(SC/BS)=p(SC\cap BS)/p(BS)=p(SC)p(BS/SC)/p(BS)=0.36$$

$$p(VC/BS)=p(VC\cap BS)/p(BS)=p(VC)p(BS/VC)/p(BS)=0.267$$

$$p(MC/BS)=p(MC\cap BS)/p(BS)=p(MC)p(BS/MC)/p(BS)=0.08$$

$$p(VR/BS)=p(VR\cap BS)/p(BS)=p(VR)p(BS/VR)/p(BS)=0.093$$

$$p(SW/BS)=p(SW\cap BS)/p(BS)=p(SW)p(BS/SW)/p(BS)=0.2$$

Note how the probability of the events SC, VC, MC, VR and SW has changed in view of the certain information BS. Without knowing any information about the failure, the most favorable troubleshooting-sequence is VR-SW-MC/SC-VC. However, if the BS possesses certain information, the most favorable troubleshooting sequence is: SC-SW-VR-VC-MC.

Exercise 19

Note that this random experiment and that of Exercise 13 are constructed using the same method, which is widely used in practical applications of probability. The random experiment consists of two phases. In the experiment of Exercise 13, the first phase involved picking a card, which could result in $KING_1$ or ACE_1. The second phase of the experiment was to pick a second card, which could result in $KING_2$ or ACE_2. For this Exercise (19), the first phase of the random experiment consists of choosing an urn, which can be U1 or U2. The second phase involves extracting a ball from the urn of our choice. The outcome can be b or n. To use theorems (9) and (10), a breakdown of the certain event E is defined using the first phase of the experiment. That is, for the situation of Exercise 13, E= "the first card is ACE or KING" = $ACE_1 \cup KING_1$; for the situation of Exercise 19, E= "urn U1 or U2"= $U2 \cup U1$. From here, we can use (9) and (10) to calculate $p(U_1/b)$ and $p(U_2/b)$:

$b=(b\cap U1)\cup(b\cap U2)$, $p(b)=p(U1)p(b/U1)+p(U2)p(b/U2)=$
$=(1/2)(9/10)+(1/2)(1/10)=1/2$

$p(U1/b)= p(U1)p(b/U1)/p(b)=(1/2)(9/10)/(1/2)=9/10$

$p(U2/b)= p(U2)p(b/U2)/p(b)=(1/2)(1/10)/(1/2)=1/10=1-p(U1/b)$

This result allows us to be certain that if the random experiment is repeated a number of times under the same conditions then we will get a white ball approximately 50% of the time. Moreover, if we count only the experiments in which a white ball has been drawn, approximately 90% of the times the urn will be U1 and the remaining 10% of the times the urn U2 was chosen.

Is it solved? Or have we missed something? Please re-read Exercise 5 and its solution. In Exercise 5 we analyzed a random experiment with four possible outcomes, R, G, A and FA. A first approach was to assume equiprobability of all of them, i.e., assume that $p(R)=p(G)=p(A)=p(FA)=1/4$. But then we thought it was unreasonable to make this assumption in that situation. Instead, we decided to work with three unknown parameters $p(R)=a$, $p(G)=b$, $p(A)=c$ and $p(FA)=1\text{-}abc$. What do Exercises 5 and 19 have in common? In the rubric of Exercise 19 it is mentioned that the urn is chosen at random. Always remember that the term "random choice" is not equivalent to "equiprobability choice". Let us imagine for example that we roll a balanced dice and that we choose the urn U1 if we get 1, and that U2 is chosen if we get any other number. In this case, the choice is random but not equiprobable. In this case, $p(U1)=1/6$, $p(U2)=5/6$. The calculations would now be:

$p(b)=p(U1)p(b/U1)+p(U2)p(b/U2)=(1/6)(9/10)+(5/6)(1/10)=7/30$

$p(U1/b)=p(U1)p(b/U1)/p(b)=(1/6)(9/10)/(7/30)=9/14$

$p(U2/b)=p(U2)p(b/U2)/p(b)=(1/2)(1/10)/(1/2)=5/14=1-p(U1/b)$

Since the statement does not explain how to choose the urn, you can assume equiprobability for U1 and U2. Write this assumption in an explicit form, i.e., write the hypothesis "Suppose that $p(U1)=p(U2)=0.5$" explicitly.

In this situation, as in Exercise 5, the best solution is to write the probabilities as functions of the unknown probability values. That would allow us to obtain the solution whatever random procedure is used to select the urn. Using p(U1)=p, p(U2)=1-p:

$$p(b) = p(U1)p(b/U1) + p(U2)p(b/U2) = 9p/10 + (1-p)(1/10)$$

$$p(U1/b) = \frac{9p/10}{9p/10 + (1-p)(1/10)} = \frac{9p}{8p+1}, \quad p(U2/b) = \frac{(1-p)(1/10)}{9p/10 + (1-p)(1/10)} = \frac{1-p}{8p+1}$$

Observe Fig. 29. We have plotted the graphs of the functions p(b), p(U1/b) and p(U2/b) as functions of the parameter p=p(U1). As you can see:

- If p=0, you can only choose the urn U2. Therefore, p(b)=p(b/U2)=0.1, p(U2/ b)=1, p(U1/ b)=0.
- If p=1, you can only choose the urn U1. Therefore, p(b)=p(b/U2)=0.9, p(U2/b)=0, p(U1/b)=1.
- If p=0.5, p(b)=0.5, p(U1/b)=0.9, p(U2/b)=0.1.
- If p=1/6=0.17, p(b)=7/30=0.23, p(U1/b)=0.64, p(U2/b)=0.36.
- p(b) increases when p grows. Why? As p grows, the most likely choice is U1, which is the urn in which there is a greater chance of getting white ball.
- p(U1/b) increases when p increases. Why? As p increases, the most likely choice is U1.
- p(U2/b) is decreasing when p increases. Why? As p grows, the less likely choice is U2.
- p(U1/b)=p(U2/b)=0.5 if p=0.1. Why? The urn U1 is chosen only in 10% of the times, but you get white ball from U1 in 90% of the cases; the urn U2 is chosen in 90% of the times, but you get white ball only from U2 in 10% of the times. The end result is that there is equiprobability between p(U1/b) and p(U2/b).

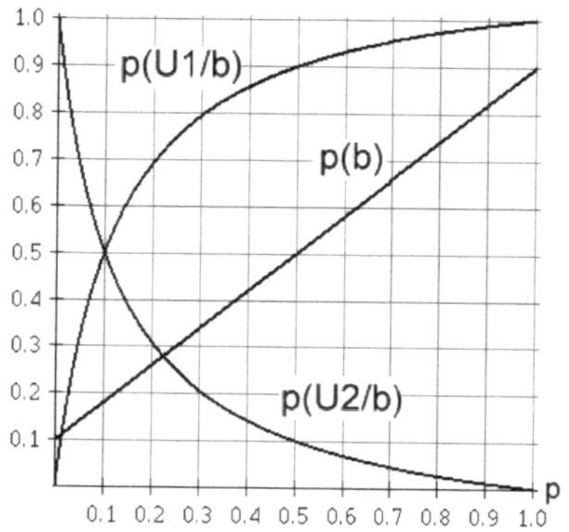

Figure 29. Solution based on p=p(U1).

Exercise 20

Let $p=p(S_1)=...= p(S_n)$. Since $p(S_1)+...+p(S_n)=1$, then $np=1$, thus $p=1/n$. If $A=\{S_1,S_2,...,S_k\}$, then $p(A)=p(S_1)+ ... +p(S_k)=1/n+...+1/n=k/n$. So to calculate the probability of an event A, just compute the *number* k of elementary events that form A and then $p(A)=k/n$. Note that no matter *what* are the elementary events that form A, it is only necessary to know *how many* they are.

Exercise 21

a) $C=\{1,2,3,4,5,6\}$, $n=6$, $m=5$. The order is relevant because the outcome is a number (e.g., 12345 and 12354 elections are different); there may be repeated digits (i.e., you can get 22233 or 44444).
b) $C=\{0,1\}$, $n=2$, $m=16$. The order in which the digits are chosen is relevant; an identical element may be chosen several times (in fact, the binary digits must be repeated because $n<m$).
c) $C=\{1,...,20\}$, $n=20$, $m=3$. Let us suppose that we elect the president first, the secretary second and finally the vocal. In this case the order with which elements are selected from C is relevant; the same element cannot be chosen more than once.
d) $C=\{1,...,10\}$, $n=10$, $m=3$. The order in which the years are chosen is irrelevant; the same year can be chosen several times.
e) $C=\{1,...,20\}$, $n=20$, $m=3$. The order in which people choose is irrelevant; the same element cannot be chosen more than once.
f) $C=\{1,...,10\}$, $n=10$, $m=10$. The order of election is relevant; the same element cannot be chosen more than once.

Exercise 22

a) $VR_{6,5} = 6^5 = 7776$ *b)* $VR_{2,16} = 2^{16} = 65536$ c) $V_{20,3} = 20{\bullet}19{\bullet}18 = 6840$

d) $CR_{10,3} = \binom{12}{3} = \dfrac{12{\bullet}11{\bullet}10}{3{\bullet}2} = 220$ e) $C_{20,3} = \binom{20}{3} = \dfrac{20{\bullet}19{\bullet}18}{3{\bullet}2} = 1140$

f) $P_{10} = 10! = 3628800$ $PR_{15}^{6,4,3,1,1} = \dfrac{15!}{6!4!3!} = 25225200$

Exercise 23

Each possible $C_{n,m}$ combinations can be ordered in P_m different ways. Thus, the value $C_{n,m}P_m$ equals the number of possible choices in order, i.e., $C_{n,m}P_{m}= V_{n,m}$.

Exercise 24

a) In the statement it is not clear whether the order in which the points are to be chosen should be taken into account. For example, are 1-3-4-5-6 and 1-4-3-6-5 results considered equal results? Perform the calculations in both cases, given that in both cases the result can be repeated.

Taking into account the order: $n=VR_{6,5}=6^5=7776$

Regardless of the order: $n=CR_{6,5}=\binom{6+5-1}{5}=252$

a1) Taking into account the order: $p(4,2,5,6,1)=\frac{P_5}{7776}=\frac{5!}{7776}=\frac{120}{7776}$

Regardless of the order: $p(4,2,5,6,1)=\frac{1}{252}$

a2) Taking into account the order: $p(4,4,5,5,5)=\frac{C_{5,2}}{7776}=\frac{C_{5,3}}{7776}=\frac{PR_5^{2,3}}{7776}=\frac{10}{7776}$.

Regardless of the order: $p(4,4,5,5,5)=\frac{1}{252}$.

If we should take the order into account, observe how you can use several approaches to calculate the number of favorable cases. We can choose without any order the two positions to place the two numbers 4 ($C_{5,2}$); or we can choose not in order the three positions where to place the three numbers 5 ($C_{5,3}$); we can also arrange the numbers 4,4,5,5,5 in every possible way ($PR_5^{2,3}$).

b) $\frac{PR_{14}^{7,7}}{VR_{2,14}}=\frac{14!}{7!7!2^{14}}\simeq 0.21.$

c1) There are five blocks where the letters ABCD can be placed. For each of the blocks, the rest of the letters may be ordered in any possible way. Thus, $p=\frac{5P_4}{P_8}=\frac{5!}{8!}\simeq 0.003.$

c2) Now, it is also possible to sort the letters A, B, C, D:

$p=\frac{5P_4P_4}{P8}=\frac{5!4!}{8!}\simeq 0.07.$

d) $p=\frac{C_{10,3}}{CR_{10,3}}\simeq 0.54.$ e) $p=\frac{1}{C_{49,6}}=\frac{1}{13983816}$.

f) The reasoning is wrong, because you have to take into account the order in which the scores of each dice appear. For example, the result

(631) can be obtained from P_3=6 possible ways, namely: 6-3-1, 6-1-3, 1-6-3, 3-6-1, 1-3-6 and 3-1-6, and the result (441) can be obtained from $PR_3^{2,1}$=3possible ways, namely: 4-4-1, 4-1-4 and 1-4-4. Using these methods, check that p(9)=25/216 and p(10)=27/216.

Exercise 25

If you flip a fair coin, p(H)=0.5. That means that on average (i.e., long-term), it will be necessary to flip the coin twice to get heads. If you roll a balanced dice, p(1)=1/6. That means that on average (long-term), you must roll the dice six times to get the number 1 (or any other number). In general, if the random experiment has k equally likely possible outcomes, it means that on an average (long-term), you must run the experiment k times to obtain a certain result. In the case of the lottery, as we have seen in Exercise 24(e), the number of possible outcomes is 13,983,816. That means that on average the combination 1-2-3-4-5-6 (or any other) is a winner every 13,983,816 draws. Even assuming that the draw is being carried out daily for 50 years, a total of 18,250 draws will be needed. The vast majority of possible combinations have not been obtained yet.

Earlier in this chapter (see Examples 1–3), we demonstrated how our intuition sometimes makes mistakes when assessing the probability of an event occurring. Well, our intuition plays tricks on us in other random situations. For example, our intuition tells us that lottery numbers with digits "ordered" such as 77777, 45454 or 76543 are less likely than other numbers whose digits are "messy", such as 49289 or 76420. In general, people have a tendency to deny the randomness of a sequence in which some kind of familiar structure can be observed. This trend is sometimes used in an egocentric way. It happens, for example, when pseudo-sciences try to explain "enigmas" such as the Egyptian Pyramids. Firstly, they show some measurements which have supposedly been taken inside the Pyramids. Surprisingly, these measurements coincide with data which only contemporary science has been able to determine (the length of the Greenwich meridian, the atomic number of cobalt, etc.). So what is the conclusion? The architects who designed the Pyramids were extra-terrestrials!

Let us examine more formally the statement. ET= "The pyramid was built by an advanced alien civilization", UD= "Some measurement of the pyramid matches data discovered by modern science." Assuming the ET event is true, it is reasonable to think that the advanced alien pyramid builders used data that only contemporary human science has been able to determine. So, we may think that the value p(UD/ET) is practically equal to unity and therefore the event UD/ET is an almost certain event. However, the relevant event is not UD/ET one, but rather ET/UD. The pseudo-scientists do not perceive the difference between the two events and assume that the event ET/UD is virtually certain instead of the event UD/ET. It is wrong to interpret the meaning of conditional probability.

An identical argument is used by parapsychologists, ufologists and fortune tellers to defend the strangest theories. They use a great mass of random data and extract some fragments which have a collective meaning of some sort. By virtue of this special meaning, they categorically reject its possible randomness categorically.[11]

In order to experiment with the idea that random sequences whether they are "ordered" or "bizarre", appear with the same frequency as "messy" sequences, just perform computer task 1 in Section 11.

Exercise 26

In the case of the drum containing 10 numbered balls, each digit is independent of the others, since each ball is returned to the drum. Thus: $p(22211)=p(2)p(2)p(2)p(1)p(1)=(0.1)^5= p(12345)=p(83056)$. As far as the second drum is concerned, each of the 100000 numbers has a probability of $1/100000=(0.1)^5$. Thus, the two situations are identical. Here again, our probabilistic intuition can be misleading (see Exercise 25). We mistakenly think that the numbers 22211 and 12345 are less likely than the number 83056, because in the first two we recognize a pattern or symmetry, "and the symmetries are less likely". Actually, the three numbers (and any other) appear with the same frequency in the long run. Similarly, we mistakenly think it is easier to get the number 12345 from the drum containing 100,000 balls than from the drum containing 10 balls. Why? This is because in the drum containing 10 balls "it is difficult to obtain exactly the sequence 1-2-3-4-5", while in the drum containing 10,000 balls "the number 12345 is already written on a ball". This intuitive reasoning is erroneous.

Exercise 27

a) and b) Both orders have the same probability p=1/5!

Here again our probabilistic intuition can be misleading (see Exercises 25 and 26). The orders 1-2-3-4-5 of question a) may seem less likely that the order 3-5-4-1-2 because the first one follows a pattern, a conventional order. Instead, it is easily accepted that the two orders of question b) have the same probability, since the sequences are formed with characters that have no conventional order. However, if you examine the font Wingdings in Windows, you can check that the characters 🗁, 🗎, 🗏, 🗐, 🗄 correspond precisely with the digits 1-2-3-4-5. This example again shows the weakness of intuitive argument when judging situations. The only way to be certain about the probability of an event is formal probability calculus. This value

[11]See Barragués & Guisasola (2009).

is the only one that will allow us to make predictions. If you run both experiments several times, you will be able to check that each of the possible arrangements shows in approximately 100/5! = 0.83% of the cases.

Exercise 28

If B1 and B3 are true, then $p(\varnothing)=p(\overline{E})=1-p(E)=1-1=0$.

If B1 and B2 are true, then $p(E)=p(\overline{\varnothing})=1-p(\varnothing)=1-0=1$.

Exercise 29

P1: $E=E\cup\varnothing \Rightarrow$ (A3) $p(E)=p(E)+p(\varnothing) \Rightarrow$ (A2) $p(\varnothing)=0$.

P2: According to (A1), then $0\leq p(A)$. The demonstration that $p(A)\leq 1$ is missing.

$E=A\cup\overline{A}\Rightarrow$ (A3) $p(E)=p(A)+p(\overline{A}) \Rightarrow$ (A2) $1=p(A)+p(\overline{A})$. If $p(A)>1$, according to (A1) then $p(A)+p(\overline{A})>1$.

P3: $E=A\cup\overline{A}\Rightarrow$ (A3) $p(E)=p(A)+p(\overline{A}) \Rightarrow$ (A2) $1=p(A)+p(\overline{A})$.

P4: $A=(A\cap\overline{B})\cup(A\cap B) \Rightarrow$ (A3) $p(A)=p(A\cap\overline{B})+p(A\cap B) \Rightarrow p(A\cap\overline{B})=$
$=p(A)-p(A\cap B)$.

$A\cup B= B\cup(A\cap\overline{B}) \Rightarrow$ (A3) $p(A\cup B)=p(B)+p(A\cap\overline{B}) \Rightarrow p(A\cup B)=$
$=p(B)+p(A)-p(A\cap B)$.

Exercise 30

$$\forall \varepsilon > 0, q \in (0,1), \exists\, n \in \mathbb{N} \,/\, p(p(A) \in (f_n(A)-\varepsilon, f_n(A)+\varepsilon))>q$$

References

Barragués, J.I., Guisasola, J. and Morais, A. 2006. Chance and probability: What do they mean to university engineering students?, International Journal of Mathematical Education in Science and Technology, 37(8): 883–900.

Barragués, J.I. and Guisasola, J. 2009. Who built the Egyptian Pyramids? The use of computational tools in solving problems to make students think about randomness. Mathematics and Computer Education, 43(1): 53–68.

Gardner, M. 1988. The Great Stone Face, in Gardner, M., The New Age, Buffalo, NY: Prometheus Books, pp. 72–78.

Kahneman, D. Slovic, P. and Tversky, A. 1982. Judgment under uncertainty: heuristics and biases. New York: Cambridge University Press.

Mlodinow, L. 2008. The Drunkard's Walk: How Randomness Rules Our Lives, New York: Pantheon books.

CHAPTER 3

Random Variables

Verônica Y. Kataoka,[1,*] *Irene M. Cazorla,*[2]
Hugo Hernandez[3] and *Claudia Borim da Silva*[4]

Phenomena is the crude data of science and they are objects of experimentation and/or simulation. For examples, we have climatic (rain, dry), geological (earthquakes), biological (metamorphosis of a butterfly) and physical (evaporation) phenomena. Some of them can be observed by sight, others require sophisticated technology (like "observing" the DNA). There is also phenomena related to mankind "that cannot be seen" but can be inferred by their manifestations, like social phenomena (migrations) or psychological phenomena (attitudes, fear). Phenomena can be deterministic or random.

A phenomenon is deterministic when we already know it and can predict its result (temperature of boiling in water). To describe those phenomena, deterministic models are proposed, indicating precise relationships between variables, where one of them is determined by the other. For example, the gravitational attraction between two objects is direct proportional to their masses and inversely proportion to the distance between them.

A phenomenon is random when we are not certain, *a priori*, which observation is going to happen, or we cannot know the result for having more than a possible outcome. To study those phenomena, probabilistic models are proposed, which aim to describe the regularity of the variables to be analyzed.

Models used in statistics are always probabilistic, and the construction of these models involves the concept of random variables.

[1] University Bandeirante of São Paulo (Brazil).
[2] Universidade Estadual de Santa Cruz (Brazil).
[3] Colegio de Ciencias y Humanidades Vallejo, UNAM (Mexico).
[4] University São Judas Tadeu (Brazil).
* Corresponding author

1. Definition

Problem 1

In order to secure some equipment on a work table, four similar bolts are needed. That type of bolt must have a specific measure and so a specific nut; they are altogether in a kit with some other pieces from a factory Y.

The factory Y buys the bolts from factory Z and the nuts from factory W. When those pieces arrive to factory Y, they are classified under the following specifications: 1st line (F), 2nd line (S) and 3rd line (T). Probabilities of having each of the types of the pieces can be noted in the table below.

Type	Probabilities	
	Factory Z—bolt	Factory W—nut
First line (F)	0.80	0.75
Second line (S)	0.15	0.15
Third line (T)	0.05	0.10

For each bolt or nut, the factory Y pays \$0.50. If the piece is made with a first line bolt and a first line nut, the manufacturer is ready to sell, at a value of \$3.00. If one of the pieces comes out as second line, it becomes necessary to repair the piece, at a cost of \$0.50. And if one of the pieces comes out as third line, the product (bolt and nut) is discarded and sold as scrap iron for \$0.50.

So, for example, the gain for a product made of a first line bolt and a second line nut would be \$1.50 = \$3.00 - \$1.00 (cost of both pieces) - \$0.50 (cost for repairing the second line nut).

If the steel mill produces 1000 pieces, what is the expected cost?

Note that a bolt is considered in the 2nd line when it is longer than specifications, and a nut is considered in the 2nd line when its diameter is less than the bolt's diameter. Repairs would be cutting the bolt to the right length and widening the nut's diameter.

For pieces in the 3rd line, the bolt would be a length shorter than specifications, impossible to repair; and the nut would have a diameter greater than that the bolt , being considered in that case as a lost piece.

To solve this problem, we need to first define the sample space (Ω) of the manufacture of the product, using a bolt and a nut: Ω = {FF, FS, FT, SF, SS, ST, TF, TS, TT}. Relating each event (ω) of that sample space to the gain (\$), we will have (Fig. 1):

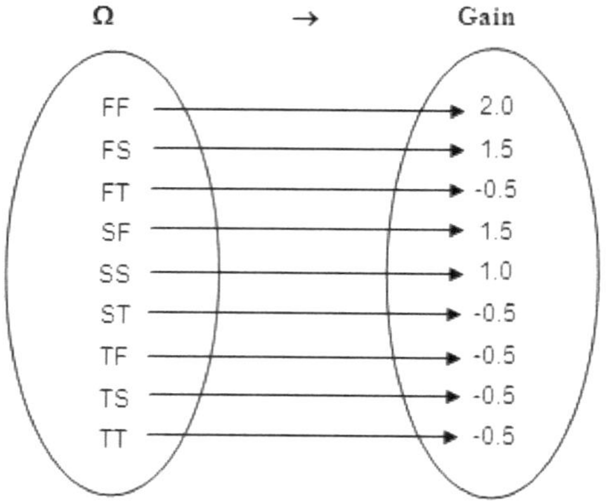

Figure 1. Relating each event (ω) of that sample space with the gain ($).

The function that relates the sample space and the set of Real Numbers is called random variable, $X(\omega):\Omega \to \mathbb{R}$. Formally, we can say that, from a space of probability $\{\Omega, Z, P[\cdot]\}$ defined by an experiment, a function $X(.)$ that associates each event of that space of probability with a real number, is called random variable.[1] The set $\mathbb{R}$ is the codomain of $X(.)$ and Ω its domain.

Usually a random variable is represented by an upper case letter, for example $X(\cdot) = X$, and the values of image set by the correspondent lower case (here, x). In our problem the image of the random variable $X(.)$ is the gain of the steel mill, this is, the image set = {-0.5, 1.0, 1.5, 2.0}.

In practice we name the image set as the random variable, so, in our problem the random variable X is gain. Here we have some other examples:

a) For the tossing of a coin three times, considering Head (H) and Tail (T), the sample space is Ω={HHH, HHT, HTH, HTT, THH, THT, TTH, TTT} and we can consider the random variable Y: number of tails, Y={0,1,2,3}.
b) For the number of siblings of the students in a class, the sample space may be Ω={0,1,2,3,4,5}, and we can consider the random variable V: number of siblings, V={0,1,2,3,4,5}.

In our problem and in the example a) the variables are classified as nominal qualitative and we can observe that the random variable transforms those variables into integer numbers. When the sample space is discrete or

[1] Meyer (2000) states that the term random variable is not accurate, since a function has been named as variable and, furthermore, the term is also inaccurate since the function P[·] is not used within its definition.

continuous, by rule, the random variable is its own identity. If the variables are qualitative or discrete quantitative (example b) they turn to be classified as discrete quantitative random variables and when the variables are continuously quantitative they can be classified as continuous random variables (example c).

Random variables assuming values within a numerable set will be denominated as discrete, and those assuming values within an interval of real numbers will be denominated as continuous.

Let's discuss further differences between discrete and continuous random variables. Let's take the spinner in Fig. 2. There are eight points labeled around it. What's the probability of turning the spinner and having the arrow pointing at number two at the end?

So, let X to be the random variable "number pointed by the arrow". Thinking that $P(X = 2) = 1/8$ would be easy, since there's only one point around the spinner labeled as 2, and there are eight points labeled, but thinking so will be wrong.

How many points are around the spinner? There are only eight of them labeled with numbers, but there is an infinite number of points. So, the probability of the spinner stopping with the arrow pointing at number two is one on an infinite, and, for practical purposes, that is zero. In other words, it's impossible to have the arrow pointing at number two, or any other number, after the spinner has stopped.

This means that $P(X = x) = 0$ for any value x for any point around the spinner. So, why are the spinners that popular if it's impossible to get a number? Now let's take the spinner in Fig. 3. What's the probability of turning the spinner and having at the end the arrow pointing at number two?

Now we have eight labeled sectors of the same size and we want to know the probability of the arrow pointing at one of them. Thinking that $P(X = x) = 1/8$ will be correct, since there's only one sector in the spinner

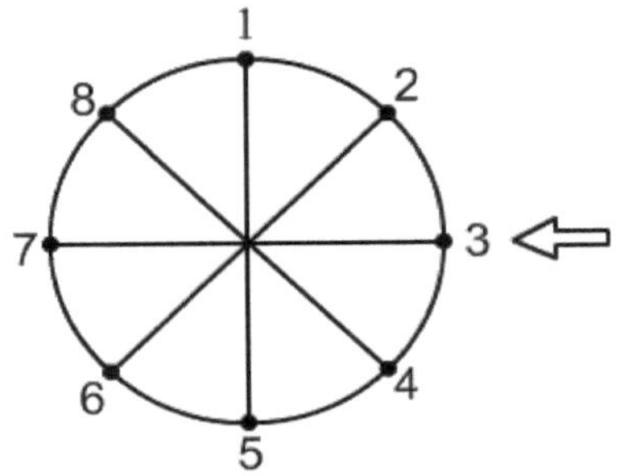

Figure 2. Spinner with eight points labeled.

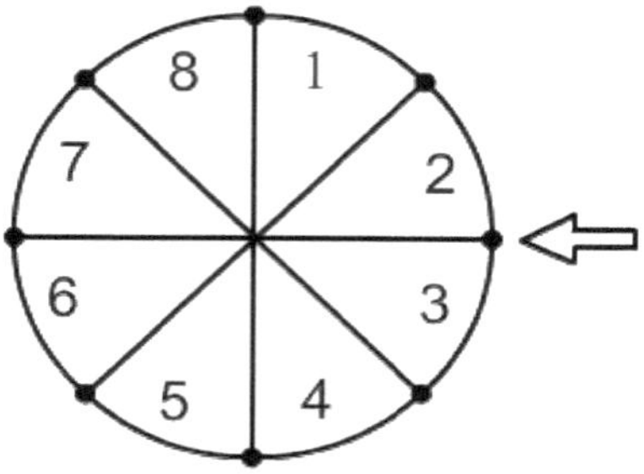

Figure 3. Spinner with eight equal sectors labeled.

labeled as 2, and there are eight equal sectors, as we can note in Fig. 4a; we are interested in just one of those sectors that represents 1/8 of the whole possible results.

Can we use this idea for the spinner in Fig. 2? We certainly can! We can take one sector of the spinner; let that be a "slice" as in Fig. 4a or an arc as in Fig. 4b, in which a new point, the hypothetical last point on the circle, has been labeled as 9, in order to cover all the sectors of 1/8 in the spinner.

If we take the arc as in Fig. 4b, we just have to adjust the probability of the sector we are calculating, from $P(X=2)$ to $P(2<X<3)$ or, more correctly, to $P(2 \le X \le 3)$, even when it doesn't matter if we use the signs "lesser than" or "lesser than or equal to", since the probability of getting exactly the point two or the point three, as we have seen, is equal to zero.

In this case, the random variable X, "number on the spinner pointed by the arrow", ranges from 1 to 9 and it can take any real value in the interval [1, 9]; so, the sample space is an infinite set and the variable cannot be counted, but just measured.

A random variable which can just be measured and take values within an interval of real numbers is known as **continuous random variable**. With a continuous random variable X, it always happens that $P(X=x)=0$.

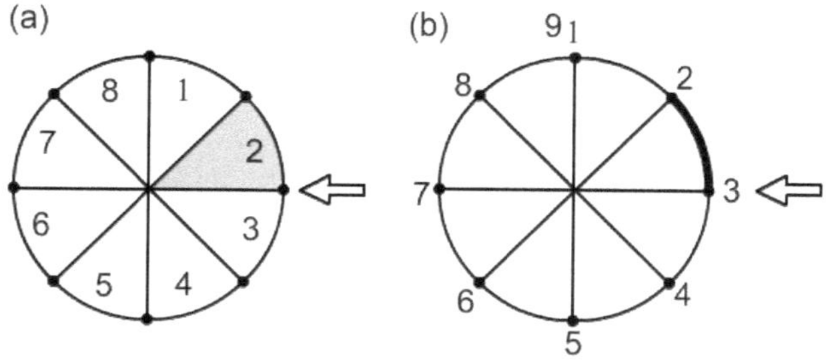

Figure 4. (a) Proportion of the spinner corresponding to X = 2; (b) Arc corresponding to values from 2 to 3.

2. Density and Distribution Functions

2.1 Density Function

Going back to the concept of random variable, it makes sense to talk about probabilities type $P[X \leq x]$, since for the set of values $X \leq x$ exist corresponding events in Ω, for which probabilities defined in the space $\{\Omega, Z, P[\cdot]\}$ exist. In this way, by definition: $P[X \leq x] = P[\omega : X(\omega) \leq x]$.

In the case of discrete random variables we can calculate $P[X = x]$, and we can check that out in our problem. For example, in order to calculate $P[X = 2]$, we have to consider probabilities for both the bolt and the nut to be first line; in the case of the bolt, that probability is 0.80 and in the case of the nut the probability is 0.75, hence, the probability of getting a product FF will be: $P[X = 2] = P[F_{bolt}].P[F_{nut}] = 0.80(0.75) = 0.60$ (Fig. 5).

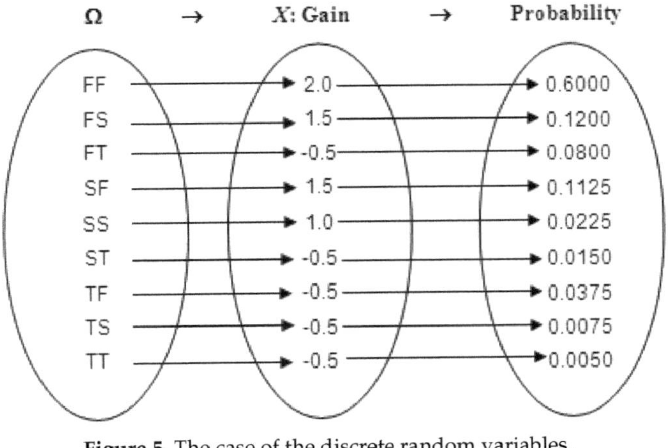

Figure 5. The case of the discrete random variables.

In the case of the discrete random variables, the function that associates each value of the random variable with its probability is known as function of probability, or discrete probability density function, which follows the next conditions:

(i) $P[X = x_i] \geq 0, \forall\ i;$

(ii) $\sum_{i=1}^{\infty} P[X = x_i] = 1$

This function is named by several authors as distribution of probability for discrete random variables. In the case of our examples, it would be:

x_i	-0.5	1.0	1.5	2.0
$P[X = x_i]$	0.1450	0.0225	0.2325	0.6000

The graphic of the density function for the discrete random variable X: gain can be observed in Fig. 6.

In the case of continuous random variables, specific values of the variable correspond to events with null probability. For example, in the population of a city, the probability of a person's height to be exactly 1.63472m is defined as null. So, there's no sense in defining a function of probability like the one for discrete random variables, but is more convenient to construct a function that gives probabilities related to the event of the variable to be contained in some interval. In this way, the probability of a single person's height to be between 1.50m and 1.70m won't be null. So, the function of density of probability (fdp), or function of density for a continuous random variable must follow this conditions:

(i) $f(x) \geq 0$, $\forall$ y;

(ii) $\int_{-\infty}^{\infty} f(x)dx = 1;$

(iii) $P[a \leq X \leq b] = \int_{a}^{b} f(x)dx, \ \forall \ a, b$

We can perceive that it's a function specifically designed for the area below it to be 1. Furthermore, areas below the curve inside an interval are probabilities.

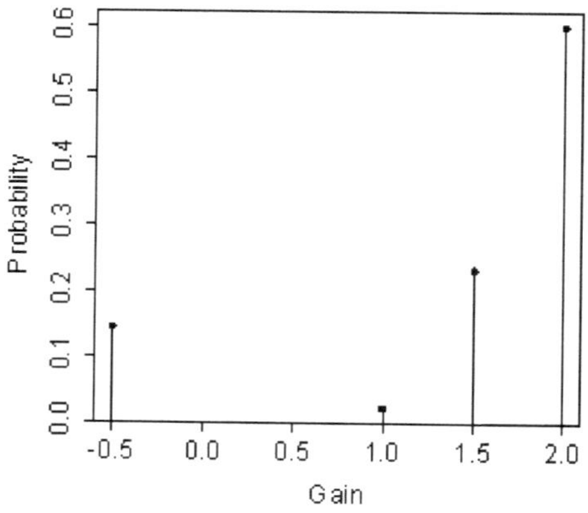

Figure 6. Graphic of the density function for a random variable X: gain.

The graphic of the density of probability can be observed in Fig. 7 for some continuous random variables.

The fdp is also known as probability distribution even when it is not a proper term in the case of the discrete random variables, since the function doesn't generate the probabilities.

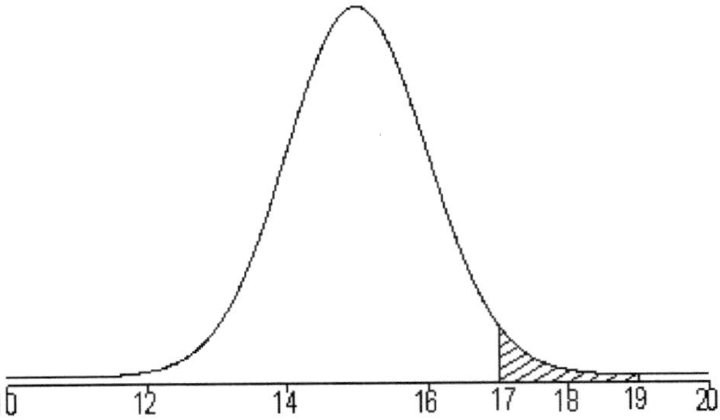

Figure 7. Example of graphic of the density probability for a continuous random variable.

2.2 Distribution Function

Despite of the random variable being discrete or continuous, we can associated an important function with it: the cumulative distribution function, also known as distribution of probability function or distribution function. With this function we can obtain the probability of the variable to be under some value. It's symbolized as $F(x)$ or $\Phi(x)$. So, $F(x) = P[X \leq x]$.

For a discrete variable we have $F(x_k) = \sum_{i=1}^{k} P[X = x_i]$. In our problem, in order to verify the probability of having a gain under \$2.00, we can calculate:

$$F(x_3) = \sum_{i=1}^{3} P[X = x_i] = P(X = x_1) + P(X = x_2) + P(X = x_3)$$

$$F(x_3) = P(X = -0.5) + P(X = 1.0) + P(X = 1.5) = 0.1450 + 0.0225 + 0.2325 = 0.4000$$

or

$$F(x_3) = 1 - P(X = x_4) = 1 - 0.6000 = 0.4000$$

The graphic of a function of distribution for a discrete random variable is represented by horizontal segments. For our problem, we have Fig. 8.

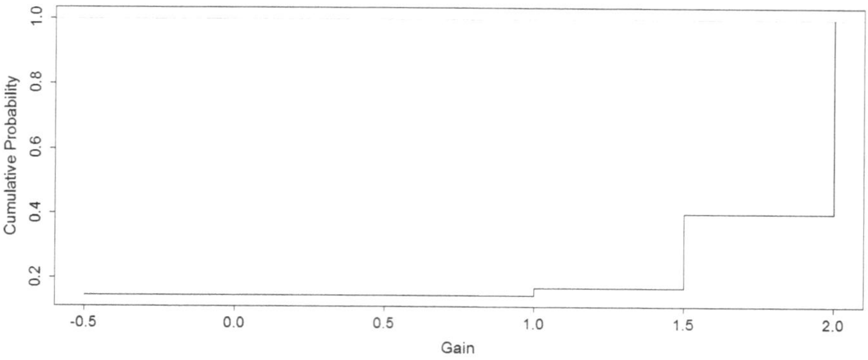

Figure 8. Graphic of a function of distribution for a random variable *X*: Gain.

If the random variable is continuous, then $F(x)$ is $F(X) = \int_{-\infty}^{x} f(z)dz$. For example, if the continuous random variable is *Y*, time of life use (in hours) of a certain electronic component of a certain brand, which function of density is $f(x) = \begin{cases} 0.02e^{-0.02y}, y \geq 0 \\ 0, y < 0 \end{cases}$ its function of distribution would be:

$$F(Y) = \int_{-\infty}^{y} 0.02e^{-0.02x}dz = \int_{-\infty}^{y} 0.02e^{-0.02z}dz = -e^{-0.02x}\Big|_{0}^{y} = -e^{-0.02x} + 1 = 1 - e^{-0.02y}$$

So, the probability of some electronic component of this brand to last 100 hours or less is given by $F(100) = 1 - e^{-0.02(100)} = 0.8650 = 86.50\%$.

For continuous random variables, the graphic for the function of distribution is a continuous curve. For the example of the electronic component, we have Fig. 9.

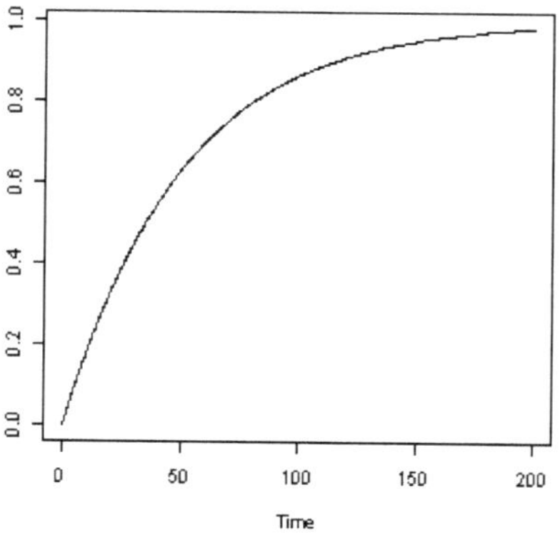

Figure 9. Graphic of the function for the continuous random variable *Y*: time.

3. Expected Value and Variance

The knowledge of the functions of density and distribution of a random variable is very important, but frequently not enough to describe the behavior of the variable. Generally, we want to know the mean value for the variable. This means, if infinity values of the variable were taken, what will be the mean? Do those values tend to concentrate around the mean or are more disperse? The answers to those questions can be obtained by the calculation, respectively, of the expected value - $E[X]$ and the variance $V[X]$, of the random variable X, which are designated as the parameters of the distribution of probability. Even when we have some other measures to describe the random variable, the expected value and the variance are the most important among them.

The expected value of a discrete random variable is given as $E(x)=\sum_{i=1}^{\infty} x_i P[X=x_i]$. In our problem, the gain for selling 1000 products is \$1,500, since:

$$E(x)=\sum_{i=1}^{4} x_i P[X=x_i]=-0.5(0.1450)+1.0(0.0225)+1.5(0.2325)+2.0(0.6000)\cong \$1.50$$

Then, the expected gain on each product will be of \$1.50, and the total expected gain of 1000 pieces will be \$1,500.00.

For continuous random variables, it's enough to take: $E(X)=\int_{-\infty}^{\infty} z.f(z)dz$. In the example of the time of life, use of the electronic component, the mean time will be:

$$E(X)=\int_0^{\infty} z.0.02e^{-0.02z}dz=-x(0.02)e^{-0.02x}\Big|_0^{\infty}+\int_0^{\infty} e^{-0.02x}dx=\frac{1}{0.02}=50 \text{ hours}.$$

The mathematical properties of the expected value are:

(i) $E(kX)=kE(X)$, with k constant;

(ii) $E(k)=k$

(iii) $E(X_1+X_2+...+X_n)=E(X_1)+E(X_2)+...+E(X_n)$, where $X_1, X_2, ...,$ X_n are random variables.

As previously stated in addition to the expected value, we may be interested in assessing if the values are concentrated or not around the mean, in order to do so, we can calculate the variance for the random variable X, which can be defined as: $V(X)=E[X-E(X)]^2=E(X^2)-[E(X)]^2$.

So, the variance for the gain in our problem can be obtained as follows:

$$E\left(X^2\right) = (-0.5)^2.(0.1450) + (1.0)^2.(0.0225) + (1.5)^2.(0.2325) + (2.0)^2.(0.6000) = 2.98$$

$$[E(X)]^2 = 1.50^2 = 2.25$$

$$V(X) = E\left(X^2\right) - [E(X)]^2 = 2.98 - 2.25 = 0.73(\$)^2$$

We must point out that the measure for the variance makes no practical sense, so we can calculate the square root, getting the standard error for the variable, which has a more immediate significance, since it's related to how much the values of the variable are away from the mean. Represented as DP, in our problem the value will be:

$$DP(X) = \sqrt{0.73} \cong \$\ 0.85$$

Variance also has certain properties:

(i) $V(X + k) = V(X)$

(ii) $V(kX) = k^2 V(X)$

Proposed exercises

1. An uneven coin has a probability for head of 0.35. For two independent tosses of that coin, study the behavior of the random variable number of heads, draw the graphs for the discrete probability density function and distribution and calculate the expected value and variance.
2. The useful life for a lamp, in hours, is modeled by a random variable X with the following discrete probability density function.

X	25	26	27	28	29	30
$P[X = x_i]$	0.05	0.10	0.25	0.35	0.15	0.10

Each lamp costs \$8.00, but if it lasts less than 25 hours, the manufacturer pays \$12.00 to the costumer. How much does the manufacturer have to sell each lamp in order to get an average gain of \$15.00?

3. Each year a factory spends a quantity represented by a random variable X with a probability density function in machinery maintenance given by:

$f_X(x) = \frac{4}{5}x - \frac{2}{5}$, if $200 < x < 400$. Obtain: $P(X < 300)$, $E[X], V[X]$.

4. Check if the following expressions are probability density functions (assume that they are null outside the specified intervals);

   a. $f_X(x) = \frac{1}{40}\left(\frac{x}{10} + 1\right)$, if $0 \le x \le 20$

   b. $f_X(x) = \frac{x^2}{2}$, if $x \ge 0$

   c. $f_X(x) = \frac{(x-3)}{2}$, if $3 \le x \le 5$

4. Discrete Probabilistic Model

Some random phenomena of discrete random variables can be adjusted to mathematical models in order to determine the probability of some value of the random variable to happen, using the very value and the parameters as the mean and the variance. Other random phenomena cannot be modeled, as it is in our problem, since for the same value of the random variable X we have different values of probability; for example, for the value -0.5 we have five different values of probability: 0.080, 0.0150, 0.0375, 0.0075 e 0.0050 (Fig. 10).

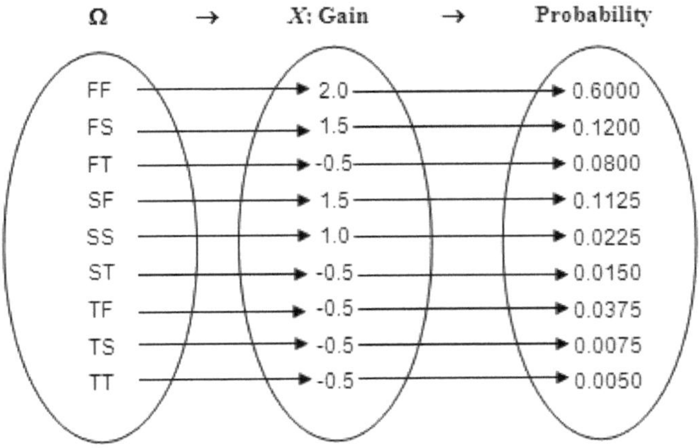

Figure 10. Graphic representation for the random variable gain with the associated probabilities.

In this section, we discussed the phenomena of discrete random variables that can be adjusted to mathematical models and have proper names for the models, like Uniform, Bernoulli, Binomial, Hypergeometric, Geometric, Negative Bonimial, Poisson Distributions.

4.1 Uniform Discrete Distribution

Problem Situation 1: Consider an urn with 100 balls labeled with numbers from 1 to 100. If you take **one** ball randomly, what's the probability of that ball to be the number 24?

In this problem situation the sample space is Ω = {1,2,...,100}. If we consider the random variable $X(.)$ the number on the ball, it can take values $x = 0,1,...,100$ and the probability of picking any of the balls is the same, this is $1/100$.

We observe that we can generalize this kind of situation, considering a random variable $X(.)$ that can take a finite number of N values, every one of them with the same probability $1/N$. So, we can say that:

A random variable X is defined as Discrete Uniform distribution if the discrete density function for X is given by:
$f_X(x) = f(x, N) = P(X = x) = \dfrac{1}{N}$ for $x = 0,1,...,N$. The parameter for the distribution is the number of elements N.

The expected value and the variance for a Discrete Uniform can be determined by the following expressions:

$$\begin{cases} E(X) = \sum_{x=1}^{N} xP[X = x] = \sum_{x=1}^{N} x\dfrac{1}{N} = \dfrac{N(N+1)}{2}\left(\dfrac{1}{N}\right) = \dfrac{(N+1)}{2} \\ V(X) = E(X^2) - [E(X)]^2 = \sum_{x=1}^{N} x^2\dfrac{1}{N} - \left(\dfrac{N+1}{2}\right)^2 = \dfrac{N(N+1)(N+2)}{6} - \dfrac{(N+1)^2}{4} = \dfrac{(N+1)(N-1)}{12} \end{cases}$$

P.S: Expression $\sum_{x=1}^{N} x$ represents the sum of N terms of an arithmetic progression with ratio 1, so $\sum_{x=1}^{N} x = \dfrac{N(N+1)}{2}$.

4.2 Bernoulli Distribution

Problem Situation 2: Consider an urn with 4 green balls and 12 yellow balls. If you pick one ball randomly and observe the color, what's the probability of that ball to be green?

In this problem situation the sample space is Ω = {ball yellow, ball green}. If we consider the random variable $X(.)$ as the outcome of a green ball in one pick, it can take values $x = 0$ or $x = 1$. So, in order to calculate the probabilities of each event we must consider the numbers of balls of each color; in this case, if we have 4 green balls (G event) out of the total of 16 balls, the probability of picking a green ball can be calculated considering $P(G) = p = \frac{4}{16} = 0.25$, and the probability of picking a yellow ball as $P(Y) = (1-p) = 0.75$. We observe that these evens are not equiprobable. We can represent this situation in the following form (Fig. 11):

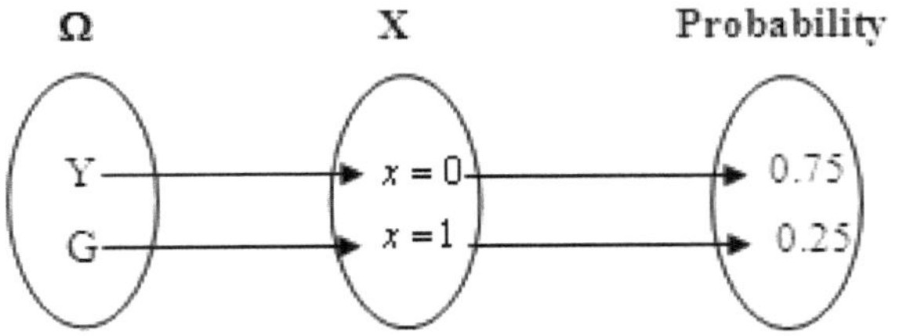

Figure 11. Graphic representation of the random variable picking of a green ball with the associated probabilities.

Can we think about a mathematical model to generalize this kind of situation depending on *x* and *p*? In this problem situation we can verify that $\begin{cases} \text{para } x=0, & P(X=0)=(1-p) \\ \text{para } x=1, & P(X=1)=p \end{cases}$. Let's have the properties: any number raised to a zero power equals one; any value raised to power one, equals itself; 1 is the neuter of multiplication, so we can represent those two expressions in the form $\begin{cases} P(X=0)=p^0(1-p)^1 \\ P(X=1)=p^1(1-p)^0 \end{cases}$.

In those expressions we verify that the power for *p* can take the value of 0 or 1, that corresponds to the occurrence of a green ball in one pick, this is, the values of the random variable X. The power of $(1-p)$ will be the complement of *x*, that means $(1-x)$. With those generalizations, we have $P(X=x)=p^x(1-p)^{1-x}$.

This kind of discrete probability distribution where the random variable is classified just from two possible outcomes, like whether the green ball is picked up or not, the plant is attacked by an insect or not, feminine or masculine, is denominated as "Bernoulli Distribution". The event where the random variable takes the value 1, is denominated as success and the probability *p* of that event to happen can be named as probability of success. So, we can say that:

A random variable X is defined as Bernoulli distribution if the discrete density function for X is given by:
$f_X(x) = f(x,p) = P(X = x) = p^x(1-p)^{1-x}$ for $x = 0$ or $x = 1$. The parameter for the distribution is the probability of success p.

The expected value and the variance for a Bernoulli distribution can be determined by the following expressions:

$$\begin{cases} E(X) = \sum_{x=0}^{1} xP[X = x] = 0.P(X = 0) + 1.P(X = 1) = p \\ V(X) = E(X^2) - [E(X)]^2 = [0^2.P(X = 0) + 1^2.P(X = 1)] - p^2 = p - p^2 = p(1-p) \end{cases}$$

4.3 Binomial Distribution

Problem Situation 3: Consider an urn with 4 green balls (G) and 12 yellow balls (Y). If you pick **three** balls randomly **with replacement**,[2] what's the probability of getting no green balls? 1 green ball? 2 green balls? 3 green balls?

In this problem situation, the sample space is conformed by 8 events ($\#\Omega = 2^3 = 8$)[3], Ω = {(YYY), (YYG), (YGY), (GYY), (YGG), (GYG), (GGY), (GGG) }. Considering that the occurrence of a success (getting a green ball) is associated with a Bernoulli distribution ($P(X = x) = p^x(1-p)^{1-x}$ for $x = 0$ or $x = 1$) we can determine a new random variable X(.) as the number of green balls obtained: $X = \{0, 1, 2, 3\}$. For example, for the event YYY $(X = 0)$, considering that each of the tries are independent, we have that $P(Y \cap Y \cap Y) = P(Y).P(Y).P(Y)$, which is meant to indicate $P(X = 0) = (0.25^0(0.75)^1).(0.25^0(0.75)^1).(0.25^0(0.75)^1) = 0.25^0(0.75)^3$; for the other events we may think in the same way (Fig. 12):

By using the discrete Bernoulli probability function, we can verify in Fig. 12 that the power of p still takes the values of the random variable X. But the power of $(1-p)$ is determined by $(3-x)$, where 3 is the number of balls taken from the urn and if we represent that number by n, the power of $(1-p)$ will be $(n-x)$.We also observe as well in Fig. 13 that an event (YYY) no green ball is picked even once; one green ball is picked three times; two green balls are picked three times; and (GGG) three green balls are picked once.

---

[2] It's worth remembering that in the experiment picking n balls with replacement, is necessary that the 1st ball is taken and then replaced in the urn before the 2nd Ball is taken, and to proceed in that way with (n-1) balls to be taken.

[3] The cardinality (#) of the sample space equals 2^n.

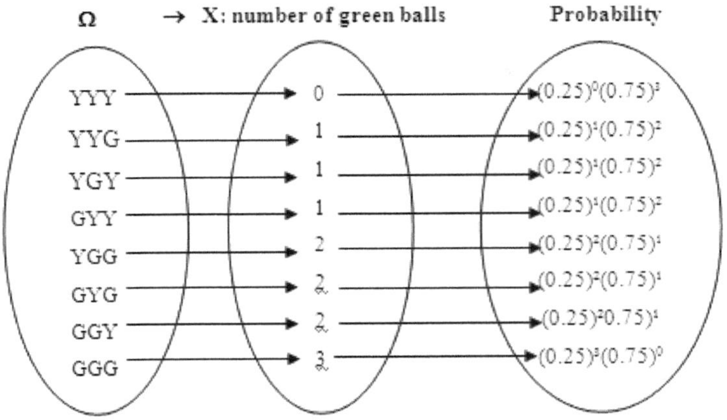

Figure 12. Graphic representation of the random variable number of green balls and the associated probabilities.

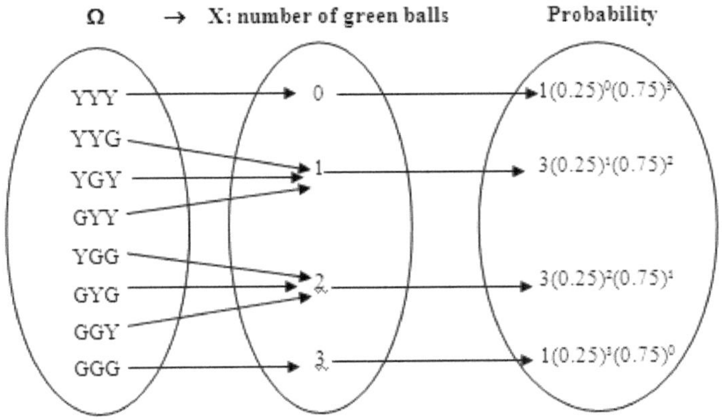

Figure 13. Graphic representation of the random variable number of green balls and the associated probabilities.

And what about the coefficient that multiplies those probabilities? This means, the number of times corresponding to each value of the random variable can be also represented in function of n and x?

If we consider that in each event the green ball can be picked in any of the three times and that independently of the position in which it happens, the value of the random variable X corresponds to the number of green balls observed, we can say that the coefficient that appears in the multiplication p^x times $(1-p)^{n-x}$ will be the combination of n balls picked x by x, this is $C_{n,x} = \binom{n}{x} = \frac{n!}{x!(n-x)!}$. With those generalizations, we have:

$P(X=x)=\binom{n}{x}p^x(1-p)^{n-x}$, that is named as "Binomial Distribution". So, we can say that:

> A random variable X is defined as Binomial distribution if the discrete density function for X is given by:
>
> $f_X(x)=f(x,n,p)=P(X=x)=\binom{n}{x}p^x(1-p)^{n-x}$ para $x=0,1,\ldots,n$. The parameters for the distribution are n and p.

This distribution is also defined as a sum of n independent variables with a Bernoulli distribution, what means that if $n=1$, the Binomial distribution turns into a Bernoulli distribution. Considering that information, the expected value and the variance for a Binomial distribution can be determined by the following expressions:

$$\begin{cases} E(X)=np \\ V(X)=np(1-p) \end{cases}$$

4.4 Hypergeometric Distribution

Problem Situation 4: Consider an urn with 4 green balls (G) and 12 yellow balls (Y). If you randomly take **three** balls **without replacement**,[4] what's the probability of getting no green balls? 1 green ball? 2 green ball? 3 green balls?

In order to make the construction of the sample space in a problem situation easier like this one, let's take a smaller number of balls in the urn, considering just five, three being yellow and two green balls, picking two balls without replacement, and that the random variable $X(.)$, number of green balls observed, can take values 0, 1 or 2.

In this new example, if we think that the three yellow balls are labeled as Y1, Y2 and Y3, and the green balls are labeled as G1 and G2, by taking two balls from the urn, we will have this sample space: Ω = {Y1Y2, Y1Y3, Y2Y3, Y1G1, Y1G2, G2Y1, Y2G2, Y3G1, Y3G2, G1G2}. We observe the order in which the balls appear does not matter, since at the end of the picking, we will have, for example, Y1 and Y2 outside the urn. In that way, if we ignore the number each ball has, we can know in many events we have two yellow balls, considering the 3 yellow balls taken 2 by 2, this means, $C_{3,2}=3$ successful events.

[4]In the experiment without replacement each ball taken out, is not taken back to the urn.

Under the same reasoning, the total number of events in that sample space can be calculated considering that the 5 balls are taken 2 by 2 as well, this is $C_{5,2} = 10$ possible events.

By remembering that observing two yellow balls is associated with the value of the random variable $X = 0$ and that by the classic definition of probability we can calculate the probability of an event by dividing the number of successful events by the number of possible events, we will have $P(X=0)=\dfrac{C_{3,2}}{C_{5,2}}=\dfrac{3}{10}$. For $X = 1$, we will have 1 green ball and 1 yellow ball taken, and then to have the total of successful events we must multiply the combination of 2 green balls taken 1 by 1, by the combination of 3 yellow balls taken 1 by 1, this is $C_{2,1}.C_{3,1}$, then $P(X=1)=\dfrac{C_{2,1}.C_{3,1}}{C_{5,2}}=\dfrac{6}{10}$. For $X = 2$, we will have $P(X=2)=\dfrac{C_{2,2}}{C_{5,2}}=\dfrac{1}{10}$.

Going back, the Problem Situation 3, with 16 balls in the urn and taking out three of them without replacement, that cardinality of the sample space can be calculated as $C_{16,3} = 560$ possible events, and for $X = 0$ we will have $C_{12,3} = 220$ successful events. Then we observe that, different from the case of just five balls in the urn, that determining the 220 associated events to $X = 0$ and the 560 events of the sample space, is not immediate, so there's the need of setting a mathematical model to calculate the probability in similar situations to the presented one.

But, before proceeding to generalize, let's consider once more the example with five balls in the urn. We noted that the probability for $X = 0$ was determined by $P(X=0)=\dfrac{C_{3,2}}{C_{5,2}}$. In order to have in that expression a combination depending on the value of *x*, we can rewrite it as $P(X=0)=\dfrac{C_{2,0}.C_{3,2}}{C_{5,2}}$, since $C_{2,0} = 1$. Then, if we represent the total number of balls, five, as *N*, the number of taken balls, two, as *n* and the number of green balls, two, as *r*, we will have $P(X=0)=\dfrac{C_{r,0}.C_{(N-r),(n-0)}}{C_{N,x}}$. Finally, considering any value for the random variable X, we can generalize the expression as $P(X=x)=\dfrac{C_{r,x}.C_{(N-r),(n-x)}}{C_{N,n}}$.

Being so, and taking back the problem situation 3, we will have:

$$P(X=0)=\frac{C_{4,0}.C_{16-4,3-0}}{C_{16,3}},\ P(X=1)=\frac{C_{4,1}.C_{16-4,3-1}}{C_{16,3}},$$

$$P(X=2)=\frac{C_{4,2}.C_{16-4,3-2}}{C_{16,3}} \text{ and } P(X=3)=\frac{C_{4,3}.C_{16-4,3-3}}{C_{16,3}}.$$

This kind of discrete density is known as "Hypergeometric Distribution" and we can say that:

> A random variable X is defined as Hypergeometric distribution if the discrete density function for X is given by:
>
> $$f_X(x)=f(x,N,n,r)=P(X=x)=\frac{C_{r,x}.C_{(N-r),(n-x)}}{C_{N,n}}, \quad \text{para } x=0,1,\dots,n.$$
>
> The parameters for the distribution are N, n and r, where N is the total of elements, n is the size of the sample or the number of observed elements of the total of N, and r is the number of successes.

The expected value and the variance for a Hypergeometric distribution can be determined by the following expressions:

$$E(X)=\sum_{x=0}^{n} x\frac{C_{r,x}.C_{(N-r),(n-x)}}{C_{N,n}}=\sum_{x=0}^{n} x\frac{\binom{r}{x}\binom{N-r}{n-x}}{\binom{N}{n}}=n\frac{r}{N}\sum_{x=1}^{n}\frac{\binom{r-1}{x-1}\binom{N-r}{n-x}}{\binom{N-1}{n-1}}$$

$$E(X)=n\frac{r}{N}\sum_{j=o}^{n-1}\frac{\binom{r-1}{j}\binom{N-1-r+1}{n-1-j}}{\binom{N-1}{n-1}}=n\frac{r}{N}$$

OBSERVATION: Using the relation $\sum_{i=o}^{m}\binom{a}{i}\binom{b}{m-i}=\binom{a+b}{m}$ we can simplyfy the expression

$$\sum_{j=o}^{n-1}\binom{r-1}{j}\binom{N-1-r+1}{n-1-j}=\binom{N-1}{n-1}, \text{ then } \sum_{j=o}^{n-1}\frac{\binom{r-1}{j}\binom{N-1-r+1}{n-x-j}}{\binom{N-1}{n-1}}=\frac{\binom{N-1}{n-1}}{\binom{N-1}{n-1}}=1.$$

$$V(X)=E\left[X^2\right]-\left(E[X]\right)^2=E\left[X(X-1)\right]+E[X]-\left(E[X]\right)^2=n(n-1)\frac{r(r-1)}{N(N-1)}+n\frac{r}{N}-n^2\frac{r^2}{N^2}$$

$$V(X)=n\frac{r}{N}\left[(n-1)\frac{r-1}{N-1}+1-\frac{nr}{N}\right]=\frac{nr}{N}\left[\frac{(N-r)(N-n)}{N(N-1)}\right]$$

OBSERVATION:

$$E\left[X(X-1)\right]=\sum_{x=0}^{n}x(x-1)\frac{\binom{r}{x}\binom{N-r}{n-x}}{\binom{N}{n}}=n(n-1)\frac{r(r-1)}{N(N-1)}\sum_{x=2}^{n}\frac{\binom{r-2}{x-2}\binom{N-r}{n-x}}{\binom{N-2}{n-2}}$$

$$E\left[X(X-1)\right]=n(n-1)\frac{r(r-1)}{N(N-1)}\sum_{y=0}^{n-2}\frac{\binom{r-2}{y}\binom{N-r}{n-2-y}}{\binom{N-2}{n-2}}=n(n-1)\frac{r(r-1)}{N(N-1)}$$

4.5 Geometric Distribution

Problem Situation 5: Consider an urn with 4 green balls (G) and 12 yellow balls (Y). If you pick a series of balls randomly **with replacement**, what's the probability of observing **the first** green ball?

In this problem situation, the random variable $X(.)$ corresponds to the number of yellow balls observed before the first green ball appears, this is $X = \{0, 1, ...\}$. For example, if the first retired ball is green (G), that means that it didn't happen to be a yellow ball and this corresponds to the event $X=0$, if the second ball were green (G) this corresponds to $X=2$, and so on. That's saying the experiment is interrupted when the first green ball appears.

If in the first pick, the ball is green, $X=0$, the probability $P(X=0)$ will be the same probability of having a green ball $p=0.25$. If we first take a yellow ball and then a green ball, $X=1$, since the picks are independent, the probability will be $P(X=1)=(1-p)p=(0.75)0.25$. If the third ball is green, 2 yellow and then 1 green, $X=2$, the probability will be $P(X=2)=(1-p)(1-p)p=(1-p)^2p=(0.75)^2 0.25$. For $X=3$, the probability will be $P(X=3)=(1-p)(1-p)(1-p)p=$ $(1-p)^3p=(0.75)^3 0.25$, and so on.

Observing those expressions of the calculated probabilities, we verify that the power for $(1-p)$ takes the value of x, and that the p value will always appears just one time, since, as stated before, the experiment is interrupted when the first green ball appears. Considering all that we can generalize: $P(X=x)=(1-p)^x p$.

A discrete probability distribution in which the random variable is the number of independent trials of a Bernoulli experiment until the first success happens, is known as "Geometric Distribution". In our problem situation, we have n–1=4 yellow balls (our failure, or F) until a green ball happens for the first time (our success, or S) on the nth=5 trial. So, we can say that:

> A random variable X is defined as Geometric distribution if the discrete density function for X is given by: $f_X(x) = f(x, p) = P(X = x) = (1-p)^x p$, para $x = 0,1,\ldots$. The parameter for the distribution is p.

The expected value and the variance for a geometrical distribution can be determined by the following expressions:

$$E(X) = \sum_{x=0}^{\infty} x(1-p)^x p = p\sum_{x=1}^{\infty} x(1-p)^x = p(1-p)\sum_{x=1}^{\infty} x(1-p)^{x-1} = \frac{p(1-p)}{p^2} = \frac{1-p}{p}$$

$$V(X) = \frac{(1-p)}{p^2}$$

OBSERVATION: Using the relation $\sum_{i=1}^{\infty} ix^{(i-1)} = \frac{1}{(1-x)^2}$ (resulting by derivation the geometric series to $\sum_{i=1}^{\infty} x^i = \frac{1}{(1-x)}$) and considering $x = 1-p$ can simplify the expression $\sum_{x=1}^{\infty} x(1-p)^{(x-1)} = \frac{1}{p^2}$.

4.6 Negative Binomial Distribution

Problem Situation 6: Consider an urn with 4 green balls (G) and 12 yellow balls (12). If you take out a series of balls randomly **with replacement**, what's the probability of observing the first two green balls?

In this problem situation the random variable $X(.)$ corresponds to the number of yellow balls taken before we observe two green balls, this is X = {0, 1, ...}. For example, we can take out two green balls in the first and second pick and in that case $X = 0$, or to have the two green balls in the second and third pick (YGG) and we will have $X = 1$. But take note that if we have the two green balls in the first and third pick (GYG), we also have $X = 1$. For $X = 2$ we could have this events: YYGG, YGYG, GYYG; for $X = 3$ the events may be: YYYGG, YYGYG, YGYYG, GYYYG.

The probability of observing two green balls, $X = 0$, considering that the picks are independent, will be $P(X = 0) = p.p = p^2 = (0.25)^2$, that can be re written as $P(X = 0) = (1-p)^0 p^2 = (0.75)^0 (0.25)^2$.

For $X = 1$ we will have 1 yellow ball and two green balls, but with 1 green ball and 1 yellow ball that can alternate their positions of occurrence and the other green ball fixed at the last pick, precisely because the experiment is interrupted when it reaches two green balls observed. In the case of $X = 1$ we can establish that the number of successful events will be obtained by the combination of the two possible places in which the yellow ball may come out -YGG or GYG- before we get two green balls; this is $C_{2,1}=2$, and the probability is going to be calculated as $P(X=1)=C_{2,1}(1-p)(p)(p)=C_{2,1}(1-p)^1 p^2=2(0.75)^1(0.25)^2$. Under the same reasoning, we will have that

$$P(X=2)=C_{3,2}(1-p)(1-p)(p)(p)=C_{3,2}(1-p)^2 p^2=3(0.75)^2(0.25)^2$$

and $P(X=3)=C_{4,3}(1-p)(1-p)(1-p)(p)(p)=C_{4,3}(1-p)^3 p^2$, and so on.

Under these considerations we may generalize, obtaining: $P(X=x)=C_{k+x-1,x}(1-p)^x p^k$. Observing the expressions for the calculated probabilities, we verify that the power of $(1-p)$ takes the value of *x*, and the power of *p* will always be the number of green balls needed to interrupt the experiment (k). About coefficients, we observe that the combination will always be of $k+x-1$ elements taken *x* by *x*, this is $C_{k+x-1,x}$.

This kind of discrete probability distribution in which the random variable corresponds to the number of failures (F), in the problem situation 6 yellow balls, until we have k successes (S), picking of green balls, in which a typical event of the random space is a sequence of length n in which we have F in $n-k$ and in the *n*-th position we have a S, is named as "Negative Binomial Distribution". So, we can say that:

> A random variable X is defined as Negative Binomial distribution if the discrete density function for X is given by:
>
> $f_X(x)=f(x,k,p)=P(X=x)=C_{k+x-1,x}(1-p)^x p^k$, for $x=0,1,\ldots$. The parameters for the distribution are *k* e *p*.

It's worth pointing out that the Geometric distribution is a particular case of the Negative Binomial, this is when $k=1$. Based on this information, the expected value and the variance for a Negative Binomial distribution can be determined by the following expressions:

$$\begin{cases} E(X)=\dfrac{k(1-p)}{p} \\ V(X)=\dfrac{k(1-p)}{p^2} \end{cases}$$

4.7 Poisson Distribution

Problem Situation 7: A poll made on a highway along several months showed the probability of severe accident is 0.5%. A person is going to use that highway in one year. What's the probability of that person to have severe accidents?

Problem Situation 8: The probability of an individual to show an allergic reaction to a serum is 0.2%. This serum was applied to a group of 1800 persons. What's the probability for two of them to show an allergic response?

Problem Situation 9: A fire station takes a mean of 3 calls in a day. What's the probability that on a certain day they take 7 calls?

To solve problem situations 7 and 8 we can use the discrete probability density function Binomial, since we have situations in which we have just two possible outcomes (severe accident or not in problem 7, allergic reaction or not in problem 8), the number of tries n (200 in problem 7, 1800 in problem 8) and the probability p of success (0.5% in problem 7, 0.2% in problem 8). The question is that calculations for the searched probabilities are extremely tedious and maybe inaccurate when using the Binomial model $\left(P(X=x)=C_{n,x}p^x(1-p)^{n-x}\right)$.

Solution for problem 7

$$P(X \le 4)=C_{200,0}0.005^0(0.995)^{200}+C_{200,1}0.005^1(0.995)^{199}+C_{200,2}0.005^2(0.995)^{198}+C_{200,3}0.005^3(0.995)^{197}+$$
$$+C_{200,4}0.005^4(0.995)^{196}$$

Solution for problem 8

$$P(X=2)=C_{1800,2}0.002^2(0.998)^{1798}$$

In problem 9 we can observe that we don't have the values of n and p, so is it possible to use the Binomial model? Could it be possible that for problems 7 and 8 we could use a model different from the Binomial model?

In 1830 Siméon Denis Poisson developed a distribution that got his name, based on the Binomial distribution. Mathematically speaking, the Poisson distribution is a limit case of the Binomial when the number of tries n ($n>0$) tends to be infinite, the probability of the event p ($0<p<1$) in each try tends to be zero and the product np tends to be a constant λ; that is, this is a Binomial distribution in which:

a) $n \to \infty, p \to 0$

b) $np \to \lambda > 0$

So, let's be X a random variable with no negative values and starting from a Binomial model, we have:

$$P[X=x]=\frac{n!}{x!(n-x)!}p^x(1-p)^{n-x}=\frac{n(n-1)...(n-y+1)}{x!}p^x(1-p)^{n-x} \qquad (1)$$

defining $\lambda = np$, we have $p = \lambda/n$ e $1 - p = (n-\lambda)/n$. Substituting the values in 1:

$$P[X=x] = \frac{n(n-1)...(n-y+1)}{x!}\left[\frac{\lambda}{n}\right]^x \left[\frac{n-\lambda}{n}\right]^{n-x}$$

$$P[X=x] = \frac{\lambda^x}{\text{x}!}\frac{n(n-1)...(n-x+1)}{n^x}\left[1-\frac{\lambda}{n}\right]^n \left[1-\frac{\lambda}{n}\right]^{-x} \tag{2}$$

Note that:

$$\frac{n(n-1)...(n-x+1)}{n^x} = \frac{(n\text{-}1)...[n\text{-}(x\text{-}1)]}{n^{x\text{-}1}} = \frac{(n\text{-}1)}{n}\frac{(n\text{-}2)}{n}...$$
$$...\frac{[n\text{-}(x\text{-}1)]}{n} = \left[1\text{-}\frac{1}{n}\right]\left[1\text{-}\frac{2}{n}\right]...\left[1\text{-}\frac{x\text{-}1}{n}\right] \tag{3}$$

And substituting (3) in (2):

$$P[X=x] = \frac{\lambda^x}{x!}\left[1\text{-}\frac{1}{\text{n}}\right]\left[1\text{-}\frac{2}{\text{n}}\right]...\left[1\text{-}\frac{\text{x-1}}{\text{n}}\right]\left[1\text{-}\frac{\lambda}{\text{n}}\right]^n\left[1\text{-}\frac{\lambda}{\text{n}}\right]^{\text{-x}} \tag{4}$$

Making $n \to \infty$, the expression of the kind $\left[1-\frac{a}{n}\right]$ tends to be 1, even $\left[1-\frac{\lambda}{n}\right]^{-x}$. The term $\left[1-\frac{\lambda}{n}\right]^n$, by the definition of neperian number, tends to be $e^{-\lambda}$ when $\text{n} \to \infty$. And so, the Poisson distribution can be presented as:

$$P[X=x] = \frac{\lambda^x e^{-\lambda}}{x!} \tag{5}$$

This model corresponds to a probability distribution, observing that:

$$\sum_{x=0}^{\infty} P[X=x] = \sum_{x=0}^{\infty}\frac{\lambda^x e^{-\lambda}}{x!} = e^{-\lambda}e^{\lambda} = 1 \tag{6}$$

The expected value of X is given by:

$$E[X] = \sum_{x=0}^{\infty} x\frac{\lambda^x e^{-\lambda}}{x!} = \sum_{x=0}^{\infty}\frac{\lambda^x e^{-\lambda}}{(x-1)!} \tag{7}$$

Defining $t = x - 1$ and substituting in 7, we have:

$$E[X] = \sum_{x=0}^{\infty} x \frac{\lambda^{t-1} e^{-\lambda}}{t!} = \lambda \sum_{x=0}^{\infty} \frac{\lambda^{t} e^{-\lambda}}{t!} = \lambda .1 = \lambda \tag{8}$$

Observe that t has a Poisson distribution. Then, the parameter λ corresponds to the mean value of the random variable. For the variance of X, we have that:

$$E\left[X^2\right] = \sum_{x=0}^{\infty} x^2 \frac{\lambda^{x} e^{-\lambda}}{x!} = \sum_{x=0}^{\infty} x \frac{\lambda^{x} e^{-\lambda}}{(x-1)!} \tag{9}$$

Once again, defining $t = x - 1$ and substituting in 9, we have:

$$E\left[X^2\right] = \sum_{x=0}^{\infty} (t+1) \frac{\lambda^{t+1} e^{-\lambda}}{t!} = \lambda \sum_{x=0}^{\infty} t \frac{\lambda^{t} e^{-\lambda}}{t!} + \lambda \sum_{x=0}^{\infty} \frac{\lambda^{t} e^{-\lambda}}{t!} = \lambda^2 + \lambda \tag{10}$$

And so:

$$V(X) = E(X)^2 - [E(X)]^2 = \lambda^2 + \lambda - \lambda^2 = \lambda \tag{11}$$

The Poisson distribution gives a realistic model for many random phenomena, and since the values for the random variable are not negative integers, random phenomena involving some kind of counting would have the Poisson distribution as a possible model.

Solved exercises

1. A factory A of equipment for cars owes \$130,000 to one provider (factory B). Each visit of the collector from factory B costs \$200. The probability for the collector to find the owner of factory A at his office is 1/3. If the collector finds the owner, he will cash the debt.
   a) What's the probability for the collector to have to pay more than 3 visits to factory A in order to cash the debt?
   b) If in the second visit to factory A the collector doesn't find the owner, what's the probability for the collector to cash the debt in the third visit?
   c) What's the expectation for the collector of factory B related to the cash value that he will have after cashing the debt?

Solution

It's an exercise of Geometrical distribution, since the random variable X(.) corresponds to the number of visits made by the collector from factory B until he aims to find the owner of factory A and cash the debt. In other words,

$P(X=0)$ is the probability of finding the owner of factory A in the first visit

$P(X=1)$ is the probability of finding the owner of factory A in the second visit

$P(X=2)$ is the probability of finding the owner of factory A in the third visit and so on.

Each visit is independent of the one before, what is similar to the replacement of the balls in the urn, on the previous experiments, since the initial conditions are the same for every try.

a) Probability for the collector has to pay more than three visits

$$P(X>3)=1-[P(X=0)+P(X=1)+P(X=2)]$$

$$P(X=x)=(1-p)^x p$$

$$P(X=0)=\left(1-\frac{1}{3}\right)^0\frac{1}{3}=\frac{1}{3}$$

$$P(X=1)=\left(1-\frac{1}{3}\right)^1\frac{1}{3}=\frac{2}{9}$$

$$P(X=2)=\left(1-\frac{1}{3}\right)^2\frac{1}{3}=\frac{4}{27}$$

Then, $P(X>3)=1-[P(X=0)+P(X=1)+P(X=2)]=1-\left(\frac{1}{3}+\frac{2}{9}+\frac{4}{27}\right)=\frac{8}{27}$

b) Probability for the collector to find the owner of factory A exactly at the third visit

This is a case of conditional probability, since the problem requires the collector to find the probability of finding the owner of factory A in the third visit, given that the former visited factory A two times already. Symbolically,

$$P(X=2\,|\,X>1)=\frac{P(X=2)\cap P(X>1)}{P(X>1)}$$

In the solution of item a9 we've got P(X=2) already. So, let's calculate P(X>1)

$$P(X>1)=1-[P(X=0)+P(X=1)]=1-\left(\frac{1}{3}+\frac{2}{9}\right)=\frac{4}{9}$$

$$\frac{P(X=2)\cap P(X>1)}{P(X>1)}=\frac{P(X=2)}{P(X>1)}=\frac{\frac{4}{27}}{\frac{4}{9}}=\frac{1}{3}$$

Observe that the probability of finding the owner of factory A at the third visit after two visits have been paid $P(X=2\,|\,X>1)$ is the same as finding the owner in the first visit $P(X=0)$. This fact comes from the characteristic of the Geometrical distribution of not having any memory.

c) Expectation of the cash value to get

Solution to this part requires to calculate the mean of the visits which is given by the expected value for the Geometric distribution.

$E(X)=\frac{1-p}{p}=\frac{\frac{2}{3}}{\frac{1}{3}}=2$ that refers to find the owner of factory 2 in the third visit.

Since the collector from factory B will have to go three times to the factory, at a cost of \$200 each time, then the expected cash value to get is 130,000 – 600 = \$129,400.

2. A company got a contract to melt 3 jet turbines. The probability of getting the accurate mold is 0.4, and being that mold destroyed when the piece is retired from the mold. The cost of each mold is \$50,000 and if the mold is inaccurate, the turbine becomes scrap with a loss of \$15,000 in materials.
   a) What's the probability of melting a maximum of 6 jet turbines in order to fulfill the contract?
   b) What's the price to be charged for the service in order to have an expected gain of \$30,000?

Solution

It's an exercise of Negative Binomial distribution, since it deals with getting three good turbines, that correspond to the random variable $X(.)$ number of lost turbines until getting 3 good turbines.

P(X = 0) is the probability of getting the three first turbines as good (not losing any turbine).

P(X = 1) is the probability of losing one turbine before getting the three good turbines.

P(X = 2) is the probability of losing two turbines before getting the three good turbines, and so on.

a) The probability of melting a maximum of 6 turbines until getting 3 good turbines means to calculate the probability of losing a maximum of three turbines.

$$P(X \leq 3) = P(X = 3) + P(X = 2) + P(X = 1) + P(X = 0)$$

$P(X = x) = C_{k+x-1,x}(1-p)^x p^k$, where k = 3, the number of good turbines needed to interrupt the experiment.

$$P(X=3) = C_{5,3}(0.6)^3(0.4)^3 = 10 \cdot 0.216 \cdot 0.064 = 0.1382$$

$$P(X=2) = C_{4,2}(0.6)^2(0.4)^3 = 6 \cdot 0.36 \cdot 0.064 = 0.1382$$

$$P(X=2) = C_{3,1}(0.6)^1(0.4)^3 = 3 \cdot 0.6 \cdot 0.064 = 0.1152$$

$$P(X=1) = C_{2,0}(0.6)^0(0.4)^3 = 1 \cdot 1 \cdot 0.064 = 0.064$$

Then, $P(X \leq 3) = 0{,}1382 + 0{,}1382 + 0{,}1152 + 0{,}064 = 0{,}4556$

b) The price to charge in order to get an expected gain of 30,000 for the contract

The cost of manufacturing those pieces is determined by the equation $C = 50{,}000(X+3) + 15{,}000X$, where X is the expected value of the random variable.

Let's calculate the expected value for the negative binomial distribution, which will give the expected number of lost turbines until we get 3 good turbines.

$E(X) = \dfrac{k(1-p)}{p} = \dfrac{3(0.6)}{0.4} = 4.5$. If the expected value for the number of lost turbines until getting three good turbines is 4.5, then the expected number of manufactured turbines is 7.5.

Then,

$$C = 50{,}000(X+3) + 15{,}000X = (50{,}000 \cdot 7.5) + (15{,}000 \cdot 4.5) = 442{,}500$$

So, they must charge $472,500.- for the service in order to obtain a gain of $30,000.-

3. The number of assistance calls received by a call center is a Poissin process with a rate of 4 each hour.

a) Calculate the probability of receiving 10 assistance calls in any term of two hours.
b) If the operators of the service take 30 minutes for lunch, what's the probability of not losing any assistance call?
c) How many assistance calls can you expect to come during lunch?

Solution

This is a case similar to that proposed in Problem 9 on section 4.7, in which we have a rate of 4 assistance calls in one hour ($\lambda = 4$) and the random variable is $X(.)$ the number of assistance calls within a certain interval of time.

a) The probability of getting exactly 10 assistance calls in any term of two hours

In a two hours term we have $\lambda = 4 \cdot 2 = 8$ and we apply Poisson to get $P(X = 10)$

$$P(X = 10) = \frac{e^{-\lambda} \cdot \lambda^x}{x!} = \frac{e^{-8} \cdot 8^{10}}{10!} = 0.0993$$

b) The probability of not losing any assistance call during lunch time, that lasts 30 minutes

In half of an hour we have $\lambda = 4 \cdot 0.5 = 2$

Calculating the probability of not losing any assistance call means to calculate the probability of zero assistance calls.

$$P(X = 0) = \frac{e^{-\lambda} \cdot \lambda^x}{x!} = \frac{e^{-2} \cdot 2^0}{0!} = 0.1353$$

4. The National Council of Engineering and Architecture is interested in knowing the engineers' opinion about legalizing drugs. Considering that the probability of someone being on favor is 0.2, what's the probability, when picking 5 engineers, to have
   a) Exactly 5 of them on favor
   b) Exactly 5 of them against
   c) A maximum of 1 on favor

Solution

Supposing that one engineer's answer is independent of some other's answer and that the probability of a positive response is always the same, the solution lays on a Binomial model, with a random variable $X(.)$ number of engineers in favor of the decriminalization of the drugs with the values for the parameters $n = 5$ e $p = 0.2$.

a) Exactly 5 of them are in favor

$$P(X=5)=C_{5,5}(0.2)^5(0.8)^0=0.0003$$

b) Exactly 5 of them are against

If the 5 engineers are against it, that means that no one is in favor and hence we are interested in the probability of zero engineers in favor.

$$P(X=0)=C_{5,0}(0.2)^0(0.8)^5=0.3277$$

c) A maximum of 1 in favor

We want to calculate $P(X \le 1)=P(X=0)+P(X=1)$

Since we have $P(X=0)$ already, we just need to calculate $P(X=1)$

$$P(X=1)=C_{5,1}(0.2)^1(0.8)^4=0.4096$$

Then, $P(X \le 1)=P(X=0)+P(X=1)=0.3277+0.4096=0.7373$

Proposed exercises

1. Taking back problem situations 2 to 6 (P2 to P6):
   a) Repeat 20 times the proposed experiments in each of the problem situations. Write down the results in tables as:

Experiment	1	2	3	4	5	6	7	8	9	10	11	12	13	14	15	16	17	18	19	20
Number of balls of green (P2, P3 and P4) or yellow (P5 and P6)																				

   b) Calculate the sample mean (or expected value) and sample variance for each problem situation.
      b.1) Compare your results with other colleagues. Write down your conclusions.
      b.2) Compare your results with $E(X)$ and $V(X)$. Write down your conclusions.
   c) Repeat 10,000 times the experiment proposed in each one of the problem situations with the R software using the following routines. Compare the simulation results with the random experimentations and write down your conclusions.

```
#Problem Situation 2
P2=matrix(rbinom(10000,1,0.25),10000,1)
mean(P2)
var(P2)
#We use the function rbinom to create the Bernoulli, once that the Binomial are n Bernoulli tries.
```

```
#Problem Situation 3
P3=matrix(rbinom(10000,3,0.25),10000,1)
mean(P3)
var(P3)
#In R size is the number of picked balls (n)
```

```
#Problem Situation 4
P4=matrix(rhyper(10000,4,12,3),10000,1)
mean(P4)
var(P4)
#In R nn is the number of experiments, m the number of green balls (r), n the number of yellow balls (N-r) and k the number of picked balls (n).
```

```
#Problem Situation 5
P5=matrix(rgeom(10000,0.25),10000,1)
mean(P5)
var(P5)
```

```
#Problem Situation 6
P6=matrix(rnbinom(10000,2,0.25),10000,1)
mean(P6)
var(P6)
#In R size is the number of green balls to interrupt the experiment (k).
```

2. Using the package Rcmdr, pick the option Distributions, Discrete Distributions and construct the graphs for the discrete density function (mass function graph) and for the distribution function (cumulative function graph) for the problem situations 2 to 6 (remembering that the Bernoulli distribution can be generated considering the Binomial distribution with a number of tries equal to 1). Observe the results and construct the graphs with different parameters (Fig. 14). Write down your conclusions.

3. Using the package Rcmdr and activating in tools RcmdrPlugin. TeachingDemos, pick the opitions Distributions, Visualize Distributions and Binomial Distributions.

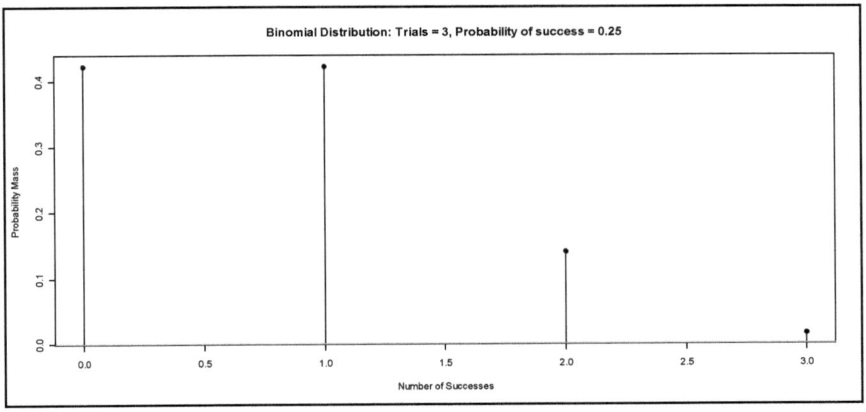

Figure 14. Example of the discrete density function for the Binomial distribution with n=3, p =0.25 generated in the package Rcmdr, R software.

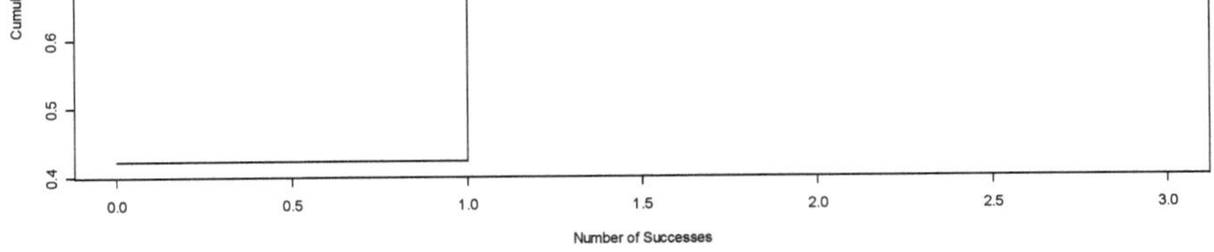

Figure 15. Example of the Binomial probability distribution function with n=3, p=0.25 generated in the package Rcmdr, R software.

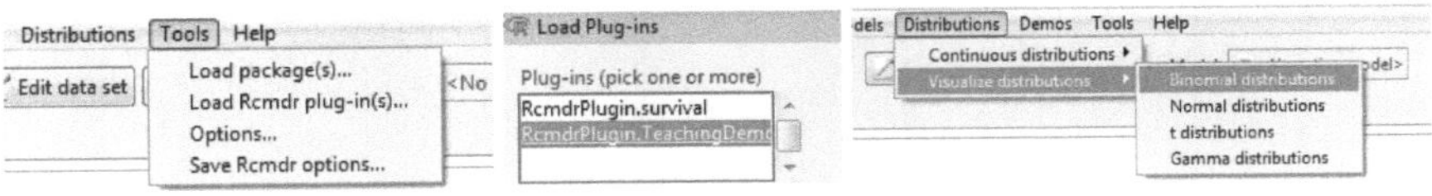

Figure 16. Steps to initialize visualization of Binomial distribution.

Color image of this figure appears in the color plate section at the end of the book.

a) Start by fixing the parameter n and varying p, and observe the graphs generated (Fig. 17). Next, turn it back, change n and fix p. Write down you conclusions.

b) Click on the option Show Poisson Approximation, verify and repeat the same procedure, fixing n and varying p, and vice-versa (Fig. 18). Write down your conclusions.

4. Because of the 100th anniversary of the Company X, the owner decided to raffle a car among his workers. One of them, John, got four tickets (41, 42, 43 and 44), and another one, Rafael, got also four numbers (3, 45, 60 and 83). Rafael says that he has a higher chance of wining the car since his numbers are not sequenced. Is his reasoning correct? Justify

5. Solve problem situations 7 and 8 using Binomial and Poisson and compare results.

6. The Council of Engineering an Architecture of a small city has just 15 associated engineers. We know that the proportion of engineers in favor of the decriminalization of the drug is 20%. A Non Governmental Organization that helps drug's dependants is hiring an engineer. If they take 4 engineers for an interview, what's the probability of all of them to be on favor of drug's decriminalization?

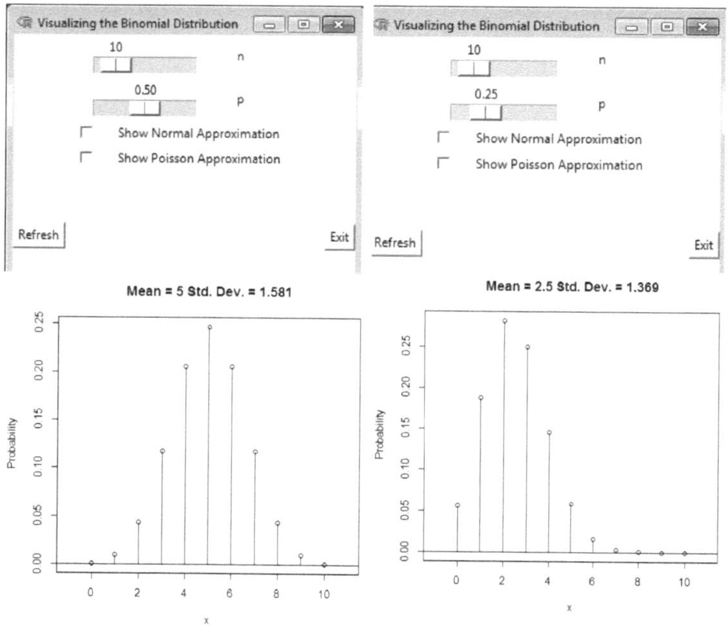

Figure 17. Examples of the graphs generated in the package Rcmdr, R software, for the binomial distributions with parameters n =10, p=0.50 (left graph) and p=0.25 (right graph).

Color image of this figure appears in the color plate section at the end of the book.

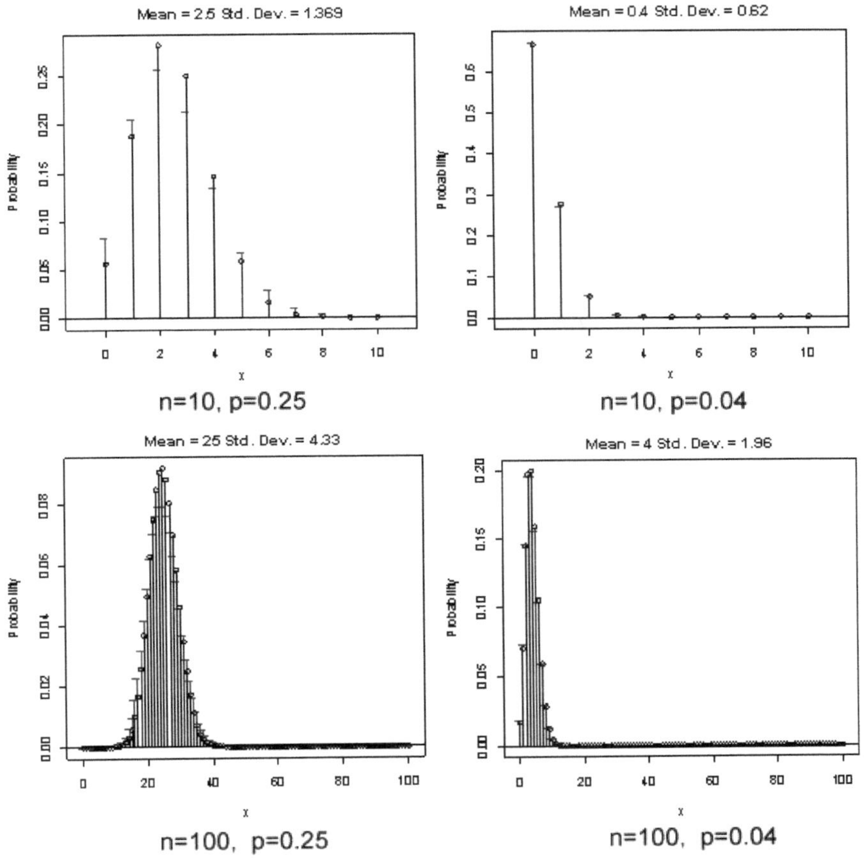

Figure 18. Examples of the graphs generated in the package Rcmdr, R software, for binomial distributions with different parameters and compared to Poisson distribution.

Color image of this figure appears in the color plate section at the end of the book.

5. Continuous Probabilistic Model

5.1 Uniform distribution

Going back to the spinner in Fig. 19, the random variable X, "number on the spinner pointed by the arrow", ranges from 1 to 9 and it can take any real value in the interval [1, 9]; so, the sample space is an infinite set, and the variable cannot be counted, but just measured.

Let's go further with the analysis of our spinner game by constructing the graph for the experiment, associating the possible values of X with their probabilities. Since we have divided the spinner in eight equal intervals, as shown in Fig. 3b, our graph will show the probabilities of each of those intervals, as shown in the Table 1:

Table 1. Distribution of probabilities for each sector around the spinner.

Interval $[x_1, x_2]$	$[1,2]$	$[2,3]$	$[3,4]$	$[4,5]$	$[5,6]$	$[6,7]$	$[7,8]$	$[8,9]$
Probability $P(x_1 \le X \le x_2)$	$1/8$	$1/8$	$1/8$	$1/8$	$1/8$	$1/8$	$1/8$	$1/8$

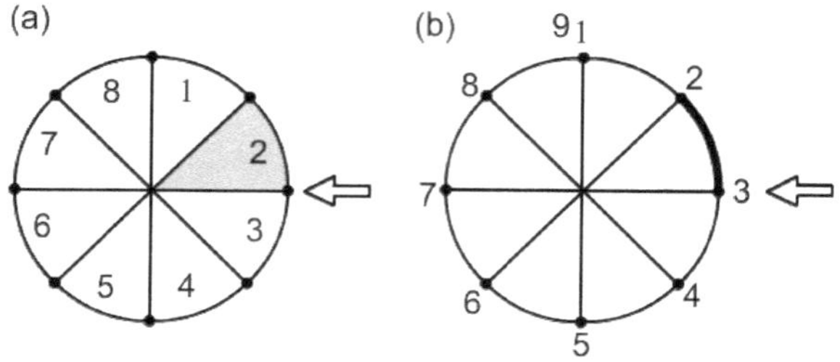

Figure 19. (a) Proportion of the spinner corresponding to X=2; (b) Arc corresponding to values from 2 to 3.

The graph then will be the histogram shown in Graph 1 (Fig. 20).

In this case the curve describing the probabilities along the sample space [1, 9] for the experiment is the constant function $f(x) = 1/8$. In order to describe this function more generally, we can say that the curve in Graph 1 is:

$$f(x) = \begin{cases} \frac{1}{8}, & x \in [1,9] \\ 0, & x \notin [1,9] \end{cases}$$

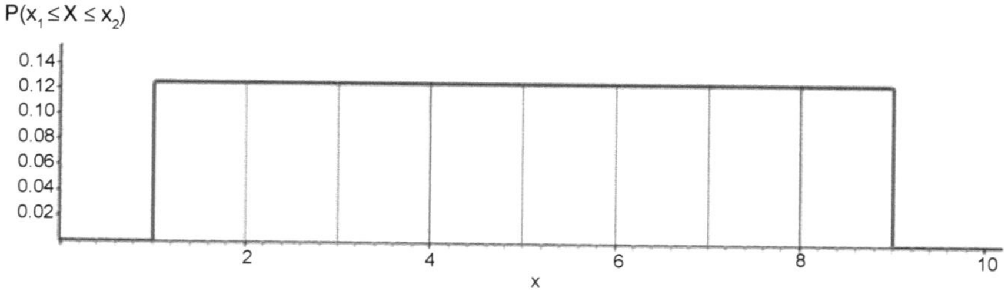

Figure 20. Graph 1 Probability distribution for the experiment of the spinner.

Color image of this figure appears in the color plate section at the end of the book.

A function describing the probabilities of a continuous random variable is known as **probability density function**, or pdf.

Can we use the graphic of this function to calculate probabilities for different intervals around the spinner? In Table 2 we can take probabilities calculated directly from the spinner to the graph.

Table 2. Comparing the probabilities around the spinner and the areas under the corresponding pdf.

Around the spinner:	In the graph:	The shaded rectangle:
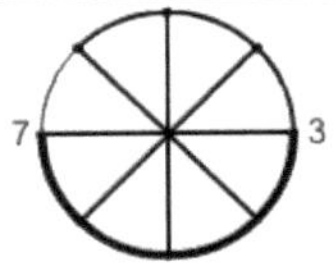 $P(3 \le X \le 7) = \frac{4}{8} = \frac{1}{2}$	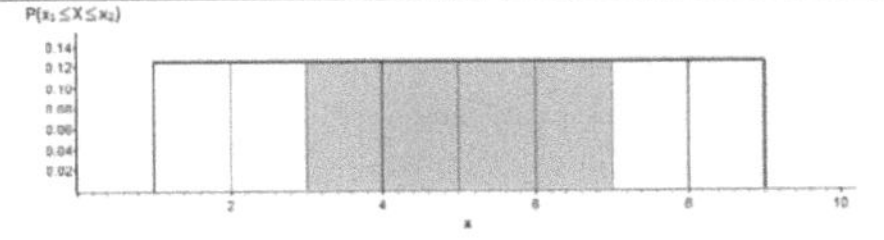	Has a base of 7-3=4. Has a height of $\frac{1}{8}$. Has an area of $(4)(\frac{1}{8}) = \frac{4}{8}$.
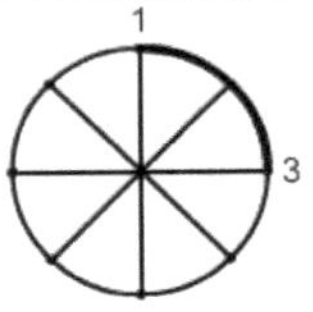 $P(X \le 3) = \frac{2}{8} = \frac{1}{4}$	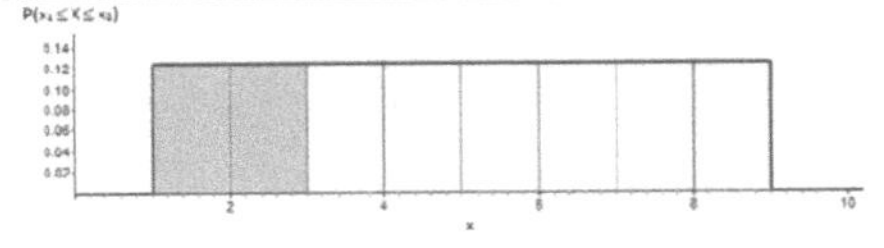	Has a base of 3-1=2. Has a height of $\frac{1}{8}$. Has an area of $(2)(\frac{1}{8}) = \frac{2}{8}$.
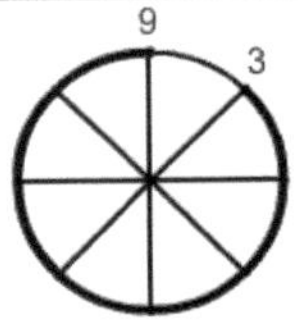 $P(X \ge 2) = \frac{7}{8}$	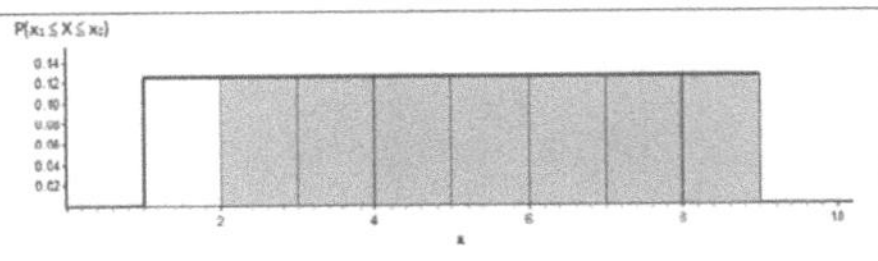	Has a base of 9-2=7. Has a height of $\frac{1}{8}$. Has an area of $(7)(\frac{1}{8}) = \frac{7}{8}$.
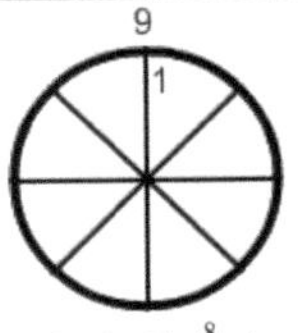 $P(1 \le X \le 9) = \frac{8}{8} = 1$	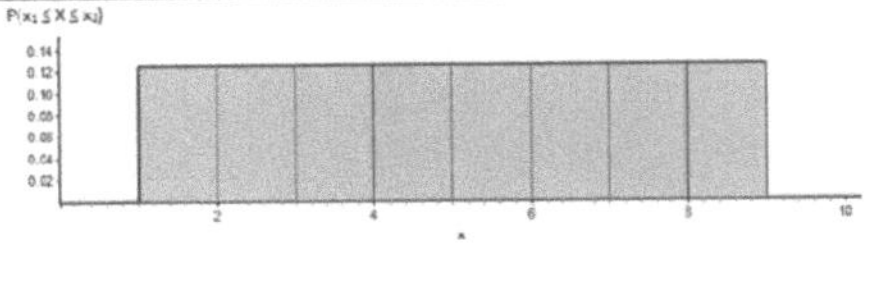	Has a base of 9-1=8. Has a height of $\frac{1}{8}$. Has an area of $(8)(\frac{1}{8}) = 1$, the probability of the sample space.

Table 2 shows that the probability of *X* getting a value within an interval, and the area under the pdf in the same interval, have the same value. In fact, the graphic representation of a pdf can be used to calculate probabilities for a continuous random variable, just by measuring the area under the curve and within the interval in which the probability is being calculated. So, for the spinner, the probabilities of Table 2 can be calculated as:

$$P(2 \le X \le 3) = \int_2^3 \frac{1}{8} dx = \frac{1}{8}(3) - \frac{1}{8}(2) = \frac{3}{8} - \frac{2}{8} = \frac{1}{8}$$

$$P(3 \le X \le 7) = \int_3^7 \frac{1}{8} dx = \frac{1}{8}(7) - \frac{1}{8}(3) = \frac{7}{8} - \frac{3}{8} = \frac{4}{8} = \frac{1}{2}$$

$$P(4 \le X \le 7) = \int_4^7 \frac{1}{8} dx = \frac{1}{8}(7) - \frac{1}{8}(4) = \frac{7}{8} - \frac{4}{8} = \frac{3}{8}$$

$$P(X \le 3) = \int_1^3 \frac{1}{8} dx = \frac{1}{8}(3) - \frac{1}{8}(1) = \frac{3}{8} - \frac{1}{8} = \frac{2}{8} = \frac{1}{4}$$

$$P(X \ge 2) = \int_2^9 \frac{1}{8} dx = \frac{1}{8}(9) - \frac{1}{8}(2) = \frac{9}{8} - \frac{2}{8} = \frac{7}{8}$$

$$P(1 \le X \le 9) = \int_1^9 \frac{1}{8} dx = \frac{1}{8}(9) - \frac{1}{8}(1) = \frac{9}{8} - \frac{1}{8} = \frac{8}{8} = 1$$

It's possible to have a function $F(x)$ calculated as $F(x) = P(X \le x) = \int_a^x f(x)dx$, where $f(x)$ is the pdf and a is the lowest value in the sample space, and which gives the cumulative probability for a x value of X. This function is known as probability distribution function, or PDF. It has the property that $F(x) = 1$ when calculated along the whole sample space.

Now let's try with a more general spinner and its pdf. Suppose that the sample space in the spinner is the interval of real numbers $[a,b]$. The pdf will be as shown in Graph 2 (Fig. 21).

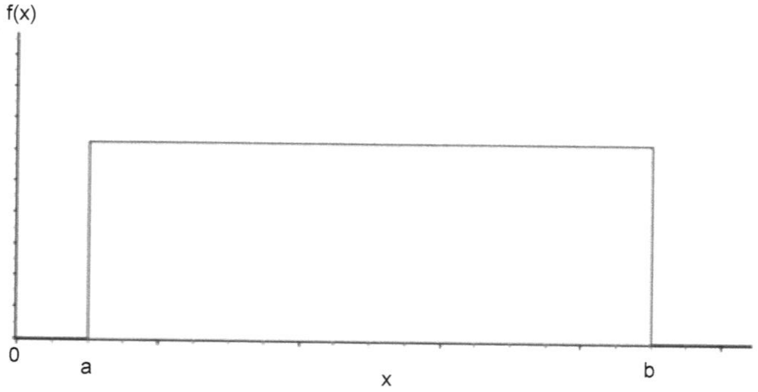

Figure 21. Graph 2: pdf for a spinner with sample space [a, b].

Color image of this figure appears in the color plate section at the end of the book.

The pdf in its algebraic form can be constructed as follows. Fig. 4 shows a rectangle with the length of the base *b–a*. Now, in order to have a total area of 1, the height must be $\frac{1}{b-a}$, and this value will define the shape of the curve. So, the pdf is

$$f(x)=\begin{cases}\frac{1}{b-a}, & x\in[a,b]\\ 0, & x\notin[a,b]\end{cases}$$

Graph 3 shows the PDF for the probability $P(c\le X\le d)$ (Fig. 22).

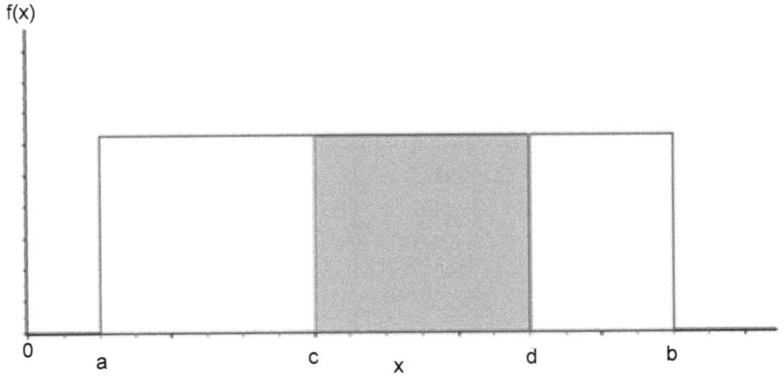

Figure 22. Graph 3: PDF for a spinner with a sample space [a,b] and interval of occurrence [c,d].

Color image of this figure appears in the color plate section at the end of the book.

So, the PDF is

$$P(c\le X\le d)=\int_c^d\frac{dx}{b-a}=\frac{1}{b-a}\int_c^d dx=\frac{1}{b-a}[x]_c^d=\frac{1}{b-a}[d-c]=\frac{d-c}{b-a}.$$

A pdf that shows a constant probability along a close interval of real numbers [a, b] is named as Uniform Distribution, which can be characterized as:

> A random variable X is defined as Uniform distribution if the density function for X is given by: $f_X(x)=f(x,a,b)=\frac{1}{a-b}$ for $\{a,b\}\in\mathbb{R}$. The parameters for the distribution are *a* and *b*.

The expected value and the variance for a Uniform distribution can be determined by the following expressions:

$$\begin{cases}E(X)=\frac{a+b}{2}\\ V(X)=\frac{(b-a)^2}{12}\end{cases}$$

Example.- Temperature of water in a public pool is uniformely distributed, from 1.3°C to 5°C. What's the probability of observing a variation greater than 3.5°C?

Our sample space is [1.3, 5], and we want to calculate P(X>3.5) (and neither clearly nor greater than 5). So, $P(X > 3.5) = \frac{5-3.5}{5-1.3} = \frac{1.5}{3.7} = 0.405.$

5.2 Exponential distribution

We've seen a continuous distribution that has the simplicity of a constant probability. But not every random experiment leads to a uniform distribution. We may find some examples which lead to a different pdf.

Let's take the next set of data, corresponding to the time, in minutes, that 100 costumers spent at a quick market:

8.7	4.9	3.4	4.5	4.5	2.1	4.6	3.2	4.8	9.2
2.6	3.7	9.8	4.4	22.9	1.9	19.9	3.7	0.4	7.5
3	6.3	2.8	8	2.2	30.3	6.1	8.2	4.8	6.9
10.4	2.1	1.8	43	1.2	21.9	0.5	7.6	0.5	9.3
18.6	32.2	2.9	16.1	15.1	1.3	3.3	3.2	12.9	29.2
0.2	5.6	1.8	11.7	13.7	18.4	5.8	13.3	5.4	2.9
10.5	51.9	4.4	1.5	17.2	5.8	4.9	2.9	15	2
22.8	2	5.7	6.2	13.5	10.2	9.1	10.2	7.9	0.1
3.2	8.8	2.4	8.7	18.9	6.7	2	2	8.2	5.9
15	2.1	4.2	1.2	3.6	2.2	2.8	9.6	14	6.1

The histogram of the distribution for the data is shown in Graph 4 (Fig. 23).

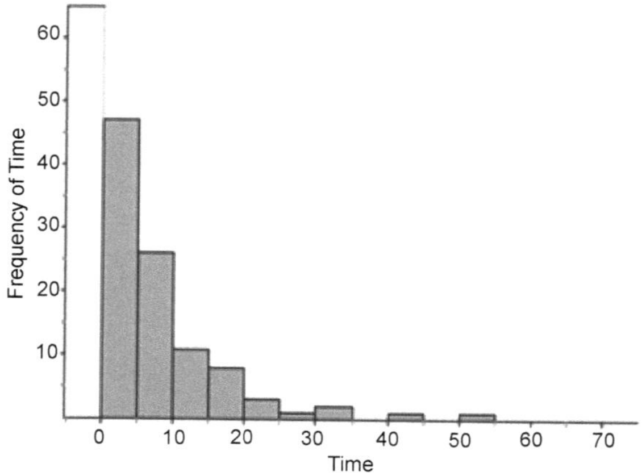

Figure 23. Graph 4: Histogram for the time in minutes spent by 100 persons at a quick market.

The distribution is clearly far from a uniform distribution. If we would want to use this data in order to calculate some probabilities related to the time that one spend at the quick market, we would need a pdf different from the uniform, but also continuous, since time is continuous. In Graph 5 it's shown a possible pdf curve for those data. It's worth pointing that the data have a mean of approximately 8.5 minutes and a standard deviation of approximately 8.8 minutes (Fig. 24).

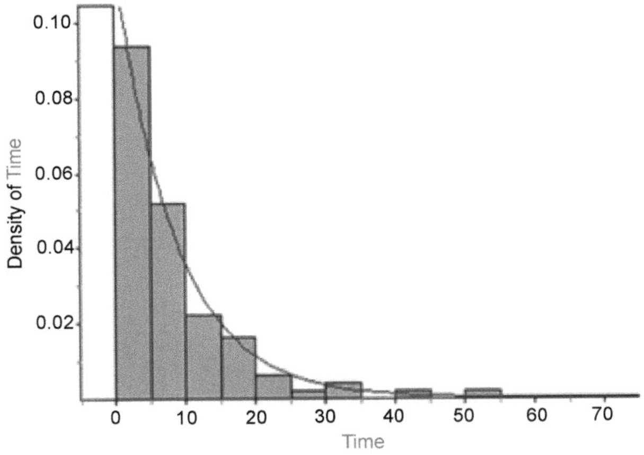

Figure 24. Graph 5: A possible pdf for the time in minutes spent by 100 costumers at a quick market.

Color image of this figure appears in the color plate section at the end of the book.

Now let's try to find the pdf in its algebraic form. First, since the possible times must be values greater than zero, so, the domain for the function is, in theory, $(0,\infty)$.

Second, the possible pdf shows an exponential decreasing, so the function has to be in the form of $f(x) = Ae^{-Bx}$. And finally, since $f(x)$ is a pdf, the condition $\int_0^\infty f(x)dx = 1$ is mandatory.

Let's try the next combination, using the mean of the set of 100 times, as $\tau = 8.5$: $f(x) = \frac{1}{8.5}e^{\frac{-x}{8.5}}$.

In Graph 6, we have this function (Fig. 25).

As we can see, this function fits well with the function proposed in Graph 5, and fulfils the domain $(0,\infty)$ and an exponential model.

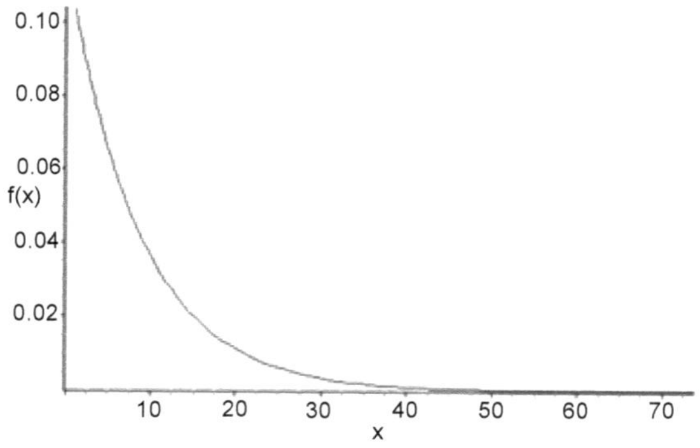

Figure 25. Graph 6: Function $f(x) = \frac{1}{8.5} e^{-\frac{x}{8.5}}$.

There's still to prove that the total area under the curve equals to 1. With $u = \frac{-x}{8.5}$, $du = \frac{-1}{8.5} dx$:

$$\int_0^\infty \frac{1}{8.5} e^{\frac{-x}{8.5}} dx = -\int_0^\infty e^u du = -e^u \Big|_0^\infty = -e^{\frac{-x}{8.5}} \Big|_0^\infty = -\left[e^{\frac{-\infty}{8.5}} - e^{\frac{-0}{8.5}} \right] = -[0-1] = 1$$

Now it seems possible to use this function as a pdf for a set of data or a situation that fits the model. For example, we may want to calculate the probabilities $P(X > 10)$ or $P(X \le 15)$, with X as the random variable that represent the time spent by a costumer at the quick market.

$$P(X > 10) = \int_{10}^\infty \frac{1}{8.5} e^{\frac{-x}{8.5}} dx = -\int_{10}^\infty e^u du = -e^u \Big|_{10}^\infty = -e^{\frac{-x}{8.5}} \Big|_{10}^\infty = -\left[e^{\frac{-\infty}{8.5}} - e^{\frac{-10}{8.5}} \right] = -[0 - 0.308] = 0.308$$

$$P(X \le 15) = \int_0^{15} \frac{1}{8.5} e^{\frac{-x}{8.5}} dx = -\int_0^{15} e^u du = -e^u \Big|_0^{15} = -e^{\frac{-x}{8.5}} \Big|_0^{15} = -\left[e^{\frac{-15}{8.5}} - e^{\frac{-0}{8.5}} \right] = -[0.171 - 1] = 1 - 0.171 = 0.829$$

So, the probability for a costumer to spend more than 10 minutes at the quick mart is 0.308, and 0.829 for the costumer to spend no more than 15 minutes. We can see both results in Graph 7a an Graph 7b (Figs. 26a e 26b).

Let's take a closer glance at both examples. In the first one, the probability of spending more than 10 minutes ended as $P(X > 10) = e^{\frac{-10}{8.5}} = 0.308$, while the probability of spending no more than 15 minutes turned as $P(X \le 15) = 1 - e^{\frac{-15}{8.5}} = 1 - 0.171 = 0.829$; we just have to keep in mind that 8.5 is the mean of the set of data corresponding to the times.

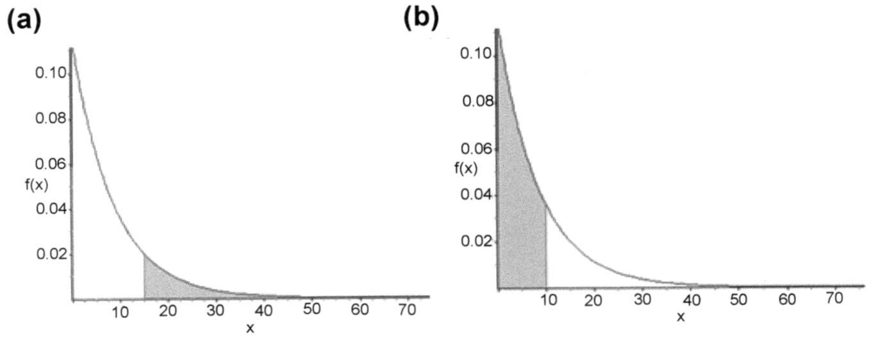

Figure 26. (a) Graph 7a: $P(X > 10) = 0.308$, (b). Graph 7b: $P(X \leq 15) = 1 - 0.171 = 0.829$.

We know that the total area under the pdf equals 1. And, by looking at both examples, it seems that $P(X > x) = e^{\frac{-x}{8.5}}$, while $P(X \leq x) = 1 - e^{\frac{-x}{8.5}}$, as the evident complement of $P(X > x)$.

Now, we can put this pdf for a random variable T in the general form:

$$f(t) = \begin{cases} \frac{1}{\tau} e^{\frac{-t}{\tau}}, & t \in (0, \infty) \\ 0, & t \in (-\infty, 0] \end{cases}$$

A pdf with the following characteristics, along a sample space ranging $(0, \infty)$ for a random variable T, is named as Exponential Distribution, which can be characterized as:

A random variable X is defined as Exponential distribution if the density function for X is given by: $f_X(x) = f(x,t) = \frac{1}{\tau} e^{-t/\tau}$ for $\tau > 0$. The parameter for the distribution is t. Furthermore, $P(T > t) = e^{-t/\tau}$ and $P(T < t) = 1 - e^{-t/\tau}$

The expected value and the variance for a Uniform distribution can be determined by the following expressions:

$$\begin{cases} E(X) = \tau \\ V(X) = \tau \end{cases}$$

Example.-The temperature in spring in Mexico City behaves under an Exponential distribution with a mean of 27°C. What's the probability of observing a temperature below 23°C in Mexico City in the spring time?

We want to calculate P(X<23) with a mean of 27 for an exponential distribution. So, $P(X<23)=1-e^{-23/27}=1-e^{-0.851852}=1-0.4266=0.5734.$

5.3 Normal Distribution

Let's develop another distribution for a continuous random variable. This time, we will start with a discrete case before passing to the continuous case.

A traditional set of dominoes has 28 pieces. A possible variable is the sum of the dots on each piece, and we can even use the pieces for constructing a bar graph for that variable, as shown in Fig. 27a. In order to have more data, let's take a not so traditional set of dominoes, like the one in Fig. 27b, which goes up to a double 18 piece, giving a total of 190 pieces.

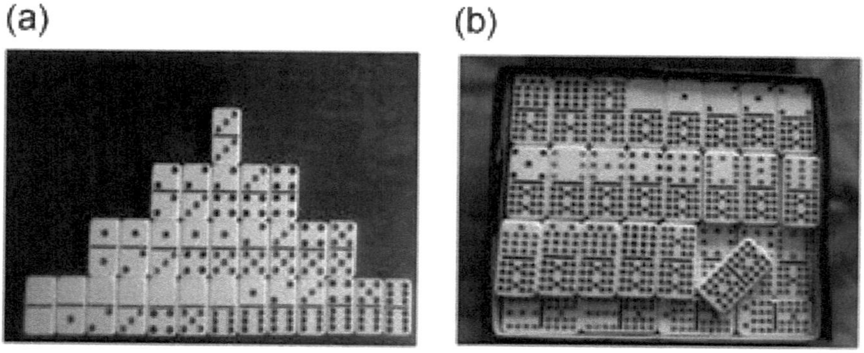

Figure 27. (a) Graph made with a set of dominoes, using the sum of the dots as the variable; (b) Set of dominoes of 190 pieces.

Color image of this figure appears in the color plate section at the end of the book.

The 190 sums from the pieces are:

0	1	2	2	3	3	4	4	4	5	5	5	6	6	6	6	7	7	7
7	8	8	8	8	8	9	9	9	9	9	10	10	10	10	10	10	11	11
11	11	11	11	12	12	12	12	12	12	12	13	13	13	13	13	13	13	14
14	14	14	14	14	14	14	15	15	15	15	15	15	15	15	16	16	16	16
16	16	16	16	16	17	17	17	17	17	17	17	17	17	18	18	18	18	18
18	18	18	18	18	19	19	19	19	19	19	19	19	19	20	20	20	20	20
20	20	20	20	21	21	21	21	21	21	21	21	22	22	22	22	22	22	22
22	23	23	23	23	23	23	23	24	24	24	24	24	24	24	25	25	25	25
25	25	26	26	26	26	26	26	27	27	27	27	27	28	28	28	28	28	29
29	29	29	30	30	30	30	31	31	31	32	32	32	33	33	34	34	35	36

The bar graph for the set of data is shown in Graph 8 (Fig. 28). The mean of the data is 18, and the standard deviation is approximately 7.9. And the distribution of the values is symmetrical, with the higher frequencies in the center of the distribution and the lower in the extremes.

In a distribution of such characteristics, the so called *empirical rule* applies. The empirical rule states that in a distribution like the one above, about 68% of the observations fall within one standard deviation around the mean, about 95% fall within two standard deviations around the mean, about 99% fall within three standard deviations around the mean, and about 100% within three and a half standard deviations around the mean. Figure 29 shows the empirical rule under a continuous curve similar with the distribution for the sum of the points on the dominoes pieces.

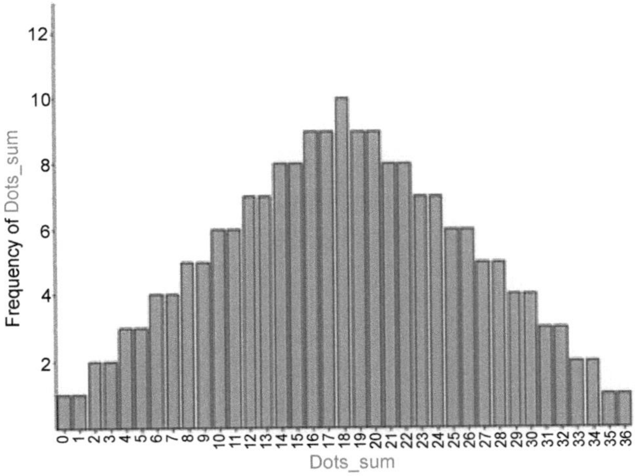

Figure 28. Graph 8: Bar graph for the sum of dots on the dominoes pieces.
Color image of this figure appears in the color plate section at the end of the book.

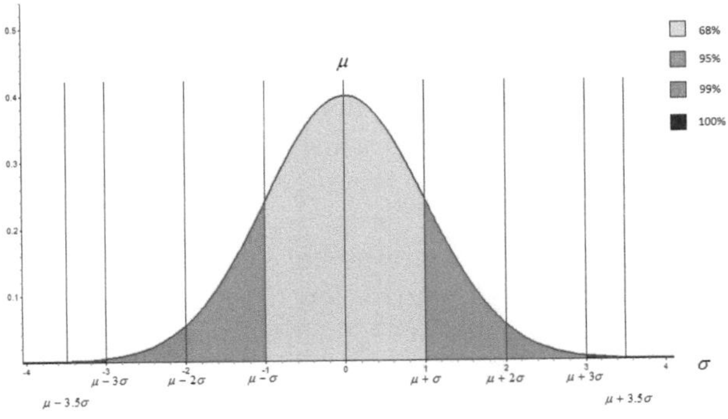

Figure 29. Empirical rule.
Color image of this figure appears in the color plate section at the end of the book.

If we apply the empirical rule to the distribution for the sums on the pieces of the dominoes, the intervals would be:

- 18 ± 7.9, [10.1, 25.9]. There are 130 pieces where the sum of the dots is between 10 and 26, being the 68.4% of the pieces.
- 18 ± 15.8, [2.2, 33.8]. There are 186 pieces where the sum of the dots is between 2 and 34, being the 97.8% of the pieces.
- 18 ± 23.7, [-5.7, 41.7]. The 100% of the pieces are within this interval, as well as it will be for the interval 18 ± 27.65.

But the empirical rule gives approximate percentages, so we can say that the rule works with our distribution.

One use that we can give to the empirical rule is calculating probabilities. For the random variable X, the sum of the points on the pieces of the dominoes shown in Fig. 27b, we may calculate, by counting pieces, that $P(X \geq 26) = \dfrac{30}{190} = 0.158 \approx 0.16$, since there are 30 pieces with a sum of their dots greater than 26.

By using the empirical rule, we know that about the 68% of the observations fall in the interval within a standard deviation around the mean, which is, more or less, [10, 26]; and we know that 26 is approximately one standard deviation away from the mean. Since the distribution is symmetrical, outside the interval should be about 32% of the observations, divided equally for the values lesser than 10 and greater than 26. This means that 16% of the observations will be lesser than ten, and 16% greater than 26. So, $P(X \geq 26) \approx 0.16$.

Now, let's try all these ideas with continuous random variables. There could be a number of phenomena that behave like the distribution shown in Fig. 7, and that are continuous. For example, the weight of adults; there could be expected that most of the adults will have an average weight while just a few of them having a weight greater or lesser than the average, and having less observation as the weight becomes ever smaller or bigger. Another example is the real quantity of juice in a can; most of the cans will actually have around the quantity of juice offered on the label, some of them will have less or more, but the number of cans with a quantity notoriously lesser or greater than the quantity offered, will be smaller than those around the average.

A third example is the consumption of electricity by a house in an urban neighborhood. We could expect just a few houses with a very small amount of consumption, as well as just a few with a very high consumption. And we could also expect as well a larger number of houses with consumption around the average.

Let's say that the consumption of electricity in some neighborhoods, which is a continuous variable, has a mean of 198 kW by month, with a standard deviation of 12 kW, and a distribution symmetrical and with higher frequencies around the mean.

By using the empirical rule, we know that about 68% of the houses will have consumption between 186 and 210 kW, about 95% will have consumption between 174 and 222 kW, about 99% will have consumption between 162 and 234 kW, and about 100% will have consumption between 156 and 240 kW.

With all this data we can calculate the next probabilities for the random variable X, consumption of electricity of a house in one month:

- $P(X > 210)$. We know that about 68% of the houses are within the interval [186, 210]. So, the remaining 32% are outside the interval, and equally divided in 16% being lesser than the inferior limit of the interval, and 16% being greater than the superior limit of the interval.

  So, $P(X > 210) = 0.16$.
- $P(X < 162)$. We know that about 99% of the houses are within the interval [162, 234]. So, the remaining 1% is outside the interval, and equally divided in 0.5% being lesser than the inferior limit of the interval, and 0.5% being greater than the superior limit of the interval.

  So, $P(X < 162) = 0.05$.
- $P(X > 150)$. We know that about 100% of the houses are within the interval [156, 240]. So, $P(X > 150) = 1$.
- $P(X = 200)$. Since the variable X is continuous, $P(X = 200) = 0$.
- $P(X > 219)$. We know that 210 kW is one standard deviation away from the mean, and 222 kW is two standard deviations away from the mean. **How many standard deviations is 219.29 kW away from the mean?**

This last question is the first step in order to calculate the requested probability using the empirical rule. The first step will be measuring the distance from 219.92 to the mean 198. This will be: 219.92 – 198 = 21.92.

The next step will be calculating how many times the standard deviation 12 fits in 21.92. This will be: $\dfrac{21.92}{12} \approx 1.83$.

This means that the value 219.92 kW is 1.83 standard deviations away from the mean. But the empirical rule does not say anything about this. We will be back to this after some general considerations.

Let's go deeper with the process of measuring the number of standard deviations that a value is away from the mean. We've taken the specific x value of X for which we were calculating the probability, and then we've subtracted the mean μ from it. Then, we've divided the subtraction by the standard deviation σ . This is $\frac{x-\mu}{\sigma}$. Let's name this value, the number of standard deviations that the specific value x of the random variable X is away from the mean, when the distribution is symmetric and with higher frequencies around the average. That means, $z = \frac{x-\mu}{\sigma}$.

A distribution that is symmetrical and with higher frequencies around the average can be modeled by the curve shown in Graph 9 (Fig. 30).

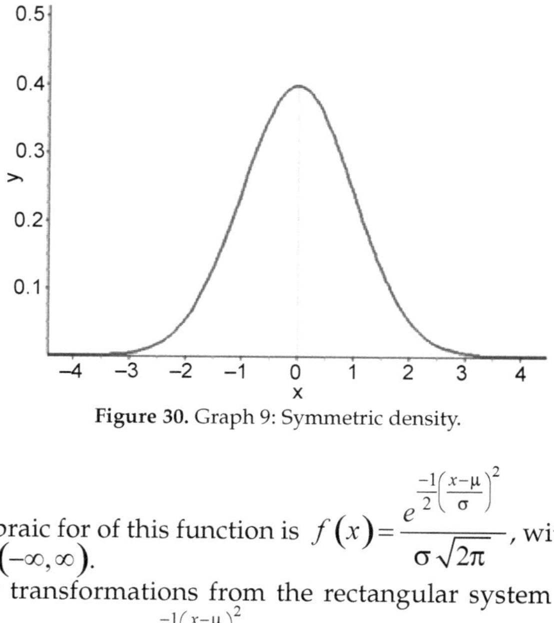

Figure 30. Graph 9: Symmetric density.

The algebraic for of this function is $f(x) = \frac{e^{\frac{-1}{2}\left(\frac{x-\mu}{\sigma}\right)^2}}{\sigma\sqrt{2\pi}}$, with $\sigma > 0$, and a domain in $(-\infty, \infty)$.

By using transformations from the rectangular system to polar, it's possible to prove that $\int_{-\infty}^{\infty} \frac{e^{\frac{-1}{2}\left(\frac{x-\mu}{\sigma}\right)^2}}{\sigma\sqrt{2\pi}} dx = 1$, what means that the function is a pdf.

This pdf is named as Normal Distribution. Its domain is $(-\infty, \infty)$, is asymptotic to the X axe and it can be characterized as:

A random variable X is defined as Normal distribution if the density function for X is given by: $f_X(x) = f(x, \mu, \sigma) = \frac{e^{-\frac{1}{2}\left(\frac{x-\mu}{\sigma}\right)^2}}{\sigma\sqrt{2\pi}}$ for $\sigma > 0$. The parameters for the distribution are μ and σ.

The expected value and the variance for a Uniform distribution can be determined by the following expressions:

$$\begin{cases} E(X) = \mu \\ V(X) = \sigma^2 \end{cases}$$

Now, it is not possible to integrate the function in an interval different from $(-\infty, \infty)$, so it's also impossible to calculate probabilities like $P(X > x)$ or $P(X < x)$ by integrating the pdf. So, we have to do something different from integrals in order to calculate areas under this curve.

The first option is to use the pdf in order to calculate $f(x)$ for different values of x in the interval where the probability is to be calculated, and then to use those points to construct rectangles under or over the curve and then approximate the area by sums of Riemann.

The second option is to calculate the z value, number of standard deviations away from the mean. This process will give standard measures under a Normal pdf despite the values of the mean, standard deviation and variable.

But, of what use can this option be? In order to simplify the calculations of probabilities under a Normal pdf, a table with values of probabilities associated to z values has been developed. In that order, any calculation of probabilities under a Normal pdf can be made by calculating the associated z value to the variable value and then by looking for it and its related probability at the table.

The process of calculating the z values can be seen as turning the parameters μ and σ of any normal situation into 0 and 1. This process is known as standardization of the Normal, the pdf with these values is known as the standardized Normal density and the PDF associated is known as standardized Normal density. So, according to tools in the package Rcmdr, software R, the probability of being further than 1.83 standard deviations away from the mean is 0.0336 (Fig. 31).

So, the probability of having a house with a consumption of electricity greater than 219.92 kW, $P(X > 219.92)$, is 0.0336.

Since we've changes or random variable X to a variable measured in z values, we can establish our problem as $P(X > 219.92) = P(Z > 1.83)$ $= 0.0336$, where Z is the random variable number of z values away from the mean, and where the z value 1.83 corresponding to the x value 219.92 kW was calculated by the expression $z = \dfrac{x - \mu}{\sigma}$.

In general, if we have a continuous random variable X that behave approximately Normal and we want to calculate the probability $P(X < x)$, we mostly standardize to z values with the expression $z = \dfrac{x - \mu}{\sigma}$ in order

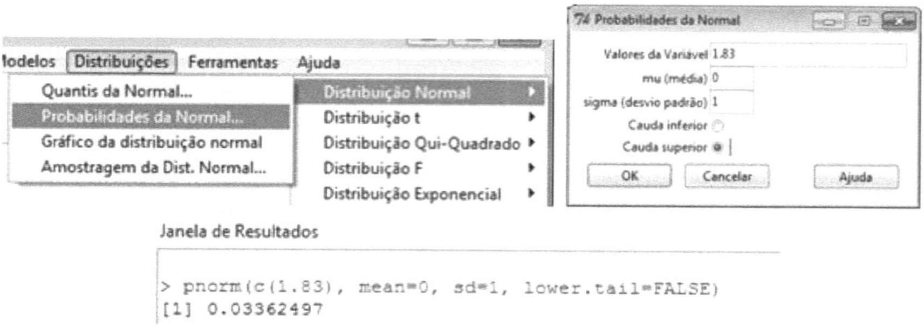

Figure 31. Outcome of package Rcmdr, software R, for the probability associated to the z value 1.83 in the Normal standardized.

to convert $P(X < x) = P(Z < z)$ and then to obtain the probability from a standardized Normal table.

Example.-The temperature in the area of Cancun behaves under a Normal distribution with a mean of 32°C and a standard deviation of 7°C along the whole year. What's the probability of having a temperature greater than 24°C in Cancun?

We want to calculate P(X>24) for a Normal distribution, with a mean of 32 and a standard deviation of 7. The first thing to do is to calculate the z value, so $\mathrm{P}(\mathrm{X} > 24) = \mathrm{P}\left(\mathrm{Z} > \dfrac{24 - 32}{7}\right) = P(Z > -1.6)$. Then, we need to check the cumulative probability for z=-1.6 in the table for the standardized Normal distribution, which is 0.9452.

Proposed exercises

1. Using package Rcmdr, pick the option Distributions, Continuous Distributions and construct the graphs for the probability density functions (mass function graph) and for the probability distribution function (cumulative probability graph) as well as for the Exponential and Normal distributions. Observe the results and construct the graphs for different parameters. Write down your conclusions.
2. Using package Rcmdr and activating, in Tools, RcmdrPlugin. TeachingDemos, pick the options Distributions, Visualize Distributions and Normal Distributions.

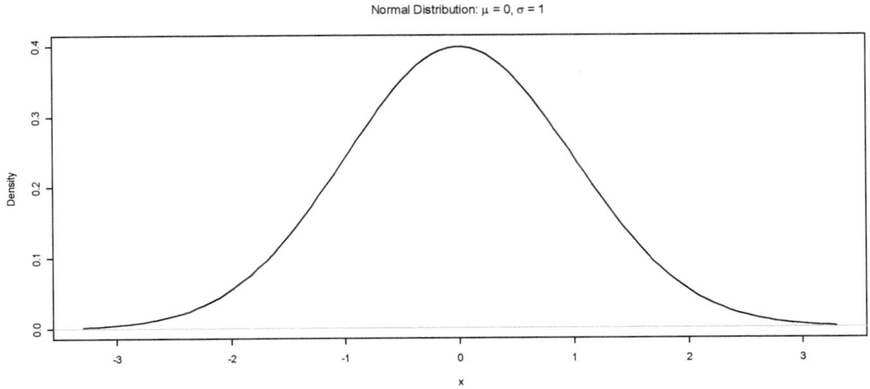

Figure 32. Example of the probability density function for standardized Normal distribution constructed in the package Rcmdr, R software.

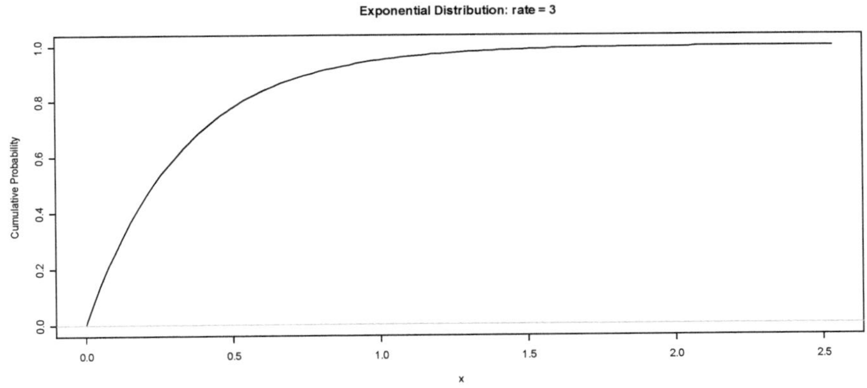

Figure 33. Example of the probability density function for exponential distribution with λ=3, constructed in the package Rcmdr, R software.

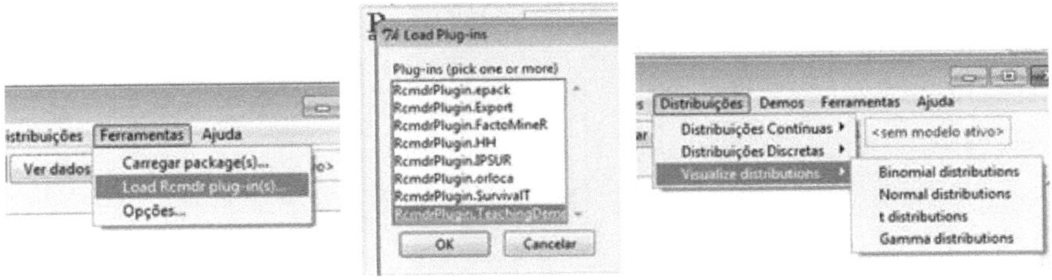

Figure 34. Steps to initialize the visualization of Binomial distribution.

a) Start by fixing the μ parameter and varying σ and observe the constructed graphs. Next, turn that around, changing μ and fixing σ. Write down your conclusions.

3. Go back to exercise 3 proposed in the section of discrete distributions (Visualize Distributions and Binomial Distributions) and click on the option Show Normal Approximation, fix n and vary p, and vice-versa, and check what happens to the binomial and normal graphs. Write down your conclusions.

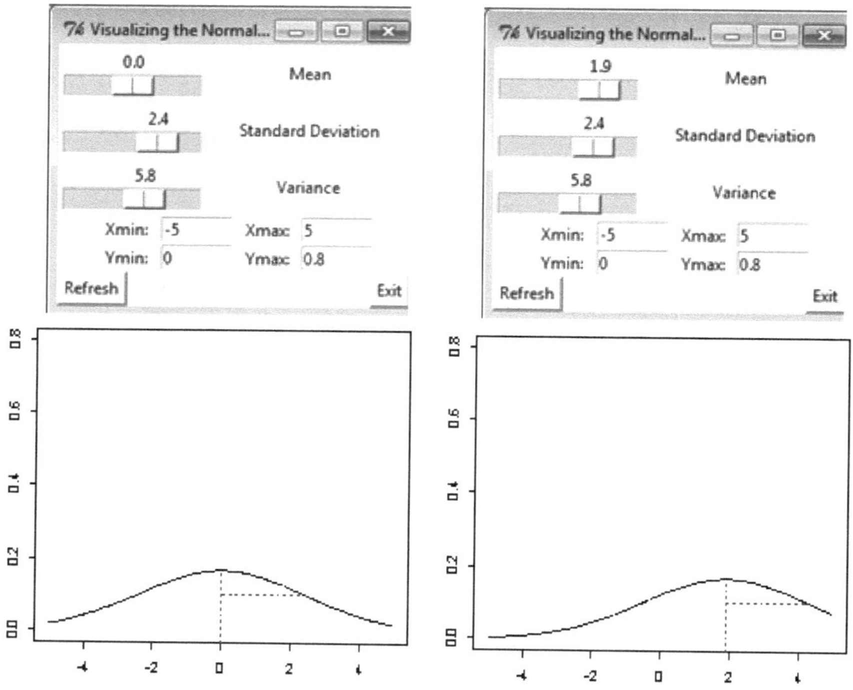

Figure 35. Example of graphs constructed with package Rcmdr, R software, for normal distributions with parameters μ = 0.0 and σ = 2.4 (left graph) and μ = 1.9 and σ = 2.4 (right graph).

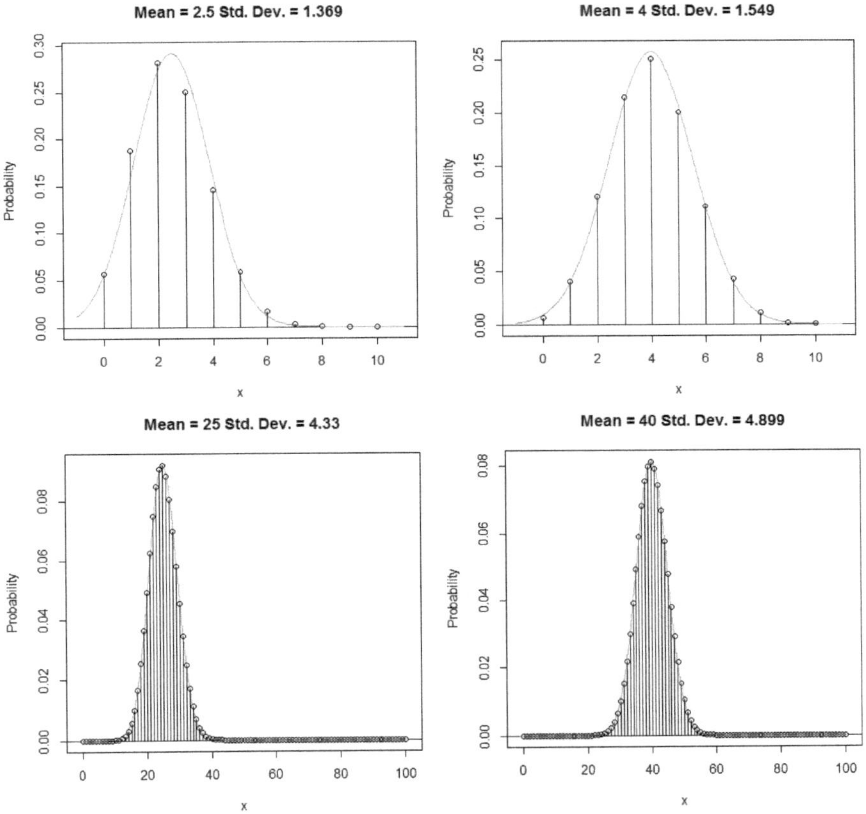

Figure 36. Examples of graphs constructed in package Rcmdr, R software, for binomial distributions with different parameters, compared to normal distributions.

Color image of this figure appears in the color plate section at the end of the book.

CHAPTER 4

Sampling

Giovanni Boscaino[1,*] and *Ornella Giambalvo*[2]

1. Introduction and First Basic Concepts

The aim of this chapter is to show how to obtain information about a phenomenon concerning a collective (population) observing just a part of it (the sample). The objective is to investigate real phenomena as they are, via a sample, and not build a sample with the aim of verifying our hypothesis at all costs. It is easier to sample badly than to sample well (Moore 2000). In fact the difficulties about getting information via a sample survey or opinion polling, concern both "ethical" and "methodological" aspects that are typical of the statistics. This chapter focuses on the characteristics and properties that a subset of the population must have to be termed a 'sample' and therefore estimating the unknown parameters of the population.

Every element of the population that is an object of the study, the elementary units or statistical units, provides useful information. The characteristics of the population can be analyzed observing just a part of the population, a subset of all statistical units: the sample. The sample has to detect a set of statistical units that can be useful to get information, conditional to the available costs and times. The chosen statistical units will affect the results (estimates) in terms of reliability (the inverse of the total error, equals bias plus estimate variance) and generalization of the whole population (the inference process, see the next chapter). The statistical selection of the group leads to the sample. All procedures adopted to select a sample represent the sampling survey.

Let us suppose that R&D staff of the hypothetical firm Sampple—a firm dealing in smartphones—wants to check the properties of a battery of a particular smartphone, following some complaints of the customers about

[1] Dipartimento di Scienze Economiche, Aziendali e Statistiche, Università degli Studi di Palermo.
[2] Università degli studi di Palermo, Facoltà di Economia, Palermo (Italia).
* Corresponding author: giovanni.boscaino@unipa.it

the duration of the charge. R&D focuses on the talktime and obviously it cannot check all the existing batteries. It has to conduct a random survey. Therefore R&D sets out to re-estimate the talk time mean, to estimate the number of batteries that are below the declared talk time mean, and the percentage of batteries that last at least a fixed number of hours. Where to start? Taking the output of batteries for a given period (one, six or more months) into account, R&D has to consider duration and costs of the survey: checking just a sample instead of the total population can reduce the costs estimating an observed statistic (the estimation of the mean talk time) of the true parameter of the population (the unknown mean talk time).

1.1 Sampling vs. *census*

The first issue regards the choice between a sample survey and a census. The first one is faster and cheaper but produces an output affected by a statistical error than the second one that considers all the possible statistical units. Is it better making a very long or expensive survey to perform a census or is it better making a shorter and cheaper sampling survey with a tolerable and known bias? And what about the destruction of the statistical units, if needed during the measurement? In the example, R&D wants to study the mean talk time of its batteries/mobiles. Theoretically, this study implies that the whole production of mobiles and batteries must be analyzed in order to measure the real mean talk time. Obviously it implies a very long and expensive study. Therefore the firm cannot analyze its own entire production. So the sampling survey is very useful in these cases. In fact it allows estimating the real value of the population performing a survey that involves just fewer units than census in a shorter time and with lower costs. Moreover, having shorter time allows managing not out-of-date data.

Taking recourse to sampling instead of census is often useful because i) a part of the error is known (the so called measurable and known 'sampling error'); ii) another part of error (the 'non-sampling error') due to human mistakes made during the acquisition and management can be reduced via a good training of the staff; iii) finally, a part of non-sampling error is tied up to the number of analyzed statistical units (the smaller the number of units, the smaller the errors due to human action).

Obviously a census has some advantages. For example, after the collection of census data, we can analyze a sub-set of the census data or select a random sample from the census data or from the subset itself.

1.2 Probabilistic sampling or not probabilistic one?

There are different ways to sample. It is basic to perform a probability sampling if the aim is to obtain an estimate of the actual value of the population parameter, measuring its reliability as well. Otherwise, you can choose a non-probability sampling. The main difference consists in the selection of the units and the measure of the error. In fact, the error is measurable and known only if we perform a probability sampling. The selection of sample units has to involve well-established selection criteria of the statistical units that will be part of the sample. Because we are just observing a subset of the population, and not the whole population, even the best settled sampling survey can be affected by bias (the distance between the result of the measurement process and the actual measure). But it also regards a "cost-benefit analysis". What is better? Performing a probabilistic sampling where the units are chosen in a random way with a non-zero and known probability of being part of the sample, or performing a non probability sampling where the units are chosen by the researcher following his own personal criteria? On one hand the probability sampling allows inference, on the other it is more expensive and longer than the non probability sampling.

This chapter focuses on probability sampling.

The probability sample is measurable: the probability of the unit being part of the sample is known. That allows more or less easily the standard errors, the likelihood functions and the sampling distribution that are basic to make an inference. More formally we define a probability sample (just sample from now on) as a set $S = \{u_1, u_2, \ldots, u_n\}$ of units u_i that are randomly drawn out from a population P with the aim to estimate one unknown parameter θ. The draw can be with or without replacements of the units. In the first situation, the drawn unit will be considered for any draw and therefore it can appear in the sample more times and consequently the sample cannot be a subset of the population (but it can be even bigger than population, also theoretically infinite). The second one is the most frequent situation: like in the lottery draw, the first drawn unit is not considered for the next draw. Moreover, another distinction regards the draw order: if the order of the draw implies different samples then the samples are called "ordered". For example, the ordered sample (3;9;10) and (10;9;3) are two different samples.

1.3 Finite vs. infinite population

The population size (finite or not) is very important for the sampling strategy. The inference (see next chapter) is based on theoretically infinite populations, and the observations are independent and identically distributed, while the survey sampling are based on finite populations.

With respect to Sampple, the output of batteries is limitless over the time, or so large that it can be considered infinite. But the production of batteries is itself finite . The finite population is a finite set of identifiable elementary units and that implies the existence of a probability that a unit is included in the sample. This aspect leads to a different approach from the classic theory.

We define: a) the inference population (object of the inference process) (*IP*); b) the target population (*TP*); c) the frame population (*FP*); and d) the surveyed population (*SP*). For example, if we consider the battery stock produced over one year by sample, that is the *IP*. Once we have set the observation period, for example three months, all the batteries produced within that time represent the *TP* by which the sample will be drawn. And once we have labelled the batteries (*FP*) we make the list of units to be drawn (*SP*). Therefore, *IP* > *TF* and *FP* > *SF*.

1.4 Sample space

If we consider an infinite population, the number of all the possible samples of size *n* is infinite. Instead, if we consider a finite population of size *N*, we can calculate that number, using the information such as the sample size and the selection method. The sample space Ω is the set of all the possible samples of size *n* (*n=1, 2, ..., N*) drawn by a population *P* of size *N*.

For example, the dimension of the sample space for non-ordered and without replacement samples is $\Omega = C_{N,n} = \binom{N}{n}$, while for non-ordered and with replacement samples is $\Omega = C^r_{N,n} = \binom{N+n-1}{n}$.

1.5 Sampling and inclusion probability

The sampling plan is a crucial aspect that affects the quality of the sampling output. It depends on costs, available information about the population, etc. Formally, let *s* be a sample and *p(s)* be its probability of being drawn. We define a sampling plan every *p(s)* function on Ω such as $p(s) \geq 0$ and $\sum_{\Omega} p(s) = 1$. The *k* order inclusion probability is the probability that *k* units belong to the same sample.

If we consider the generic unit *i* of the population and the set A_i as the set of all the samples of Ω that include the *i*-unit, the first order inclusion probability is equal to the sum of the probabilities of the samples belonging to A_i: $\pi_i = \sum_{A_i} p(s)$. In the same way, the second order inclusion probability is $\pi_{ij} = \sum_{A_{ij}} p(s)$, where A_{ij} is the set of all the samples that include the two units i and j. If $\pi_i = c \ \forall i$ then the sampling plan is self-weighting.

1.6 The Sample design

The sample design concerns sample plan and estimator too. Once a sample plan is set, there are a lot of estimators that can be chosen. The choice of the 'best' sample design depends on estimator properties, and the object of the survey. For example, the estimators chosen by the R&D of sample are the mean talk time of the batteries, the number (total) of batteries that are below the declared talk time mean, and the percentage of batteries that last at least a fixed number of hours.

The aims of the survey plan are measured in terms of precision and accuracy. Precision is equal to the inverse of the variance of the estimator, while the accuracy is the inverse of the total error (bias plus variance of estimator) and both depend on sample size (see next chapter).

1.7 Design effect

Let $\hat{\theta}$ be an unbiased estimator of θ, the effect of the design (*Deff*) is:

$$Deff = \frac{V\left(\hat{\theta}\right)}{V_0\left(\hat{\theta}\right)}$$

where $V(\hat{\theta})$ is the variance of $\hat{\theta}$ for a chosen sampling plan and $V_0(\hat{\theta})$ is the variance of $\hat{\theta}$ for a benchmark sampling plan (simple sampling random, typically), for the same n. If $Deff < 1$ the chosen sampling plan is more precise then the benchmark one.

2. Notation

a) Population
P = population
N = population size
1, 2, ... , N units' labels for population P
Y = variable of interest
$(Y_1, Y_2, \ldots, Y_N)$ = values of Y over the units of the population P
$\bar{Y}$ = mean of Y over the population:

$$\overline{Y} = \begin{cases} \dfrac{1}{N}\sum_{i=1}^{N} Y_i & \text{if } Y \text{ is metric} \\ \dfrac{1}{N}\sum_{i=1}^{N} Y_i = \dfrac{N_A}{N} & \text{if } Y \text{ is dichotomous:} \\ & Y_i = 1 \text{ if } i \text{ belongs to the set } A \text{ of units with} \\ & \text{a specific characteristic and } \sum Y_i = N_A \end{cases}$$

$\dddot{Y}$ = total of Y over the population:

$$\dddot{Y} = N\overline{Y} = \begin{cases} \sum_{i=1}^{N} Y_i & \text{if } Y \text{ is metric} \\ \sum_{i=1}^{N} Y_i = N_A & \text{if } Y \text{ is dichotomous:} \\ & Y_i = 1 \text{ if } i \text{ belongs to the set } A \text{ of units with} \\ & \text{a specific characteristic and } \sum Y_i = N_A \end{cases}$$

V or σ^2 = variance of Y over the population: $\sigma^2 = \dfrac{1}{N}\sum\left(Y_i - \overline{Y}\right)^2$

As a matter of notation, later result is presented in terms of a slightly different expression, in which the divisor N-1 is used instead of N (Cochran 1977). So the variance of Y is:

$$S^2 = \frac{\sum_{i=1}^{N}\left(Y_i - \overline{Y}\right)^2}{N-1} = \frac{N}{N-1}\sigma^2$$

Its advantage is that most results take a simpler form (above all in sampling for finite population case).

b) Sample

s = sample

n = sample size

$(y_1, y_2, \ldots, y_n)$ = observations of Y over the sample s

$\bar{y}$ = mean of Y over the sample:

$$\bar{y} = \begin{cases} \dfrac{1}{n}\sum_{i=1}^{n} y_i & \text{if } Y \text{ is metric} \\ \dfrac{1}{n}\sum_{i=1}^{n} y_i = \dfrac{n_A}{n} & \text{if } Y \text{ is dichotomous: } Y_i = 1 \text{ if } i \text{ belongs to the set } A \text{ of units with a specific characteristic and } \sum y_i = n_A \end{cases}$$

$\ddot{y}$ = total of Y over the sample s:

$$\ddot{y} = n\bar{y} = \begin{cases} \sum_{i=1}^{n} y_i & \text{if } Y \text{ is metric} \\ \sum_{i=1}^{n} y_i = n_A & \text{if } Y \text{ is dichotomous: } Y_i = 1 \text{ if } i \text{ belongs to the set } A \text{ of units with a specific characteristic and } \sum y_i = n_A \end{cases}$$

S_s^2 = variance of Y over the sample:

$$S_s^2 = \frac{1}{n-1}\sum (y_i - \bar{Y})^2$$

s^2 = estimate of S_s^2 if $\bar{Y}$ is unknown:

$$s^2 = \frac{1}{n-1}\sum (y_i - \bar{y})^2 . \tag{1}$$

3. The Sampling from Infinite Populations

The sample R&D sets down that the production of batteries can be considered as an infinite population. Therefore, the only sampling plan is the simple random sampling with replacement.

3.1 The simple random sampling with replacement (SRSR)

The simple random sampling is known to be the simplest sampling plan. Practically, the simple random sampling technique consists in labelling each unit of the population numerically and then generating random values corresponding to the sampled units. Every unit has the same probability of being part of the selected sample, at each draw.

The simple random sampling is rarely used in practice because it does not take into account additional information about the units of the population and because it is often expensive. On the contrary, for its simplicity it is considered the benchmark sampling plan (see *Deff* in the previous paragraph).

The simple random sampling with replacement (SRSR) is a sampling plan where every sample, among all the possible N^n samples with-replacement, has the same probability of being drawn:

$$p(s)=\frac{1}{N^n} \quad \forall s \in \Omega .$$

Even if the population is finite, the replacement process makes the population theoretically infinite, as it is useful for inference. SRSR consists in drawing the units one by one, considering the drawn units also during the other draws: at each draw the population size remains N. It is like a box with N labelled balls from which we draw a ball randomly and, before the next draw, we replace the ball in the box. Therefore, at each draw, every unit has the same probability of being part of the sample that is equal to $1/N$.

The first and second order inclusion probability (π_i e π_{ij} respectively) are:

$$\pi_i = 1-\left(1-\frac{1}{N}\right)^n$$

$$\pi_{ij} = 1-2\left(\frac{N-1}{N}\right)^n+\left(\frac{N-2}{N}\right)^n$$

PROOF: Let s be a simple random sample of size n with-replacement, and i_1 be the unit drawn at first draw, i_2 at second draw and so on. The probability of being drawn with-replacement for i_1 is:

$$\begin{aligned}\pi_i &= \Pr(i \in s) = \Pr\left[(i_1 = i)\cup(i_2 = i)\cup\ldots\cup(i_n = i)\right] = \\ &= 1-\Pr\left[(i_1 \neq i)\cap(i_2 \neq i)\cap..\cap(i_n \neq i)\right]\end{aligned}$$

$i_1, i_2,..i_n$ are independents and $\Pr(i_j \neq i) = 1 - \Pr(i_j = i) = 1 - P_i$ occurs n times, therefore:

$$\Pr(i, j \neq i) = 1 - \Pr(i_j = i) = 1 - (1 - P_i)^n$$

In particular, if $P_i = 1/N \quad \forall i,$

$$\pi_i = 1 - (1 - P_i)^n = 1 - \left(1 - \frac{1}{N}\right)^n = 1 - \left(\frac{N-1}{N}\right)^n = 1 - \left(1 - \frac{1}{N}\right)^n$$

that is the probability of being drawn with-replacement for unit i.

The second order inclusion probability, for the i and j units, is:

$$\pi_{ij} = \Pr\left[(i \in s) \cap (j \in s)\right] = 1 - \Pr\left[(i \notin s) \cup (j \notin s)\right] =$$
$$= 1 - \Pr(i \in s) - \Pr(j \in s) + \Pr\left[(i \notin s) \cap (j \notin s)\right]$$

and, as before, we obtain: $\pi_{ij} = 1 - (1 - P_i)^n - (1 - P_j)^n + (1 - P_i - P_j)^n$, therefore:

$$\pi_{ij} = 1 - 2\left(\frac{N-1}{N}\right)^n + \left(\frac{N-2}{N}\right)^n = 1 - 2\left(1 - \frac{1}{N}\right)^n \left(1 - \frac{2}{N}\right)^n$$

It is better to consider the expected frequency of inclusion φ, because the draw is with-replacement. Let τ_{ij} be a random variable that takes 1 if the i unit is drawn at *jth* draw, and 0 otherwise; and $\gamma_i = \sum_{j=1}^{n} \tau_{ij}$ and $\varphi_i = E(\gamma_i)$. Therefore, $\varphi_i = \sum_{j=1}^{n} E(\tau_{ij})$ and, since $E(\tau_{ij}) = 1/N$, $\varphi_i = n/N$.

That is the only sampling plan that produces samples with *i.i.d.* observations.

The estimator[1] for the mean of Y is the mean computed over the sample (sample mean):

$$\bar{y} = \frac{\sum_{i=1}^{n} y_i}{n}. \tag{2}$$

For the Sampple example, y_i is the number of minutes of talk time for the battery i and n is the number of the batteries analyzed.

[1] In the sampling by infinite population case, estimators are random variables.

The variance of the random variable $\overline{y}, V(\overline{y})$, is taken over all the $C^r_{N,n}$ samples. It can be proved that (for the proof see the next paragraph):

$$V_{SRSR}(\overline{y}) = E(\overline{y} - \overline{Y})^2 = \frac{S^2}{n}$$

The inverse of the $V(\overline{y})$ is a measure of the precision of the output of the sample design. The squared root of $V(\overline{y})$ is the standard error (*s.e.*), and it measures the average gap between the sampled value and the actual parameter of the population:

$$s.e._{SRSR}(\overline{y}) = \sqrt{V_{SRSR}(\overline{y})} = \sqrt{\frac{S^2}{n}} = \frac{S}{\sqrt{n}}.$$

Moreover, S^2 is often unknown, therefore it is taken into account its estimate (1): $v_{SRSR}(\overline{y}) = \frac{s^2}{n}$.

Example 1

R&D wants to estimate the mean talk time over 87 batteries drawn from a theoretically infinite population, adopting an SRSR. We have to observe the 87 talk times y_i, to add together and to divide that sum by 87. That is the estimation of the mean talk time, for example $\overline{y} = 8.65$. If we suppose that the estimate of the variance of Y over the sample is $s^2 = 1.3$, the estimate of the variance of $\overline{y}$ is $v_{SRSR}(\overline{y}) = 1.3/87 = 0.0149$. Therefore, the standard error is $\sqrt{v_{SRSR}(\overline{y})} = \sqrt{1.3/87} = 0.122$ and the precision of the estimator is $1/v_{SRSR}(\overline{y}) = 1/0.0149 = 67.1$.

The estimator for the total is:

$$\dddot{y} = \frac{N}{n}\sum_{i=1}^{n} y_i = N\overline{y} \quad (3)$$

And the variance of the total and its estimate are:

$$V_{SRSR}(\dddot{y}) = E(\dddot{y} - Y)^2 = \frac{N^2 S^2}{n} \quad v_{SRSR}(\dddot{y}) = \frac{N^2 s^2}{n}.$$

The estimate for proportion (or relative frequency) of a character of the population is quite similar to the mean and total cases, if we consider Y as a dichotomous variable. Let Y' the new variable that takes 1 if its observation over a unit of the population is the wanted characteristic, 0 otherwise. For Sampple, Y'=1 if the battery talk time lasts at least a fixed amount of hours. $A = \sum_{i=1}^{N} Y_i'$ stands for the absolute frequency of the number of batteries which talk times exceed a fixed amount. The mean of Y'_i is the relative frequency, or proportion, P:

$$P = \bar{Y} = \frac{\sum_{i=1}^{N} Y_i}{N} = \frac{A}{N} \tag{4}$$

and, with respect to the sample, if a is the number of unit with the wanted characteristic, the estimate p of P is:

$$p = \bar{y} = \frac{\sum_{i=1}^{n} y_i}{n} = \frac{a}{n}. \tag{5}$$

The variance of Y' is:

$$S^2 = \frac{N}{N-1} P(1-P) = \frac{N}{N-1} PQ \tag{6}$$

where $Q=1-P$. Its sample estimate is

$$s^2 = \frac{n}{n-1} pq \tag{7}$$

where p is the proportion computed over the sample and $q=1-p$.

Since the population is infinite, $N/(N-1) \rightarrow \infty$, the variance of p and its estimate are:

$$V_{SRSR}(p) = \frac{S^2}{n} = \frac{PQ}{n} \quad v_{SRSR}(p) = \frac{s^2}{n} = \frac{pq}{n}.$$

Following the (4), $A= NP$, therefore the absolute frequency estimate is

$$\hat{A} = Np. \tag{8}$$

Its variance and the variance estimate are:

$$V_{SRSR}\left(\hat{A}\right) = N^2 V_{SRSR}(P) = N^2 \frac{PQ}{n}$$

$$v_{SRSR}\left(\hat{A}\right) = N^2 v_{SRSR}(p) = N^2 \frac{pq}{n}.$$

4. The Sampling from Finite Populations

For the Sampple example, keeping the same R&D staff aims, it is more likely to consider the production of batteries as finite than infinite, if we limit it to a definite period. There are several sampling plans for that case and the option depends on aims, costs, timing and available information about population.

4.1 The simple random sampling without replacement (SRS)

In paragraph 4.3.1 we have shown the simple random sampling as the simplest sampling plan. In particular, the without replacement sampling is the simplest one. The Simple Random Sampling without replacement (SRS) is based on the draw of the units so that a drawn unit can not be considered in the next draws. Therefore, the number of the balls in the box reduces of one ball at every draw. The sample space consists in all the samples that are non ordered and whose elements are all different from each other; and its size is $C_{N,n}$. The sampling plan is such that $p(s)=\frac{1}{\binom{N}{n}}$.

PROOF: first unit i_1 is drawn with a probability equal to n/N. SRS is without replacement, therefore second unit i_2 is drawn with a probability equal to $(n-1)/(N-1)$, and so on still the *n-th* draw for the unit i_n, with a probability equal to $1/(N-n+1)$. Following the theorem of composite probability for independent events, the probability of the sample $\{i_1,\ldots.i_n\}$ is:

$$p(s)=\frac{n}{N}\frac{n-1}{N-1}\cdots\frac{1}{N-n+1}=\frac{1}{\binom{N}{n}}.$$

In the SRS plan, each unit has the same probability of being part of the drawn sample, at each draw. Moreover, the first order inclusion probability is constant $\forall i$ and equals to:

$$\pi_i=\frac{n}{N}. \tag{9}$$

PROOF: the probability of drawing the unit *i* at first draw is n/N. Let *A* be the event "drawing unit *i* at second draw" and *B* be the event "the unit *i* did not drawn at first draw". Following the Bayes theorem:

$$P(A\mid B)=\frac{P(A\cap B)}{P(B)}$$

then

$$P(A\cap B)=P(B)P(A\mid B)=\frac{N-1}{N}\frac{1}{N-1}=\frac{1}{N}$$

where $P(B)=\frac{N-1}{N}$ and $P(A|B)=\frac{1}{N-1}$ are the probability of *B* and of being drawn at second draw conditional to the event B, respectively, for the unit *i*. Now, let us consider the third draw. *A* and *B* events become: "drawing of unit *i* at third draw" and "the unit *i* was not drawn at first and second draw". Therefore $P(B)=\frac{N-1}{N}\frac{N-2}{N-1}$ and $P(A|B)=\frac{1}{N-2}$, and following again the Bayes theorem:

$$P(A\cap B)=\frac{1}{N-2}\quad\frac{N-1}{N}\frac{N-2}{N-1}=\frac{1}{N}.$$

It is possible to repeat the procedure *n* times so that we have the total probability or rather the probability of drawing at one of *n* draws the unit *i*:

$$\frac{1}{N}+\frac{N-1}{N}\frac{1}{N-1}+\frac{N-1}{N}\frac{N-2}{N-1}\frac{1}{N-2}+....=\frac{1}{N}+\frac{1}{N}+\frac{1}{N}+....\frac{1}{N}=\frac{n}{N}.$$

By definition, the SRS is a self-weighting plan.

The second order inclusion probability for the units *i* e *j* is:

$$\pi_{ij}=\frac{n(n-1)}{N(N-1)}\quad\forall i,j. \tag{10}$$

The estimator for the mean coincides with (2). For SRS the variance of the estimator is quite different from SRSR case:

$$V_{SRS}(\bar{y})=\frac{S^2}{n}\frac{N-n}{N}=\frac{S^2}{n}(1-f) \tag{11}$$

where $f=n/N$ is the sampling fraction and it coincides with the first order inclusion probability. Moreover, $1-f$ is called finite population correction factor (fpc), that measures the gap between the (11) and the V_{SRSR}. Fcp can be ignored if $f \le 0.05$, that is the case of very big size of population with respect to the sample size, so big that the population can be considered infinite (compared to the sample).

PROOF: since every unit belongs to the same number of sample of Ω

$$\binom{N-1}{n-1}=\frac{(N-1)!}{(n-1)!(N-n)!}$$

$E(y_1+y_2+\ldots+y_n)$ will be a multiple of $(Y_1+Y_2+\ldots+Y_N)$, that is:

$$E(y_1+y_2+...+y_n)=k(Y_1+Y_2+...Y_N)$$

where $k=n/N$. Therefore we have:

$$n(\bar{y}-\bar{Y})=(y_1+y_2+..+y_n)-n\bar{Y}=(y_1-\bar{Y})+(y_2-\bar{Y})+..+(y_n-\bar{Y}). \quad (12)$$

Let us consider the quantity:

$$E\left[(y_1-\bar{Y})^2+..+(y_n-\bar{Y})^2\right]=\frac{n}{N}\left[(Y_1-\bar{Y})^2+..+(Y_N-\bar{Y})^2\right], \quad (13)$$

and

$$\begin{aligned}&E\left[(y_1-\bar{Y})(y_2-\bar{Y})+(y_1-\bar{Y})(y_3-\bar{Y})+..+(y_{n-1}-\bar{Y})(y_n-\bar{Y})\right]=\\&\frac{n(n-1)}{N(N-1)}\left[(Y_1-\bar{Y})(Y_2-\bar{Y})+(Y_1-\bar{Y})(Y_3-\bar{Y})+..+(Y_{N-1}-\bar{Y})(Y_N-\bar{Y})\right]\end{aligned} \quad (14)$$

In (14) the sum of products has been extended to all the couples y_i e y_j of the sample and Y_i e Y_j of the population, that are respectively $n(n-1)/2$ for the sample and $N(N-1)/2$ for the population. For both, the products of each values by itself have not been considered. Let us square (12):

$$n^2(\bar{y}-\bar{Y})^2=\left[(y_1-\bar{Y})+(y_2-\bar{Y})+..+(y_n-\bar{Y})\right]^2$$

and, following (13) and (14), applying the expected value, we get:

$$\begin{aligned}&n^2E(\bar{y}-\bar{Y})^2=\frac{n}{N}\left[(Y_1-\bar{Y})+(Y_2-\bar{Y})+...+(Y_N-\bar{Y})\right]^2=\\&\frac{n}{N}\left\{(Y_1-\bar{Y})^2+...+(Y_N-\bar{Y})^2+2\frac{(n-1)}{N-1}\left[(Y_1-\bar{Y})(Y_2-\bar{Y})+...+(Y_{N-1}-\bar{Y})(Y_N-\bar{Y})\right]\right\}\end{aligned}$$

Adding and subtracting the same quantity:

$$\frac{n-1}{N-1}(Y_1-\bar{Y})^2-...-\frac{n-1}{N-1}(Y_N-\bar{Y})^2$$

we get:

$$\begin{aligned}&n^2E(\bar{y}-\bar{Y})^2=\frac{n}{N}\left\{(Y_1-\bar{Y})^2+...+(Y_N-\bar{Y})^2-\frac{n-1}{N-1}(Y_1-\bar{Y})^2-...-\frac{n-1}{N-1}(Y_N-\bar{Y})^2+\right.\\&+\frac{n-1}{N-1}(Y_1-\bar{Y})^2+...+\frac{n-1}{N-1}(Y_N-\bar{Y})^2+2\frac{n-1}{N-1}(Y_1-\bar{Y})(Y_2-\bar{Y})+...\\&\left....+2\frac{n-1}{N-1}(Y_{N-1}-\bar{Y})(Y_N-\bar{Y})\right\}=\end{aligned}$$

$$=\frac{n}{N}\left(1-\frac{n-1}{N-1}\right)\left[\left(Y_1-\bar{Y}\right)^2+\ldots+\left(Y_N-\bar{Y}\right)^2\right]+\frac{n-1}{N-1}\left[\left(Y_1-\bar{Y}\right)+\ldots+\left(Y_N-\bar{Y}\right)\right]^2$$

where the second member on the right side of the equation is equal to 0 due to first property of the arithmetic mean. Then:

$$\frac{n}{N}\left(\frac{N-1-n+1}{N-1}\right)\left[\left(Y_1-\bar{Y}\right)^2+\ldots+\left(Y_N-\bar{Y}\right)^2\right]=$$

$$=\frac{n}{N}(N-n)\frac{\left(Y_1-\bar{Y}\right)^2+\ldots+\left(Y_N-\bar{Y}\right)^2}{N-1}$$

and last fraction on the right side of the equation is S^2, then we have: $\frac{n}{N}(N-n)S^2$, and therefore:

$$V_{SRS}\left(\bar{y}\right)=E\left(\bar{y}-\bar{Y}\right)^2=\frac{n(N-n)}{Nn^2}S^2=\frac{S^2}{n}\frac{N-n}{N}.$$

and its estimate is $v_{SRS}\left(\bar{y}\right)=\frac{s^2}{n}\frac{N-n}{N}$.

The standard error of the estimate is:

$$s.e._{SRS}\left(\bar{y}\right)=\sqrt{V_{SRS}\left(\bar{y}\right)}=\frac{S}{\sqrt{n}}\sqrt{\frac{N-n}{N}}=\frac{S}{\sqrt{n}}\sqrt{1-f}, \qquad (12)$$

Example 2

Following Example 1 but referring to the SRS and a population size N=2320, the estimate for the variance of the sample mean is

$$v_{SRS}\left(\bar{y}\right)=\frac{s^2}{n}\frac{N-n}{N}=\frac{1{,}3}{87}\frac{2320-87}{2320}=0{,}00144,$$

where $(2320-87)/2320=0{,}9625$ is the fcp.

Self Evaluation Exercise – SEE – 1

Following the Example 2, one has to estimate how big must n be to discard the fcp.

If the aim is to estimate the number of batteries whose talk time is less than the one declared, R&D can refer to the estimator for the total (3) also for SRS. In this case, just the variance of the total changes (with respect to the SRSR):

$$V_{SRS}(\ddot{y})=V_{SRS}(N\bar{y})=N^2V_{SRS}(\bar{y})=\frac{N^2S^2}{n}\frac{N-n}{N}=\frac{N^2S^2}{n}(1-f) \qquad (15)$$

and the standard error is $s.e._{SRS}(\ddot{y})=\sqrt{V_{SRS}(\ddot{y})}=\frac{NS}{\sqrt{n}}\sqrt{\frac{N-n}{N}}=\frac{NS}{\sqrt{n}}\sqrt{1-f}$.
If S^2 is unknown, following its estimator (1), the estimator of the variance of the total is: $v(\ddot{y})=\frac{N^2s^2}{n}(1-f)$.

Equation (5) is the estimator for the proportion, but the variance of p, following the (6), changes in:

$$V_{SRS}(p)=\frac{S^2}{n}\frac{N-n}{N}=\frac{N}{N-1}\frac{PQ}{n}\frac{N-n}{N}=\frac{PQ}{n}\frac{N-n}{N-1} \qquad (16)$$

and if S^2 is unknown, following the (7), the estimator of the variance of the proportion is:

$$v_{SRS}(p)=\frac{s^2}{n}\frac{N-n}{N}=\frac{n}{(n-1)}\frac{pq}{n}\frac{N-n}{N}=\frac{N-n}{N(n-1)}pq=\frac{1-f}{n-1}pq. \qquad (17)$$

If N is big compared to n, f is small and then $v_{SRS}(p)\approx\frac{pq}{n-1}$.

As well as in the paragraph 4.3.1, the estimator for the absolute frequency A follows the (8), and its variance is:

$$V_{SRS}(\hat{A})=N^2V_{SRS}(p)=N^2\frac{PQ}{n}\frac{N-n}{N-1} \qquad (18)$$

If S^2 is unknown, the estimator for the variance of $\hat{A}$ is

$$v_{SRS}(\hat{A})=N^2v_{SRS}(p)=N^2\frac{N(N-n)}{n-1}pq\,. \qquad (19)$$

4.2 The stratified random sampling (STS)

Sampple R&D decides to consider the plants that produce the batteries for it. In fact, R&D supposes that the bad performance of the batteries can be due to plants: likely, there is more homogeneity about performance among the batteries produced by the same plant and more variability among plants. Therefore, R&D wants to take account of that information, so a suitable sampling plan is the stratified random sampling (STS).

Supplementary information can divide the population into L internally homogeneous strata,[2] that is each stratum has a low variability of Y. Then, S^2 is divided into h (h=1,...,L) components $S_1^2, S_2^2, \ldots, S_L^2$. If each S_h^2 is low, the

[2]It is obvious that the reference population is finite: it is impossible to stratify the population, because the strata would be infinite too.

estimator component linked to each stratum is more precise, therefore the complete variance of the estimator will be more precise. That homogeneity is 'technical', in opposition to the 'logical', indicates the plausibility of the stratification variable. The choice process is basic: differentiating the batteries by plant, the homogeneity is both logical and technical. If R&D discriminates on the basis of the colour of the batteries, it will be not a logical stratification and also it will not produce more homogeneity in the technical sense. Moreover, the stratification can be originated by more than one characteristic. Then we can have L_1 strata for variable 1 and L_2 strata for variable 2, originating a stratification grid of $L_1 \times L_2$ dimension.

Formally, the population *P* of size *N* is divided in *L* sub-populations (or strata) of size N_h (h=1, …, *L*), where the units will be homogeneous according to that criterion (correlated to *Y*). Then, *L* independent SRS samples, n_h sized, are drawn from each stratum. There are several advantages in STS:

a) it is possible to adopt different methods and procedures for different strata (e.g., different administration of questionnaires for each stratum);
b) it is possible to set the precision of the estimator via oversampling or undersampling procedures that allow variation of the sizes of the strata according to different aims;
c) it allows one to obtain information for sub-populations or sub-domains of analysis. For example, R&D can estimate the mean talk time for each plant;
d) all the characteristics of the strata are reported in the sample. In Fig. 1 (left side) a population which units have different shapes is showed. If the SRS sample is drawn, it is possible that not all the shapes will be in the drawn sample (e.g. we can obtain a sample with all star units). If we consider the stratified population (right side of Fig. 1) at least 1 unit per shape type will be part of the sample. That provides information from all the shape types of the population;
e) if the logical stratification leads to a technical homogeneity, we have a sampling plan that usually produces more precise estimators or we are able to obtain the same efficiency with a smaller sample.

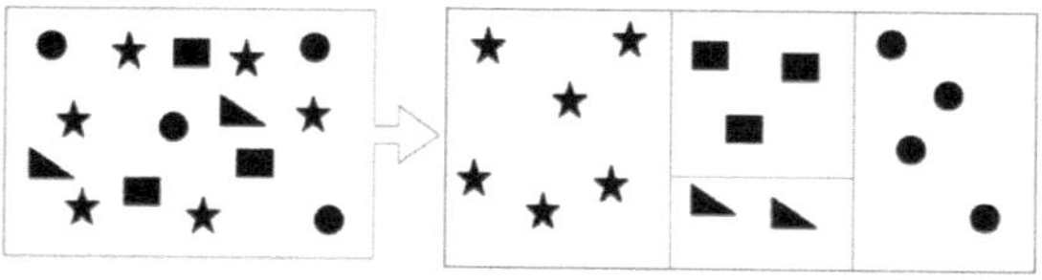

Figure 1. Stratification effect (right side) on the population (left side).

Some notations:

L number of strata

h generic stratum indicator ($h = 1, \ldots, L$)

N_h number of units of the population in the stratum h

$N = \sum_{h=1}^{L} N_h$ population size

n_h number of units of the sample for the stratum h

$n = \sum_{h=1}^{L} n_h$ sample size

$W_h = \frac{N_h}{N}$ weight of the stratum for the population

$w_h = \frac{n_h}{n}$ weight of the stratum for the sample

$f_h = \frac{n_h}{N_h}$ sampling fraction for the stratum h

$\overline{Y}_h = \frac{\sum_{i=1}^{N_h} Y_{hi}}{N_h}$ mean of Y for the stratum h

$\overline{y}_h = \frac{\sum_{i=1}^{n_h} y_{hi}}{n_h}$ mean of Y over the sampled units in stratum h

$S_h^2 = \frac{\sum_{i=1}^{N_h} \left(Y_{hi} - \overline{Y}_h\right)^2}{N_h - 1}$ variance of Y in stratum h

$$s_h^2 = \frac{\sum_{i=1}^{n_h} \left(y_{hi} - \overline{y}_h\right)^2}{n_h - 1} \text{ estimated variance of } Y \text{ in stratum } h. \tag{20}$$

Example 3

Let $N_1=10$, $N_2=13$, $N_3=12$ be the population size divided into three strata, and let $n_1=5$, $n_2=9$, $n_3=6$ be the SRS sample sizes drawn from those strata. How large is the size of Ω? Since Ω is the number of all the possible samples that can be drawn from that population and since a SRS sample is drawn from each stratum, for STS the dimension of Ω is the product of the sample space related to each stratum:

$$\Omega=\binom{N_1}{n_1}\times\binom{N_2}{n_2}\times.....\times\binom{N_h}{n_h}=\binom{10}{5}\times\binom{13}{9}\times\binom{12}{6}=166,486,320$$

SSE – 2

Calculate the first order inclusion probability.

With respect to the second order inclusion probability, we have to consider two different situations: i) both the units *i* and *j* belong to the same stratum, and ii) *i* and *j* belong to two different strata. Case i) is the simplest one: we draw a SRS without replacement 2 units and the draws are independents. Then:

$$\pi_{hij}=\frac{n_h}{N_h}\frac{n_h-1}{N_h-1} \qquad \forall\ i\neq j.$$

For case ii) we have to consider that unit *i* belongs to the stratum h and unit *j* belongs to the stratum h' (for example, the two strata are two different plants), because the two draws are independent:

$$\pi_{(hi)(h'j)}=\frac{n_h}{N_h}\frac{n_{h'}}{N_{h'}}.$$

In STS plan some basic decisions must be taken by the researcher. The most important are:

a) detecting the stratification criterion, via one or more auxiliary variable;
b) detecting the number of strata and, if the stratification variable is metric, the interval extremes for each stratum;
c) once the sample size is set, detecting a criterion to determine the sample size for each stratum: that procedure is called allocation.

The allocation can be equal if n is divided equally among strata, that is $n_h=n/L=k \quad \forall h=1,...,L$.

The allocation is proportional if n is divided among the strata on the basis of the weight of the stratum for the population, that is $n_h\propto W_h$ or

$$\frac{n_h}{n}=\frac{N_h}{N}\Rightarrow n_h=n\times\frac{N_h}{N}.$$

For both allocation, each unit in each stratum has the same inclusion probability. Therefore, the sampling plan is self-weighting.

Finally, adopting the Neyman optimum allocation (Neyman, 1934), sample sizes of the strata are set following the aim to maximize the precision of the estimator, considering the variability of Y in each stratum. It can be proved (see paragraph 4.4.2.1) that for each stratum h:

$$n_h = n\frac{W_h S_h}{\sum W_h S_h}. \tag{21}$$

Comparing the three types of allocation, it can be noticed that they are all 'proportional'. In fact, for equal allocation n_h is proportional to L/n; while in Neyman optimum allocation n is divided among strata proportionally to $W_h S_h / \sum W_h S_h$.

The allocation type is so basic as to determine the estimator and its variance that we can affirm that there are three different STS plans, one for each allocation.

With respect to the mean of Y for a population divided in L strata, according to the associative property of the arithmetic mean, it is:

$$\overline{Y}_{STS} = \frac{\sum_{h=1}^{L} N_h \overline{Y}_h}{N} = \sum_{h=1}^{L} W_h \overline{Y}_h.$$

Therefore, the weighted mean of the $\overline{y}_h$ could be an estimator for the mean of Y:

$$\overline{y}_{STS} = \frac{\sum_{h=1}^{L} N_h \overline{y}_h}{N} = \sum_{h=1}^{L} W_h \overline{y}_h. \tag{22}$$

Weights (W_h) take account of the population size of the stratum compared to the population size. The mean of Y over the sample is different from the (22), in fact it is:

$$\overline{y} = \frac{\sum_{h=1}^{L} n_h \overline{y}_h}{n} = \sum_{h=1}^{L} w_h \overline{y}_h, \tag{23}$$

where w_h is the weight, that is the ratio between the sample size in the stratum and the sample size

SEE – 3

When do (22) and (23) coincide?

Since the samples are independent, the variance of the estimator for the mean is:

$$V(\overline{y}_{STS}) = \sum_{h=1}^{L} W_h^2 V(\overline{y}_h), \tag{24}$$

where $V(\bar{y}_h)$ is the variance of the estimator for the mean for the stratum h.

PROOF: Since $\bar{y}_{ST} = \sum_{h=1}^{L} W_h \bar{y}_h$, that is a linear combination of $\bar{y}_h$ with weights W_h, it is possible to write the variance in terms of linear combination (variance of an addition):

$$V(\bar{y}_{STS}) = \sum_{h=1}^{L} W_h^2 V(\bar{y}_h) + 2\sum_{h=1}^{L}\sum_{j>h}^{L} W_h W_j \operatorname{cov}(\bar{y}_h \bar{y}_j).$$

Since the draws are independents, the covariation is null, then the (24).

Equation (24) shows that the variance of the estimator depends on just the L variances computed for the L estimators for the mean in each stratum. Since in each stratum a SRS sample is drawn, the variance of the estimator of the mean for the stratum h is:

$$V(\bar{y}_h) = \frac{S_h^2}{n_h}\frac{N_h - n_h}{N_h}. \tag{25}$$

If it could possible to stratify the heterogeneous population in a such way that it is very homogenous with respect to Y in each stratum, we could estimate the Y mean without bias, because every S_h^2 tends to zero. The variance of the estimator for the mean is obtained putting (25) in (24):

$$V(\bar{y}_{STS}) = \sum_{h=1}^{L} W_h^2 \frac{S_h^2}{n_h}(1 - f_h). \tag{26}$$

If *fcp* can be ignored, the (26) becomes:

$$V(\bar{y}_{STS}) = \frac{1}{N^2}\sum_{h=1}^{L}\frac{N_h^2 S_h^2}{n_h} = \sum_{h=1}^{L}\frac{W_h^2 S_h^2}{n_h}.$$

Most of the times S_h^2 is unknown, so it is necessary to estimate it. If we draw at least two units per stratum, an unbiased estimator for S_h^2 is the (20). Then, the estimator for the variance of the estimator for the mean is:

$$v(\bar{y}_{STS}) = \sum_{h=1}^{L} W_h^2 \frac{s_h^2}{n_h}(1 - f_h) = \sum_{h=1}^{L}\frac{W_h^2 s_h^2}{n_h} - \sum_{h=1}^{L}\frac{W_h s_h^2}{N}, \tag{27}$$

Where the last term of the very right side is the *fcp*.

With respect to the equal allocation, (24) becomes:

$$V_{eq}(\bar{y}_{STS}) = \frac{L}{n}\sum_{h=1}^{L} W_h^2 S_h^2 - \frac{1}{N}\sum_{h=1}^{L} W_h S_h^2,$$

while if we consider the proportional allocation, since $f_h = \frac{n_h}{N_h} = \frac{n}{N} = f$ and then $n_h = n\frac{N_h}{N}$, we get:

$$V_{prop}\left(\bar{y}_{STS}\right) = \frac{1-f}{n}\sum_{h=1}^{L} W_h S_h^2. \tag{28}$$

SEE – 4

If the variances of Y in each stratum are equals and the allocation is proportional, which is the variance of the estimator for the mean?

With regard to the Neyman optimum allocation, it can be proved that by using (21) in (26) it becomes:

$$V_{\min}\left(\bar{y}_{STS}\right) = \frac{\left(\sum W_h S_h\right)^2}{n} - \frac{\sum W_h S_h^2}{N}, \tag{29}$$

where the last term is the *fcp*. Equation (29) which is called 'minimum' variance because it is the minimum variance that can be obtained for the estimator for the mean conditional to a set sample size n, to W_h, and to S_h^2.

If the aim of sampling is to estimate the total $\dddot{Y}$, the estimator for the total is

$$\dddot{y}_{STS} = N\bar{y}_{STS}, \tag{30}$$

and the its variance is:

$$V\left(\dddot{y}_{STS}\right) = \sum_{h=1}^{L} N_h\left(N_h - n_h\right)\frac{S_h^2}{n_h}. \tag{31}$$

SEE – 5

Prove the (31), following the proof of the (24).

Example 4

Sampple R&D wants to estimate the number of batteries of which talk time is lower than the declared time. Then it considers a STS where the stratification variable is the size of the plants (big, medium, little), setting n=185. The estimate of the mean using the following information is:

Plant h	$\bar{y}_h$	N_h	n_h	s_h^2
Big	9,5	820	62	2,5
Medium	10,4	546	41	1,9
Little	12,9	1084	82	0,8
Total		2450	185	

$$\bar{y}_{STS} = \sum_{h=1}^{3} W_h \bar{y}_h = \frac{820}{2450} \times 9.5 + \frac{546}{2450} \times 10.4 + \frac{1084}{2450} \times 12.9 = 11.2,$$

the estimate of the total is $\ddot{y}_{ST} = N\bar{y}_{ST} = 2450 \times 11.2 = 27,440.$

The estimate of its variance is:

$$v(\ddot{y}_{STS}) = \sum_{h=1}^{3} N_h (N_h - n_h) \frac{s_h^2}{n_h} =$$

$$= 820(820 - 62)\frac{2.5}{62} + 546(546 - 41)\frac{1.9}{41} + 1,084(1,084 - 82)\frac{0,8}{82} = 48,437.4$$

Therefore, the standard error is $s.e.(\ddot{y}_{STS}) = \sqrt{v(\ddot{y}_{STS})} = \sqrt{48,437.4} = 220.1.$

SEE – 6

How do the (31) reduce in the equal, proportional and Neyman optimum allocation cases?

As well as SRS, the estimator for the proportion can be thought of as the estimator for the mean of the dichotomous variable Y', and the estimator for the frequency (the numerator of the proportion) can be thought of as the estimator for the total. Following the notation used for the SRS, let A_h be the total of units that have the characteristic for the stratum h, and $P_h = A_h / N_h$ the proportion of units in the same stratum ($Q_h = 1 - P_h$), its estimator is $p_h = a_h / n_h$ ($q_h = 1 - p_h$). Then, the estimator for the proportion is:

$$p_{STS} = \sum_{h=1}^{L} \frac{N_h p_h}{N},$$

and the estimator for the absolute frequency is:

$$a_{STS} = \sum_{h=1}^{L} N_h p_h$$

Considering the (6) and the (16) for each stratum, the variance of the estimator for the proportion is:

$$V(p_{ST}) = \sum_{h=1}^{L} W_h^2 V(p_h) = \frac{1}{N^2} \sum_{h=1}^{L} N_h^2 \frac{(N_h - n_h)}{N_h - 1} \frac{P_h Q_h}{n_h} \qquad (32)$$

where $(N_h - n_h)/(N_h - 1)$ is the *fcp*.

When $1/N_h$ is negligible, $S_h^2 = P_h Q_h$ and then the (32) reduces:

$$V(p_{STS}) = \sum_{h=1}^{L} \frac{W_h^2 P_h Q_h}{n_h} (1 - f_h) \qquad (33)$$

and if *fcp* is negligible too, the (33) becomes:

$$V\left(p_{STS}\right)=\sum_{h=1}^{L}W_h^2\frac{P_hQ_h}{n_h}.$$

SEE – 7

How do the (32) reduce in the equal allocation case?

If the allocation is optimum in Neyman way, replacing S_h of the (29) with those of the (6), we get:

$$V_{\min}\left(p_{STS}\right)\simeq\frac{1}{n}\left(\sum_{h=1}^{L}W_h\sqrt{P_hQ_h}\right)^2-\frac{1}{N}\sum_{h=1}^{L}W_hP_hQ_h \quad (34)$$

where $\frac{1}{N}\sum_{h=1}^{L}W_hP_hQ_h$ is the *fcp*. Usually, P_h e Q_h are unknown and to estimate the (32) $(P_hQ_h)/n_h$ is replaced by $(p_hq_h)/(n_h\text{-}1)$. Another way is set P_h=Q_h=0.5 because it is the maximum presumable variability for the proportion of Y'.

Example 5

Following the Example 4, the aim is to estimate the percentage of batteries that last at least 10 hours. With respect to a Neyman optimum STS, to determine n_h we use the (21):

$$n_1=185\times\frac{0.335\times\sqrt{2,5}}{0.335\times\sqrt{2,5}+0.223\times\sqrt{1,9}+0.442\times\sqrt{0,8}}=185\times\frac{0,529}{1,232}=79$$

$$n_2=185\times\frac{0,307}{1.232}=46$$

$$n_3=185\times\frac{0,396}{1.232}=60.$$

For example, if the totals of the batteries that last at least 10 hours are 12, 5, and 10 for each stratum, the proportions are:

$$p_1=\frac{12}{185}=0,065 \qquad p_2=\frac{5}{185}=0,027 \qquad p_3=\frac{10}{185}=0,054.$$

Then, the estimate of the percentage of the batteries is:

$$p_{STS}\times100=\sum_{h=1}^{3}\frac{N_hp_h}{N}\times100=\left(\frac{820\times0.065}{2450}+\frac{546\times0.027}{2450}+\frac{1084\times0.054}{2450}\right)\times100=51.6\%$$

with an estimate of the standard error (following the (34)) equals to:

$$\left[v_{\min-opt}\left(p_{STS}\right)\right]^{1/2}=\left[\frac{1}{n}\left(\sum_{h=1}^{3}W_h\sqrt{p_h q_h}\right)^2-\frac{1}{N}\sum_{h=1}^{L}W_h p_h q_h\right]^{1/2}=\left[0.000278\right]^{1/2}=0.017$$

4.2.1 Some remarks on sample size

In the previous paragraph we have shown the variances for the three estimators, as per different allocations. All those formulae depend on n_h and it depends on n. Often the sample size is just a matter of opportunities. For example, costs play an important role. Sampple should pay to collect the batteries from the plants, or for labelling the batteries, or for overtime. Practically, there is a trade-off between hoped precision and costs for survey (for both SRS and STS). There are several solutions that involve sample size, variances and costs. With respect to n_h, if we want to consider the bearable costs, we have to define them and typically you can refer to a linear function:

$$C=C_0+\sum_{h=1}^{L}C_h n_h, \tag{35}$$

where:

- C_0 are the fixed costs, those that are strictly linked to the survey (finding information, training, planning, etc.).
- C_h are the costs related to gather information from each sampled unit in the stratum h: it is a variable cost because it can be different among the strata (different costs to collect the batteries from different plants).

Once the cost function is set, the aim of the researcher will be (as alternative):

a) minimizing the variance of the estimator conditionally to costs;
b) minimizing the costs conditionally to a precision level of the estimator;
c) minimizing the variance of the estimator conditionally to a fixed sample size n.

All these alternative choices can be handled via a strategy for finding the local minimum of a function subject to equality constraints. Therefore, it is possible to find the solution using the Lagrange multipliers method, with a single constraint. We refer to the estimator of the mean case and therefore to the (22).

a) The n_h that minimize the variance (26), fixing the costs to C', can be derived from the Lagrangian function:

$$\Lambda(n_h,\lambda)=V(\bar{y}_{ST})+\lambda(C-C')=\sum\frac{W_h^2S_h^2}{n_h}-\frac{\sum W_hS_h^2}{N}+\lambda\left(\sum c_hn_h+c_0-C'\right).$$

Setting the gradient equal to zero, we get:

$$\frac{\partial\Lambda}{\partial n_h}=-\frac{W_h^2S_h^2}{n_h^2}+\lambda c_h=0 \tag{36}$$

$$\frac{\partial\Lambda}{\partial\lambda}=\sum c_hn_h+c_0-C'=0 \tag{37}$$

From (36) we can find n_h:

$$n_h^2=\frac{W_h^2S_h^2}{\lambda c_h}\quad\Rightarrow\quad n_h=\frac{W_h\,S_h}{\sqrt{\lambda}\sqrt{c_h}}$$

that, put in (37), implies:

$$\sum c_h\frac{W_h\,S_h}{\sqrt{\lambda}\sqrt{c_h}}+c_0-C'=0\quad\Rightarrow\quad\frac{1}{\sqrt{\lambda}}\sum\sqrt{c_h}W_h\,S_h=C'-c_0\quad\Rightarrow\quad\frac{1}{\sqrt{\lambda}}=\frac{(C'-c_0)}{\sum\sqrt{c_h}W_h\,S_h}$$

Therefore, the n_h that minimize the (26) with the constraint of fixed costs C', using the (35), are:

$$n_h=\frac{(C'-c_0)}{\sum\sqrt{c_h}W_h\,S_h}\frac{W_h\,S_h}{\sqrt{c_h}}. \tag{38}$$

The n_h allocation, obtained taking account the fixed costs, the weights W_h of strata, and the variability of Y in the strata, maximizes the variance of the estimator. Such allocation is known as optimum allocation with fixed costs (Cochran 1977).

b) The n_h that minimizes costs (34), fixing the variance V' of the estimator, can be derived from the Lagrangian function, too:

$$\Lambda(n_h,\lambda)=C+\lambda\left(V(\bar{y}_{ST})-V'\right)=\sum c_hn_h+c_0-C+\lambda\left(\sum\frac{W_h^2S_h^2}{n_h}-\frac{\sum W_hS_h^2}{N}-V'\right).$$

Setting the gradient equal to 0, we get:

$$\frac{\partial\Lambda}{\partial n_h}=c_h+\lambda\left(-\frac{W_h^2S_h^2}{n_h^2}\right)=0 \tag{39}$$

$$\frac{\partial\Lambda}{\partial\lambda}=\sum\frac{W_h^2S_h^2}{n_h}-\frac{\sum W_hS_h^2}{N}-V'=0. \tag{40}$$

From (39) we can find n_h:

$$n_h^2 = \lambda \frac{W_h^2 S_h^2}{c_h} \quad \Rightarrow \quad n_h = \sqrt{\lambda} \frac{W_h S_h}{\sqrt{c_h}}$$

that, put in (40), implies:

$$\sum \frac{W_h^2 S_h^2}{\sqrt{\lambda} \frac{W_h S_h}{\sqrt{c_h}}} = \frac{\sum W_h S_h^2}{N} + V' \quad \Rightarrow \quad \frac{1}{\sqrt{\lambda}} \sum W_h S_h \sqrt{c_h} = \frac{\sum W_h S_h^2}{N} + V' \quad \Rightarrow$$

$$\Rightarrow \sqrt{\lambda} = \frac{\sum W_h S_h \sqrt{c_h}}{\frac{\sum W_h S_h^2}{N} + V'}.$$

Therefore, the n_h that minimize the (26) with the constraint of fixed variance V', are :

$$n_h = \frac{\left(\sum W_h S_h \sqrt{c_h}\right)}{\frac{\sum W_h S_h^2}{N} + V'} \frac{W_h S_h}{\sqrt{c_h}} \tag{41}$$

The previous allocation considering the fixed variance, W_h, and the variability of Y in the strata, minimizes the costs. Such allocation is known as optimum allocation with fixed variance.

c) The n_h that minimize the variance (26), fixing the sample size n', can be derived from the Lagrangian function:

$$\Lambda(n_h, \lambda) = V(\bar{y}_{ST}) + \lambda(n - n') = \sum \frac{W_h^2 S_h^2}{n_h} - \frac{\sum W_h S_h^2}{N} + \lambda\left(\sum n_h - n'\right).$$

Setting the gradient equal to zero, we get:

$$\frac{\partial \Lambda}{\partial n_h} = -\frac{W_h^2 S_h^2}{n_h^2} + \lambda = 0 \tag{42}$$

$$\frac{\partial \Lambda}{\partial \lambda} = \sum n_h - n' = 0. \tag{43}$$

From (42) we can find n_h:

$$n_h^2 = \frac{W_h^2 S_h^2}{\lambda} \quad \Rightarrow \quad n_h = \frac{W_h S_h}{\sqrt{\lambda}}$$

that, put in (43), implies:

$$\sum \frac{W_h S_h}{\sqrt{\lambda}} - n' = 0 \quad \Rightarrow \quad \frac{1}{\sqrt{\lambda}} = \frac{n'}{\sum W_h S_h}.$$

Therefore, the n_h that minimize the (26), with the constraint of fixed sample size n', are :

$$n_h = n \frac{W_h S_h}{\sum W_h S_h}. \tag{44}$$

This allocation obtained fixing n', considering W_h and the variability of strata, optimizes the variance of the estimator. Such allocation is the Neyman optimum one: (29) is derivable from (26) if $c_h = c \ \forall h$.

4.3 Outline on other sampling techniques

The aim of the sample can be achieved in several other ways. For example, R&D staff can get a sample of battery plants and measure the talk time of all the present batteries. That sampling plan is known as cluster sampling (CLS). The rationale is that the population is divided in "natural" and homogeneous groups (but internally heterogeneous) that collect the elementary units. Formally, a CLS is a SRS where the units are the groups: therefore, N is the number of groups in the population, and n is the sample size of groups. For the sample, the groups are the plants and the elementary units are the batteries. The Heterogeneity within the clusters allows homogeneity between the clusters and the sampling of a few clusters.

CLS is a simple sampling plan, but it becomes harder if the groups differ for the number of elementary units. Therefore, it is impossible to know *a priori* the number of the sampled elementary units that have repercussions on estimators and their variances.

Even if the CLS and STS seem to be quite similar (because both groups and the units), the draw scheme is very different: STS divides the population in different strata and, from each stratum, draws the elementary units. CLS considers the population divided in several groups (like strata) but only the elementary units belonging to the drawn groups are analyzed. Moreover, in STS within the strata, the homogeneity is desiderable; while in CLS within the group the heterogeneity is needed.

Another different sampling plan is the systematic sampling (SYS). Let us suppose that the Sampple has just one plant that produces the batteries, and imagine that there is a conveyor belt for the batteries. SYS consists of selecting one battery at random (for example the battery that is on the belt at a fixed moment) and then select one battery every k (known as sampling interval) batteries. Implicitly, SYS is an equal-probability sampling. Main advantages are the easiness and the quickness of the sampling method. SYS is often regarded as pseudo-random sampling because just the first unit is selected at random.[3] If the order of the population units is at random, SYS is equal to SRS. In other situations the population can be ordered (for example alphabetically) and the researcher must ensure that the chosen sampling interval does not hide a pattern (in respect to Y), because any pattern would lose randomness.

Let us suppose that the Sampple wants to take account of the different plants but that it can analyze just a fixed number of batteries. A solution can be the merging of different sampling plan: for example it is possible to draw a sample of plants and then, from each sampled plant, drawing a sample of elementary units. Formally, we are taking account of two nested units of population: the first one is composed of the primary units (or first stage units, e.g., the plants), and the second one of the secondary units (or second stage units, e.g., the batteries). Once the sample of first units is drawn, the set of elementary units that belongs to that sample becomes the population for the second draw. The sampling plan that considers just two nested units of population and one or two sampling plans is called two stages sampling (TSS). It consists of drawing, without replacement, a random sample of primary units and at the second step (stage), drawing at random the secondary units from the sampled primary ones. For example, Sampple R&D staff can choose a CLS of plants for first stage and a SRS of batteries, belonging to the selected plants, for the second one. TSS can be generalized to r stages, with a different sampling plan for each stage.

Finally, Sampple can be interested in taking into account both the plants and their size, for example in terms of batteries produced per time unit or in terms of the numbers of employees. Moreover, it wants that most of the sampled units come from big plants. STS and CLS do not assure that, and obviously not even the SRS and SIS. In such cases it is possible to refer to the family of variable probability sample, known as probability proportional to size samples (PPS). In PPS the units are not equal-probability to be drawn. So it is possible to assign different weights (probabilities) to a group of units so that their draw is more likely. For example, the weights can be the normalized sizes of the plants. PPS is a very common sampling plan, and in literature more than 60 different methods exist, classified according to selection method, inclusion probability, etc. (Brewer and Hanif 1983, Chaudhuri and Vos 1988).

---

[3]For further close examination see, among others, Ken Black (2004), *Business Statistics for Contemporary Decision Making* (Fourth (Wiley Student Edition for India) ed.), Wiley-India; and Madow, G.W. and Madow L.H. 1944, On the Theory of Systematic Sampling, I. Annals of Math Stats 25: 1–24.

Solution SEE – 1

Fcp can be ignored in (11) or in the estimator for the (11) if $n/N \leq 0{,}05$. Therefore, since $n \leq 0{,}05N = 0{,}05*2320=116$.

Solution SSE – 2

The first order inclusion probability is that the battery *i* of the stratum *h* is included in the sample drawn, following the (9), is:

$$\pi_{hi} = \frac{n_h}{N_h} = f_h \quad \forall i = 1,\dots,N$$

It is equal to the sampling fraction for each stratum.

Solution SEE – 3

(22) and (23) are equals just in proportional allocation case, that is if $\frac{n_h}{n} = \frac{N_h}{N}$ or $\frac{n_h}{N_h} = \frac{n}{N}$ or $f_h = f$.

Solution SEE – 4

With respect to the proportional allocation, if the variances of *Y* in each stratum are equals, $S_h^2 = S_w^2$ and then (28) reduces $V_{prop}\left(\overline{y}_{ST}\right) = \frac{S_w^2}{n}\frac{N-n}{N}$.

That variance is equal to (9), therefore the stratification process does not improve the precision of the estimator.

Solution SEE – 5

Taking account of (30) and (26), we get:

$$V\left(\ddot{y}_{ST}\right) = V\left(N\overline{y}_{ST}\right) = N^2 V\left(\overline{y}_{ST}\right) = N^2 \sum_{h=1}^{L} W_h^2 \frac{S_h^2}{n_h}\left(\frac{N_h - n_h}{N_h}\right) = \sum_{h=1}^{L} N_h^2 \frac{S_h^2}{n_h}\left(\frac{N_h - n_h}{N_h}\right) =$$

$$= \sum_{h=1}^{L} N_h \left(N_h - n_h\right)\frac{S_h^2}{n_h}$$

Solution SEE – 6

If the allocation is equal, $n_h = n/L$ modify the (31) in such way:

$$V_{eq}\left(\ddot{y}_{STS}\right) = \frac{L}{N}\sum_{h=1}^{L} N_h^2 S_h^2 - \sum_{h=1}^{L} N_h S_h^2$$

If the allocation is proportional, putting $n_h = nW_h$ in (31), we get:

$$V_{prop}\left(\ddot{y}_{STS}\right) = \sum N_h \left(N_h - n\frac{N_h}{N}\right)\frac{N}{nN_h} S_h^2 = \sum N_h \left(\frac{N}{n} - 1\right) S_h^2 = \frac{N-n}{n}\sum N_h S_h^2$$

Same results can be obtained if you consider $V_{prop}(\ddot{y}_{ST}) = V_{prop}(N\bar{y}_{ST})$.

If the allocation is optimum in the Neyman sense, the variance of the estimator for the total is derived putting (29) in (30):

$$V_{\min}(\ddot{y}_{ST}) = V_{\min}(N\bar{y}_{ST}) = N^2 V_{\min}(\bar{y}_{ST}) = \frac{\left(\sum N_h S_h\right)^2}{n} - \sum N_h S_h^2$$

Solution SEE – 7

If allocation is equal, $n_h = n/L$ modify the (32) in such a way:

$$V_{eq}(p_{ST}) = \frac{1}{n}\sum_{h=1}^{L} W_h^2 \frac{(LN_h - n)}{N_h - 1} P_h Q_h.$$

If allocation is proportional, putting $n_h = nW_h$ in (33), we get:

$$V_{prop}(p_{ST}) = \frac{1}{N^2}\sum_{h=1}^{L} N_h^2 \frac{\left(N_h - n\frac{N_h}{N}\right)}{N_h - 1} \frac{P_h Q_h}{n\frac{N_h}{N}} = \frac{1}{N^2}\sum_{h=1}^{L} N_h^2 \frac{\frac{NN_h - nN_h}{N}}{N_h - 1} \frac{NP_h Q_h}{nN_h} =$$

$$= \frac{1}{N^2}\sum_{h=1}^{L} \frac{N_h^2}{N_h - 1} \frac{N_h(N-n)}{N} \frac{NP_h Q_h}{nN_h} = \frac{N-n}{nN^2}\sum_{h=1}^{L} \frac{N_h^2}{N_h - 1} P_h Q_h =$$

$$= \frac{1-f}{n}\sum_{h=1}^{L} \frac{N_h}{N}\frac{N_h}{N_h - 1} P_h Q_h \simeq \frac{1-f}{n}\sum_{h=1}^{L} W_h P_h Q_h$$

Solved Problems

Problem) Proof the (10)

Answer) Following the rationale of the proof of the (9), if we consider the units i and j, all the samples that contains them are $\binom{N-2}{n-2}$; while, the number of all samples is $\binom{N}{n}$. As in the proof of the (9), the second order inclusion probability is the ratio of that two quantities:

$$\pi_{ij} = \frac{\binom{N-2}{n-2}}{\binom{N}{n}} = \frac{n(n-1)}{N(N-1)}.$$

P) Find the first order inclusion probability for units 1 and 4, belonging to the population {1,2,3,4}, taking into account the not ordered and without replacement scheme. Moreover, the samples *s* and their probability *p(s)* are:

s	(1;2)	(1;3)	(1;4)	(2;3)	(2;4)	(3;4)
p(s)	0.15	0.10	0.20	0.15	0.20	0.20

A) The first order inclusion probability for unit 1 is $\pi_1 = 0.15+0.10+0.20 = 0.45$; and for the unit 4 is $\pi_4 = 0.20+0.20+0.20 = 0.60$

P) Let *P* be a population of *N=5* different elements {1;2;3;4;5}. In the table are reported all the samples drawn with a not ordered and without replacement scheme, and *n*=4, along with their probabilities:

p4	(1;2;3;4)	(1;2;3;5)	(1;2;4;5)	(1;3;4;5)	(2;3;4;5)
p(s)	0.15	0.25	0.10	0.20	0.30

Find the second order inclusion probability for all the couples of units.

A) Let π_{ij} be the generic second order inclusion probability, we get:
$\pi_{12}= 0.15+0.25+0.10 = 0.50$; $\pi_{13}= 0.15+0.25+0.20 = 0.60$;
$\pi_{14}= 0.15+0.10+0.20 = 0.45$; $\pi_{15}= 0.25+0.10+0.20 = 0.55$;
$\pi_{23}= 0.15+0.25+0.30 = 0.70$; $\pi_{24}= 0.15+0.10+0.30 = 0.55$;
$\pi_{25}= 0.25+0.10+0.30 = 0.65$; $\pi_{34}= 0.15+0.20+0.30$;
$\pi_{35}= 0.25+0.20+0.30$; $\pi_{45}= 0.10+0.20+0.30$.

P) Following the Example 4, how big is the stratum sample size if the sample size *n* is the 10% of the population size *N*?

A) The first step consists of identifying the type of allocation: Exercise 4 data of shows that the allocation is proportional. In fact: n_h/N_h=62/820=41/546=82/1084=*n/N*=185/2450=0.075. Since *n* is the 10% of *N*, *n=0,1N=245*, n_h will be:

$$n_h = n\frac{N_h}{N} = 0.10N\frac{N_h}{N} = 0.10N_h$$

$$n_1 = 0.10 \times 820 = 82 \quad n_2 = 0.10 \times 546 = 55 \quad n_3 = 0.10 \times 1084 = 108$$

Proposed Problems

1) Estimate of sample size in SRS

Estimate sample size *n* with respect to the estimator for the mean, in *SRS* plan, fixing a variance of the estimator (that is the bearable error for the estimator). Starting from the absolute error $d = |\bar{y} - \bar{Y}|$ (or the relative error $r = \left|\frac{\bar{y} - \bar{Y}}{\bar{Y}}\right|$), the aim is minimizing the probability that sampling error is larger than the bearable one, that is $\Pr\left\{\left|\bar{y}-\bar{Y}\right| \geq d\right\} = \alpha$, with α small. You have to make some assumptions over the distribution of the mean (14).

$$\left[n = \frac{\dfrac{t^2 S^2}{d^2}}{1 + \dfrac{1}{N}\dfrac{t^2 S^2}{d^2}} ; n = \frac{\dfrac{t^2 S^2}{r^2 \bar{Y}^2}}{1 + \dfrac{1}{N}\dfrac{t^2 S^2}{r^2 \bar{Y}^2}} \right]$$

2) Estimate of sample size in STS

In STS, with respect to the estimator for the mean, determine the sample size n, when the variance of the estimator is fixed and in proportional, Neyman optimum and for fixed costs optimum allocation cases.

$$\left[n = \frac{\sum W_h S_h^2}{V' + \dfrac{\sum W_h S_h^2}{N}} ; n = \sum n_h = \frac{\left(\sum W_h S_h \sqrt{c_h}\right)}{\dfrac{\sum W_h S_h^2}{N} + V'} \sum \frac{W_h S_h}{\sqrt{c_h}} ; n = \sum n_h = \frac{(C' - c_0)}{\sum \sqrt{c_h} W_h S_h} \sum \frac{W_h S_h}{\sqrt{c_h}} \right]$$

3) Comparison among different sampling plans

Fixing the sample size, and referring to the estimator for the mean, compare the precisions of the estimators among STS with proportional allocation, STS with minimum variance, and SRS plans. Try to decompose the variance of Y into two components: the variance between and the variance within groups (strata):

$$S^2 = \sum_{h=1}^{L} W_h S_h^2 + \sum_{h=1}^{L} W_h \left(\bar{Y}_h - \bar{Y}\right)^2$$

$$\left[V\left(\bar{y}_{SRS}\right) \geq V_{prop}\left(\bar{y}_{ST}\right) \geq V_{\min}\left(\bar{y}_{ST}\right) \right]$$

Several Youtube-videos related to sampling are available on the webpage.

Acknowledgement

The article is the result of the productive collaboration between the authors. In particular, paragraphs from 4 to 4.1 can be ascribed to Ornella Giambalvo, and paragraphs 4.2 and 4.3 can be ascribed to Giovanni Boscaino.

References

Brewer, K. and Hanif, M. 1983. Sampling with Unequal Probabilities, Springer- Verlag, New York, 2.

Chaudhuri, A. and Vos J.W.E. 1988. Unified theory and strategies of survey sampling. North-Holland.

Cochran, W.G. 1977. Sampling Techniques, 3Rd Edition. Wiley India Pvt. Limited.

Moore, D.S. 2000. Statistics: Concepts and controversies, 5Th Edition. Freeman Publishers.

Neyman, J. 1934. On the two different aspects of the representative method: the method of stratified sampling and the method of purposive selection. Journal of the Royal Statistical Society 97, 4: 558–625.

CHAPTER 5

Point Estimation and Statistical Intervals

Martin Griffiths

1. Introduction to Point Estimation

Introductory Problem

Treatments of many illnesses require patients to stay in hospital for a numbers of days, weeks or even months. Owing to the large number of inpatients that hospitals are expected to deal with, there is always an issue over the availability of hospital beds. In a particular large hospital an administrator is concerned that she will not have enough beds to cope with all the patients that are scheduled to be admitted over the coming months. In order to assess the current situation and put in place any contingency plans that might be necessary, she decides that it would be useful to know what the 'typical' length of stay was of an inpatient in her hospital. How might the administrator go about finding this typical value?

Initial Discussion

An initial, and certainly pertinent, question here is: "What do we mean by a 'typical' value, and how might we go about estimating it?" (Watier et al. 2011). Well, the notion of a 'typical' or an 'average' value of a set of data is something that is introduced in mathematics and statistics lessons at high school. In fact, we learn that three types of average are commonly considered; the *mean*, the *mode* and the *median*. Each of these gives a way

Mathematical Institute, University of Oxford (United Kingdom).

of summarizing the location of the data in some sense. They are sometimes termed *measures of location* or *measures of central tendency*. A key point is that the mean, mode and median each summarize the entire data set using just one number.

For example, suppose that the lengths of stay (in days) of the last 10 patients to leave the hospital were:

25, 28, 1, 3, 32, 24, 27, 3, 137, 29.

The mean of this set of data is given by

$$\frac{25+28+1+3+32+24+27+3+137+29}{10} = 30.9,$$

while 3 is the mode since it is the only piece of data to occur more than once, and the median is obtained by ordering the data to give 1, 3, 3, 24, 25, 27, 28, 29, 32, 137 and then finding the middle value in this ordered list; note that the median is 26 here since in the case of an even number of data items, we take the mean of the two middle values.

Computer Comment

Throughout the two following chapters we will make frequent use of a free piece of software entitled R (R Development Core Team). It is an extremely powerful language and environment for statistical computing. The short series of commands below calculate the mean, median and mode of the data given above.

```
> x=c(25,28,1,3,32,24,27,3,137,29)
> mean(x); median(x); names(sort(-table(x)))[1]
[1] 30.9
[1] 26
[1] "3"
```

There are several points to note here:

i) The symbol > is a prompt that appears at the beginning of each line of input when using R; it is not part of the code.
ii) It is possible to enter several commands and routines on the same line by using semicolons. Whether we use semicolons or start on a new line will be largely determined by visual clarity here.
iii) The command `x=c(25,28,1,3,32,24,27,3,137,29)` stores the data as a vector x.
iv) The R command `mode` does not do what you would expect. In other words, it does not calculate the mode of a set of data. It is thus necessary

to make up your own routine; the code `names(sort(-table(x)))[1]` is just one way of obtaining the mode.

Let us now consider how well each of these measures of average would appear to 'represent' the set of 10 data items. On glancing at this data set, it soon becomes evident that the mode gives a somewhat false picture of the general location since most of the data seems to be clustered between 24 and 32 days. Both the mean and the median give us a better idea of where the majority of the data is located, although the mean of 30.9 would seem to be unduly affected by the 137 data item and is consequently possibly a little on the high side to be regarded as truly representative of the data. We might thus argue that the median gives the best measure of location for this particular set of data.

Incidentally, before carrying out any serious analysis on the data set, it is worth bearing in mind the fact that it may contain erroneous values. In our case, there would not appear to be anything particularly suspicious in this regard; the person who was purportedly in hospital for one day might have had a simple key-hole operation on his/her knee, while the person who apparently spent 137 days there may have been recovering from a serious motorbike accident. However, there could be scenarios in which these extremes are highly unlikely to occur, in which case these data items may have been recorded in error.

Of course, it will not always be the case that the median gives us a better picture of the location of data then the mode or mean; this will depend very much on the nature of the data set. In practice, we tend to use the mean as a measure of location because its mathematical properties ensure that any ensuing calculations remain tractable. In particular, we know, by way of the central limit theorem (Grimmett and Stirzaker 2001), that the mean of a set of n identically and independently distributed random variables is approximately normal for sufficiently large n. Indeed, this is something we return to at various points in both the current and the following chapter.

Self-evaluation Problems 1.1

1. Consider the following set of ordered data:

$$4, 5, 8, 12, 12, 12, 16, 19, 20.$$

Note that it is in some sense symmetrical about the middle value.

i) Suppose that a new item of data, 17, is inserted into this data set. What other number would have to be inserted in order to maintain the symmetry?
ii) Is it true that in symmetrical data sets the mean is always equal to the median?

iii) In the above data set, the mode is equal to both the mean and the median. Will this be the case for all symmetrical data sets?

2. The diagram below shows the section of a number line (or axis) with a series of dots representing the numerical values of some items of data. Which of the mean, median or mode would you use as a measure of location in this instance? (Note that the missing scale on the axis is deliberate.)

3. The data displayed in Table 5.1 summarizes the number of pets owned by the children in a secondary-school class:

Table **5.1**

Number of pets	0	1	2	3	4
Frequency	6	14	4	1	5

i) Find the mean, median and mode of the number of pets owned.
ii) A new pupil joins the class. She is a beekeeper, and has 20,000 bees. If each of these bees is counted as a pet, calculate the new mean number of pets owned by pupils in this class. State, with a reason, which of the mean or median would be deemed the best measure of location in this particular case.

Self-evaluation Problem Solutions 1.1

1. i) The symmetry about the central data item, 12, means that we would need to insert a number that is as far below 12 as 17 is above it. The required number is thus 7.
   ii) Yes, this is true. You might like to think about why this will always be the case.
   iii) On considering 1, 1, 3, 3, we see that the mode is not in general equal to the mean or the median for symmetric data sets.
2. On the face of it, the apparent presence of the outlier might indicate that the median would be the best measure of location to use in this instance. However, in order to make a firm decision, we would need to see the scale on the axis. It could be that the relative differences of all these data items are extremely small, in which case we might as well use the mean. There is also the possibility that the axis had a non-linear scale (such as a logarithmic one), in which case we would have be careful in deciding which data items might be deemed outliers.
3. i) The mean is given by

$$\frac{6\times0+14\times1+4\times2+1\times3+5\times4}{30}=1.5.$$

The median and mode are both equal to 1.

ii) The new mean is

$$\frac{6\times0+14\times1+4\times2+1\times3+5\times4+1\times20{,}000}{31}\approx647.$$

This is clearly not representative of the majority of the data! It would be best to use the median, which is not affected by outliers to the extent that the mean is. In fact, in this case the median remains at 1.

Computer Comment

The mean of the data in the frequency table above is easily calculated by way of the following code:

```
> x=c(0:4); f=c(6,14,4,1,5); sum(x*f)/sum(f)
[1] 1.5
```

Population Parameters

The notion of a *population parameter* is very important here. By way of an example, let us consider the statistical population comprising the heights of all 18-year old females in the United Kingdom. In theory at least, we could measure the height of every 18-year old female in the United Kingdom and thus obtain the population mean μ. We call μ a population parameter since it provides us with a numerical summary of some aspect of the population; in this case that aspect is the location of the females' heights.

Is it, however, feasible in practice to measure the height of every single 18-year old female in the United Kingdom? A moment's thought tells us that the answer is indeed a very firm "No". There are approximately half a million 18-year old females in the United Kingdom and they are scattered all over the country. Thus, not only is there a large number of heights to record, but there would be the additional problem of locating all the members of the population.

From the previous paragraph we see that, unless the population is extremely small in number and easy to locate, it is simply not practical to obtain the exact value of the population parameter μ, representing the mean of the population. It is therefore the case that if we want to know something about its true value we will have to be content with estimating μ. This will involve taking a sample from the population.

Returning to our featured example discussed in the previous section, the statistical population might consist of the lengths of stays of all inpatients at all hospitals in a particular country over the last 10 years, say. For the purposes of forward planning, our hospital administrator would like to know the population mean μ. Once more, it would seem an almost impossible task to obtain μ precisely. To gather all the appropriate data would be expensive and extremely time-consuming. Furthermore, some patients' records may have been lost in the intervening period. She will therefore have to make do with an estimate for μ. However, the administrator will clearly want in some sense to obtain as 'good' an estimate as possible, and this is something that we discuss in the forthcoming sections.

Discussion

A point worth noting here is that the task the administrator wishes to carry out is called *point estimation* since she uses the data to calculate one value, the sample mean, that is used to estimate another, the population mean μ. On its own, this might be regarded as a somewhat primitive method of estimating a population parameter, since much of the information in the original data set has been 'lost' in summarizing it by just one number.

The key concepts at this stage are the notions of *population parameter* and *point estimate*, and it important to gain an intuitive appreciation of the ideas behind point estimation by discussing possible solutions to this problem in an informal manner (Lavy and Mashiach 2009, Kaplan et al. 2009). It is best initially to focus very much on the meaning of these concepts, and to contextualize them via our introductory problem; the importance of using real-life examples to teach abstract statistical concepts is discussed in (Mvududu and Kanyongo 2011). This helps circumvent issues arising from misconceptions associated with definitions or meanings. Indeed, statistical ideas are paramount here, rather than the memorization of formulas (O'Brien 2008). The dangers of over-mathematizing course content or over-emphasizing the teaching of formulas without having much concern for the underlying ideas are spelt out in (Batanero 2004, Schau and Mattern 1997).

The Relevance of Sampling

The notions of estimation and sampling are closely linked. Before considering this in detail, it is worth generating, possibly from an intuitive or naive perspective, some practical ideas for obtaining an estimate of μ, the mean length of stay of an inpatient at this hospital. For example, we could look at the hospital records of all the people discharged on a particular day, find the length of stay for each of them and then work out the sample mean $\bar{x}$. Here are some questions to ponder over in this regard:

i) Would we expect this to give us a good estimate of μ? Is it likely to be representative of the population taken as a whole?

ii) Are you able to devise a sampling procedure that is likely to provide us with better estimates?
iii) What are some of the fundamental issues associated with the taking of samples?

Further Population Parameters

In essence, we can think of a population parameter as something that captures one particular aspect of the population by way of a single number. In addition to the population mean μ, you will also already have encountered the population variance σ^2, which gives an indication of the degree of spread of the data. Indeed, you may well have seen how parameters can be used to define random variables. Suppose, for example, that X is the random variable representing the length of stay of a patient in the hospital. Then, if X is normally distributed, we may write

$$X \sim N(\mu,\sigma^2)$$

for population parameters μ and σ^2, representing the population mean and variance, respectively. In this case, from an intuitive point of view we might think of μ as specifying the 'position' of the distribution and σ^2 as specifying its 'shape'.

For example, in Fig. 1 we see the probability density function of the normal random variable X when $\mu = 26$ and $\sigma^2 = 64$. Changing μ and σ^2 to 31 and 36, respectively, results in the transformed distribution shown in Fig. 2. The resultant graph is 5 units further to the right and is also taller and

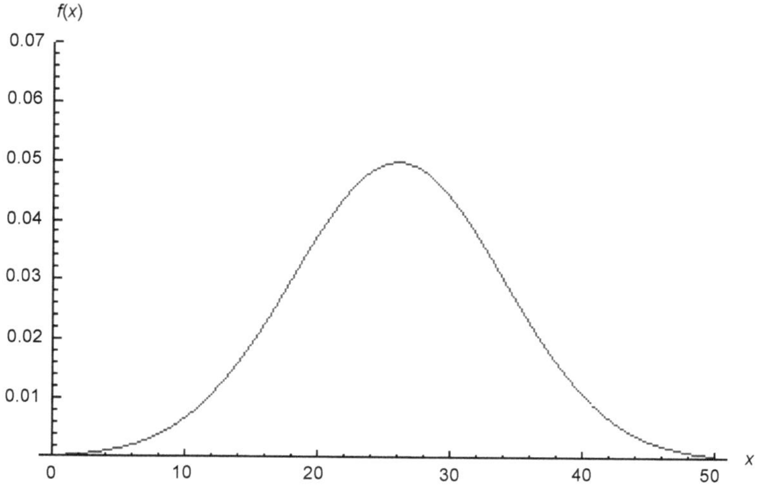

Figure 1. The probability density function for $X \sim N(26,8^2)$.

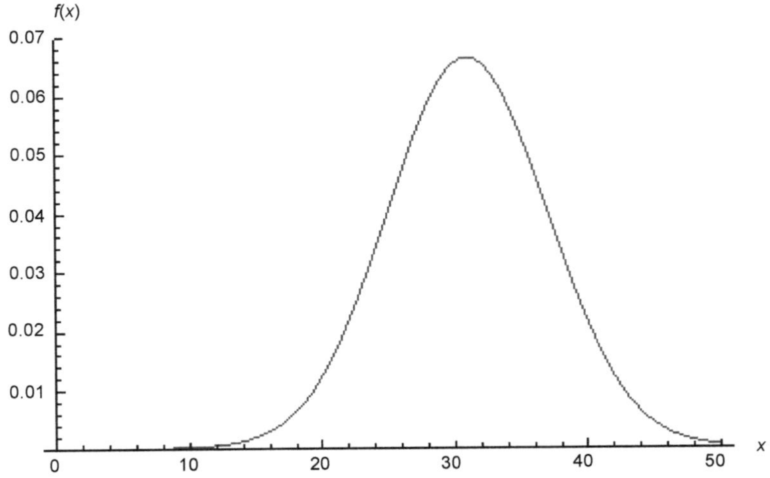

Figure 2. The probability density function for $X \sim N(31, 6^2)$.

thinner than the original distribution (the latter because of the decrease in population variance), remembering that the area under both of the curves is equal to 1.

Another common parameter is the population proportion. For example, p could denote the proportion of females in a particular population. You will have met this notion whilst studying the binomial distribution.

Problems 1.1

1. A student interested in basketball wanted to estimate the mean height of the 4,000 professional basketball players in the world (1,200 of which are women). In order to do so, he selected a sample of 10 male and 5 female basketball players at random. The mean height of the males was 205 cm and that of the females was 193 cm. He estimated the mean height of the 4,000 players by calculating the mean of 205 cm and 193 cm, which is 199 cm. Is this likely to be a good estimate?
2. Let p denote the population parameter representing the proportion of adults in the United Kingdom that are left-handed. A sample of size n is taken in order to obtain an estimate for p. What is wrong with the following statement?

   "The larger n is, the more accurate the estimate for p."

3. We have thus far met five possible population parameters: mean, median, mode, variance and population proportion. Find at least four more, and describe what information each conveys about the population.

Solutions to Problems 1.1

1. It is very unlikely to be a good estimate of the mean height. First, notice that 199 cm is not the mean of the sample. The sample mean is given by
$$\frac{10\times 205+5\times 193}{10+5}=201 \text{ cm}.$$
Second, because the mean height of the males will be significantly different to that of the females, it would have better either to take a random sample from the population as a whole, or to take a stratified sample (28 men and 12 women, for example). Finally, the size of the sample might be considered a little on the small side.
2. Although we often hear statements such as this, it is not strictly true. Suppose, for example, that the true proportion is 0.1. If we take a sample of size 10 then there is the possibility that exactly one of these people will be left-handed. In this case our estimate would then be exactly equal to true population proportion. If we now increased the size of the sample, to 20 say, there is of course the possibility that none of the next 10 will left-handed. Our estimate for p at this point would be 0.05. Thus, in increasing the sample size, we have obtained an inferior estimate. It is better to say something like "The larger n is, the more accurate we would expect the estimate for p to be." Such a statement can be expressed in more rigorous terms concerning probabilistic statements, but we are not concerned with such matters here.
3. Here are some further examples: skewness, kurtosis, range, inter-quartile range, maximum value of the population and minimum value of the population. For any of these that you have not heard of before (possibly skewness and kurtosis), carry out some research to find out as much about them as you can; particularly with respect to what information they give you about the population.

2. Estimates and Estimators

Initial Discussion

Our aim in this section is to give at least a partial explanation as to what is meant by a 'good' estimate for a population parameter such as μ, the mean length of stay of inpatients in hospitals nationally. We also consider how such an estimate might be obtained.

In order to be able to plan for the future demand on hospital beds, the administrator would like to be in a position to obtain an estimate for μ that was in some sense accurate. We have already pointed out that it is simply not a viable proposition to obtain the population parameter μ exactly, and consequently that some sort of sampling procedure will need to be carried out to enable us to estimate it. A question we may now ask ourselves is "Does it matter how this sampling procedure is carried out?"

The answer to this question is a resounding "Yes." There are two aspects of sampling to be considered initially; the size of the sample and the way in which the sample is obtained. Dealing with the first of these, it is an intuitive notion that somehow the larger the sample size n, the more accurate the estimate for μ. This is admittedly a little vague, and of course it does not necessarily follow that larger sample sizes will always lead to more accurate estimates for μ than will smaller ones, as we saw in Problem 1.1(3). However, it is the case that this somewhat imprecise statement may be sharpened up and subsequently proved rigorously (in a probabilistic sense) by way of the Law of Large Numbers (Grimmett and Stirzaker 2001).

The notion of sampling, covered in an earlier chapter, is a key one with respect to the process of estimating parameters. For this reason, we reinforce several points here. The following example emphasizes the point that larger samples do not guarantee more accurate estimates.

Example 2.1

Suppose that there are four siblings in a family, aged 5, 7, 12 and 19. Let us consider the estimation of the mean age μ of these siblings. For this example we can easily compute μ, the population mean, to be equal to 10.75. Now say we estimate μ by taking a sample of size two. If 7 and 12 were in the sample, then the estimate, assuming we use the mean of these values to estimate μ, would be 9.5. On the other hand, the sample of size three given by 5, 7 and 12 would give an estimate of 8. Thus in this case the smaller sample gives the better estimate for μ. (Of course, for this particular example, the size of the population is extremely small and there is no real need to obtain samples of size two of three to estimate μ; however, we wished merely to illustrate a point.)

Let us now move on to the second aspect of sampling that needs to be considered; the method by which the sample is obtained. We would like the sampling procedure to be fair in the sense that each of the data items is equally likely to be in the sample. If a sample is carried out under these conditions then it is termed a *random sample*. The selection of the sample might be carried out by compiling a list of all the data items in the population (this is known as a *sampling frame*) and then using random numbers generated on a computer to select the sample from this list. Alternatively, for smaller populations, we might devise some mechanical procedure for obtaining a sample, such as writing each of the data items on a small piece of paper, folding them up and then placing them in a hat. Of course, to ensure that obtaining a sample by picking numbers out of a hat was fair, we would need to consider the conditions of the process very carefully; for example, have all the pieces of paper been folded in such a way that they are all of equal size and are hence equally likely to be picked out?

Random sampling may be carried out either *with* or *without* replacement. For the former, if a particular data item has been selected as part of the sample, this does not preclude it from being selected again. With respect to the numbers-out-of-a-hat scenario, this corresponds to folding the paper back up each time and replacing it in the hat. Thus, at each stage of the sampling process, every one of the k data items in the population is equally likely to be selected. On the other hand, for random sampling without replacement, once a data item has been taken from the hat, it stays out. In this case, if the sample is of size n then we would wish this to be random in the sense that each possible sample of size n is equally likely to be selected. If the sample size is very small compared to the population size, we may essentially ignore this subtle difference in the sampling procedure. Indeed, we shall make the assumption throughout this chapter that the sample size is considerably smaller than the population size.

Computer Comment

To make concrete some of the simple ideas discussed above, we used R to obtain random samples from a uniform discrete distribution with mean μ = 50.5. In the first simulation 10 samples of size 5 were drawn, whilst in the second the 10 samples were of size 20. By comparing the outputs of the two runs, it may be seen that the means of the samples of size 20 provide generally superior estimates of μ, which is certainly what we might expect intuitively.

```
> w=c(1:100)
> for(k in 1:10){y[k]=mean(sample(w,5,replace=FALSE))}
> y[1:10]
 [1] 34.6 26.2 42.8 48.4 72.6 83.6 48.4 38.2 50.2 39.0
```

```
> w=c(1:100)
> for(k in 1:10){y[k]=mean(sample(w,20,replace=FALS
E))}
> y[1:10]
 [1] 42.55 48.80 54.15 50.75 46.15 48.05 37.45 54.65
46.35 51.80
```

As discussed earlier, we are unlikely to have a complete sampling frame. For the example we are considering here, this would consist of the lengths of stays of all inpatients in all hospitals over the last 10 years. It would be totally unrealistic to expect to be able to acquire this. In practice, therefore, any sample of waiting times that we may obtain will not be a truly random one since it is not the case that all data items have an equal chance of being selected; indeed, for those items that do not appear on our sampling frame, the probability of being selected is zero.

If a sample is not truly random then there is the possibility that it may not be fully representative of the general population. This leads to the notion of *bias*, which will be discussed at some length in this chapter. Before moving on, let us take an informal and intuitive look at this concept. Suppose that we obtain a sample to estimate the population parameter μ. Although we might use the sample mean $\bar{x}$ to estimate μ, we would not necessarily expect these two quantities to be actually equal. However, it would not be an unreasonable expectation that this procedure gives us the correct value of μ 'on average'.

Let us give another illustration as to how bias may be introduced at the data collection stage. Suppose, for example, that a particular disease had recently swept through the country. The symptoms of this disease were highly regular from one patient to another, and sufferers were required to stay in hospital for a particular period of time; something in the region of 40 days. Say, for the sake of convenience, the administrator decides to use data only of the last six months to estimate μ. Then, due to the high preponderance of lengths of stay of around 40 days due to this particular disease, the data may not be at all representative of the data throughout the 10-year period (and may in fact lead to an overestimate of μ).

Estimation and Random Variables

Now that we have considered, in a somewhat informal and intuitive way, some of the underlying notions associated with point estimation, this is an appropriate point at which to introduce random variables into our problem and to make things a little more rigorous (Crawshaw and Chambers 1994, Grimmett and Stirzaker 2001). Students require an intuitive appreciation of random variables, random sampling and the variance of the sample mean (Lecoutre et al. 2006) in addition to a robust understanding of what is meant by statistical variation (Peters 2011, delMas and Liu 2005) before proceeding much further.

Let the random variable X represent the length of stay in the hospital for a randomly-chosen patient. If we use a realisation x of X to estimate μ of then we call X an estimator for μ. (Incidentally, at this point we do not need to know the exact distribution of X.)

It is very important to highlight the distinction between an *estimate* of a population parameter and an *estimator* for this parameter. Indeed, many students find it difficult to appreciate the difference between an estimate and an estimator initially. The estimate is a numerical value that gives us an approximate value of μ (this might be thought of as a calculated 'guess' for μ, given the available data), while an estimator is a random variable that may be used to obtain the estimate. An estimator essentially dictates both the sample size and how the sample is used to obtain the estimate for μ. If the estimator is X we would normally denote the estimate by x, so x is actually a realisation of X.

It has already been mentioned that we would, at the very least, hope that our method for estimating μ leads to the right answer on average. Stating this in a slightly more precise manner, we would require the expectation of any estimator we use to be equal to μ. In such cases it is known as an *unbiased* estimator. On the other hand, if the expectation of the estimator was not equal to μ then it would be termed a *biased* estimator (it is not, in general, desirable to utilize such an estimator). These ideas are best illustrated via an example.

Example 2.2

Let us consider the estimation of μ by way of four simple estimators.

1. The simplest and most obvious choice for our estimator is to use X itself. We have $\mathrm{E}(X) = \mu$.
2. Next, let the random variables X_1, X_2 and X_3 be independently and identically distributed as X. We will consider the sample mean $\bar{X}$ given by

   $\bar{X} = \frac{1}{3}(X_1 + X_2 + X_3)$.

   By the linearity property of the expectation operator,

   $\mathrm{E}(\bar{X}) = \frac{1}{3}\mathrm{E}(X_1) + \frac{1}{3}\mathrm{E}(X_2) + \frac{1}{3}\mathrm{E}(X_3) = \frac{1}{3}\mu + \frac{1}{3}\mu + \frac{1}{3}\mu = \mu.$

   Note that the estimate associated with the estimator $\bar{X}$ is usually written $\bar{x}$.
3. Let U be given by

$U = \frac{1}{3}(2X_1 + X_2).$

Then

$$E(U) = \frac{2}{3}E(X_1) + \frac{1}{3}E(X_2) = \frac{2}{3}\mu + \frac{1}{3}\mu = \mu.$$

4. Finally, V is defined by

$V = \frac{1}{3}(2X_1 + X_2 + X_3),$

and

$$E(V) = \frac{2}{3}E(X_1) + \frac{1}{3}E(X_2) + \frac{1}{3}E(X_3) = \frac{2}{3}\mu + \frac{1}{3}\mu + \frac{1}{3}\mu = \frac{4}{3}\mu.$$

Let us now try to figure which of the estimators X, $\bar{X}$, U and V might be considered in some sense the 'best' estimator for μ. On looking at the above expectation calculations, we see that V may be discounted straight away; this is because it is the only biased estimator of these four random variables. So, is there any way in which we may distinguish between X, $\bar{X}$ and U? We know that they all give the right answer on average, so we might be interested in the estimator that gives the most accurate answers on average; in other words, let us look for the one with the smallest variance. To this end, we denote the population variance by σ^2 (noting that this is another population parameter). Then we have

1. $\text{Var}(X) = \sigma^2.$
2. $\text{Var}(\bar{X}) = \frac{1}{3^2}\text{Var}(X_1) + \frac{1}{3^2}\text{Var}(X_2) + \frac{1}{3^2}\text{Var}(X_3) = \frac{1}{3}\sigma^2.$
3. $\text{Var}(U) = \frac{2^2}{3^2}\text{Var}(X_1) + \frac{1}{3^2}\text{Var}(X_2) = \frac{5}{9}\sigma^2.$

Thus, with respect to the criteria considered previously, we may regard $\bar{X}$ as the best estimator for μ. In fact, more generally, for a sample of size n, the sampling mean given by

$$\bar{X} = \frac{1}{n}(X_2 + X_1 + \cdots + X_n)$$

satisfies

$$E(\bar{X}) = \mu \quad \text{and} \quad \text{Var}(\bar{X}) = \frac{\sigma^2}{n}.$$

So $\bar{X}$ an unbiased estimator for μ. Furthermore, it is the best such estimator in that it has the least variance of all similarly-constructed random variables based on a sample of size n. In fact, because of this, $\bar{X}$ is known as an *efficient estimator* of the mean. An interesting point to make here is that in our informal introductory discussion in this chapter we were intuitively using the sample mean to estimate μ (refer back to Example 2.1). It is important also to realize that the calculations in Example 2.2 are in fact distribution-independent. The fact that $\overline{X}$ has the same expectation as X yet a smaller variance means that it might be regarded as a better estimator for μ than is X, since 'on average' it will be closer to μ.

The larger the value of n the better the estimator $\overline{X}$ is for μ, in the sense that variance of $\overline{X}$ decreases as n increases (Kozack 2009, Mulekar and Siegel 2003). Of course, in reality we cannot make n indefinitely large. Indeed, there are practical limitations associated with any statistical enquiry; it is a compromise between accuracy on the one hand and time and money on the other. This is certainly another point for discussion.

Computer Comment

We show, via the R code and subsequent output below, that $\overline{X}$ would indeed appear to be a more efficient estimator than U for the particular case of the standard normal distribution.

```
> for(k in 1:10){C[k]=(rnorm(1)+rnorm(1)+rnorm(1))/3}; C
[1]  0.53372326  -0.10500105  -0.37569105  0.66300589
-0.05528143 -0.78194773
[7] -0.73870192 0.68248223 0.15571681 -0.22357169

> for(k in 1:10){C[k]=(2*rnorm(1)+rnorm(1))/3}; C
 [1] -0.5315980 -0.1399744 0.5402137 1.3912576 1.7592367
-0.2403519
[7] -1.2478048 -1.6963734 0.6116141 0.3855511
```

Example 2.3

We now consider an example for which the parameter being estimated is not the population mean.

Suppose that Jennifer wants to estimate the number n of taxis in her home town. All she knows about them is that they are numbered consecutively from 1 to n, with the numbers clearly visible next to the registration plate. (So n may be regarded as our population parameter here.) Jennifer writes down the numbers of the first three numbers she sees (ignoring any repeats). Compare, for the case n = 5, the utility of the estimators for n given by

i) $2\overline{X}-1$;
ii) $\frac{4}{3}X_{max}-1$;
iii) $2M-1$.

where $\overline{X}$ denotes the mean of the sample (of size 3 in this case), M is the median and X_{max} represents the largest number in the sample. A list of all possible outcomes is given below:

Sample	$2\overline{X}-1$	$\frac{4}{3}X_{max}-1$	$2M-1$
1, 2, 3	3	3	3
1, 2, 4	$3\frac{2}{3}$	$4\frac{1}{3}$	3
1, 2, 5	$4\frac{1}{3}$	$5\frac{2}{3}$	3
1, 3, 4	$4\frac{1}{3}$	$4\frac{1}{3}$	5
1, 3, 5	5	$5\frac{2}{3}$	5
1, 4, 5	$5\frac{2}{3}$	$5\frac{2}{3}$	7
2, 3, 4	5	$4\frac{1}{3}$	5
2, 3, 5	$5\frac{2}{3}$	$5\frac{2}{3}$	5
2, 4, 5	$6\frac{1}{3}$	$5\frac{2}{3}$	7
3, 4, 5	7	$5\frac{2}{3}$	7

First, as you should check, each of these is an unbiased estimator for n in this case. The variances now need to be calculated in order to ascertain which of them is the most efficient. On carrying this out you will find that the most efficient estimator for n in this particular case is $\frac{4}{3}X_{max}-1$.

Computer Comment

As n increases it becomes increasingly long-winded to list all possible samples of size 3. Indeed, when n=40, there are 9,880 possible outcomes. In this case it is easier to carry out a series of simulations in order to estimate n. The following piece of R code generates samples of size 3 for the case when n=40. It does this 30 times and takes the mean and variance of these realizations for each of the random variables given previously. Note that the means are relatively close to 40, and that the smallest variance is achieved by $\frac{4}{3}X_{max}-1$.

```
> x=c(1:40); A=c(1:30); B=c(1:30); C=c(1:30)
> for(k in 1:30) {y=sample(x,3,replace=FALSE);
A[k]=2*mean(y)-1; B[k]=(4*max(y)/3)-1;
C[k]=2*median(y)-1}
> mean(A); mean(B); mean(C); var(A); var(B); var(C)
[1] 41.97778
[1] 39.84444
[1] 42.66667
[1] 194.26
[1] 75.70677
[1] 294.9195
```

Suppose now that the sample size is greater than 3. Is it still the case that is $\frac{4}{3}X_{max}-1$ is an unbiased estimator for n? Use R to run a series of simulations that will enable you to answer this question fairly conclusively.

A Note of Caution

There is a little more to some statistical concepts than might meet the eye initially. We have already seen that the sample mean $\bar{X}$ is an unbiased estimator for the population mean μ. We now give two examples which show that it is not always a good idea to use the sample version of a statistic as an estimator of the population version of the same statistic (Sanqui and Arnholt 2011) because such estimators may be biased. These examples might be considered counter intuitive, and it is important to dispel misconceptions in this regard.

Example 2.4

As a particularly straightforward initial example, suppose that three students had taken a mental arithmetic test, and that they scored 7, 1 and 3. Of course, we know that the exact value of the population median, m say, is 3 merely by glancing at the data. Let us, however, consider estimating it by taking a sample of size 2, and then using the median M_2 of the sample as an estimator for m. Is this an unbiased estimator?

Well, the possible samples of size 2 and their respective medians are given by:

$1, 3 \rightarrow 2;$

$1, 7 \rightarrow 4;$

$3, 7 \rightarrow 5.$

The expected value of M_2 is thus given by

$$E(M_2) = \frac{1}{3}\times 2 + \frac{1}{3}\times 4 + \frac{1}{3}\times 5 = \frac{11}{3} \neq 3,$$

from which we that of M_2 is a biased estimator for m.

Admittedly, the population size is very small here (as in Example 1), and we do not in practice actually need to take a sample of size 2 in order to estimate m, but the point is that this demonstrates, in general, that using the sample median as an estimator for the population median does not on an average give us the correct answer. Incidentally, the same would be true if we took a sample of size 1.

Problems 2.1

1. Consider a population having four members given by 1, 3, 5, 10. Using the median M_3 of a sample of size 3 as an estimator, show that, for this particular set of data, it is an unbiased estimator for the population median. Would this result be true for any set of four integers? Explain your answer. What about the case in which the median of a sample of size two is used as an estimator?
2. **Computer task:** Consider the data set consisting of the first n consecutive positive integers 1,2,3,...,n. Suppose you use the median M_k of a sample of size $1 \leq k \leq n$ to estimate the median of the populations. For what values of k is M_k an unbiased estimator? Use R to help.

Solutions to Problems 2.1

1. The possible samples of size 3 and their respective medians are given by:
   $1, 3, 5 \rightarrow 3;$
   $1, 3, 10 \rightarrow 3;$
   $1, 5, 10 \rightarrow 5;$
   $3, 5, 10 \rightarrow 5.$
   From this it is clear that the expected value of M_3 is 4, which is the same as the median of the population. This will in fact always be the case for a population with four members; see if you can provide your own

explanation of this fact. It might help to work through another example in order to see exactly what is happening. You will find that M_2 is in general a biased estimator for populations with four members.

2. You will find that M_k is an unbiased estimator for the median for all values of k with $1 \le k \le n$. In fact, more generally, we may state that this will be the case for any population that has a symmetric distribution.

Example 2.5

Since the variance of a set of n pieces of data $x_1, x_2, \ldots, x_n$ is given by

$$\frac{1}{n}\sum_{i=1}^{n} x^2 - \bar{x}^2,$$

it is only natural to presume that the random variable V defined by

$$V = \frac{1}{n}\sum_{i=1}^{n} X_i^2 - \bar{X}^2$$

is an unbiased estimator for the population variance σ^2. We show here, however, the rather non-obvious, and possibly surprising, fact that V is actually a biased estimator for σ^2. In order to do this we need to demonstrate that $\mathrm{E}(V) \neq \sigma^2$. To this end, note first that

$$\mathrm{E}(V) = \mathrm{E}\left(\frac{1}{n}\sum_{i=1}^{n} X_i^2 - \bar{X}^2\right) = \frac{1}{n}\left(n\mathrm{E}(X^2)\right) - \mathrm{E}(\bar{X}^2) = \mathrm{E}(X^2) - \mathrm{E}(\bar{X}^2). \quad (1)$$

Then, since $\mathrm{Var}(X)=\mathrm{E}(X^2)-\mu^2$ by definition, we have

$$\mathrm{E}(X^2) = \sigma^2 + \mu^2. \quad (2)$$

Similarly, $\mathrm{Var}(\bar{X})=\mathrm{E}(\bar{X}^2)-\mu^2$ leads to

$$\mathrm{E}(\bar{X}^2) = \frac{\sigma^2}{n} + \mu^2. \quad (3)$$

Now, using (1), (2) and (3) we obtain

$$\mathrm{E}(V) = \mathrm{E}(\bar{X}^2) = \sigma^2 + \mu^2 - \left(\frac{\sigma^2}{n} + \mu^2\right) = \frac{n-1}{n}\sigma^2 \neq \sigma^2,$$

as required. In fact, from this result we may obtain an unbiased estimator for σ^2 that we denote by S^2:

$$S^2 = \frac{n}{n-1}V = \frac{1}{n-1}\left(\sum_{i=1}^{n} X_i^2 - n\bar{X}^2\right).$$

Example 2.6

Suppose that an engineer wants to check the long-term reliability of a particular component. He puts a sample of 200 components through a rigorous testing regime that is equivalent to 10 years normal usage, and 30 of them had failed by that point. How may we obtain an unbiased estimate for the proportion p of these components that fail within 10 years? Intuitively, we might feel that an unbiased estimate is simply $\frac{30}{200} = 0.15$, but is this in fact correct? (Our intuition cannot always be relied on; consider Example 2.4, for example.) Note that the number of components C in the sample that fail within 10 years is distributed as

$$C \sim B(200, p).$$

Thus, with P_s denoting the random variable representing the proportion of components in the sample that fail within 10 years, we have

$$P_s = \frac{C}{200}.$$

Therefore,

$$E(P_s) = E\left(\frac{C}{200}\right) = \frac{1}{200}E(C) = \frac{200p}{200} = p,$$

on remembering that a random variable distributed as $B(n,p)$ has expectation np. It is thus the case that the proportion in the sample that fail within 10 years does in fact give us an unbiased point estimate of the population proportion p.

Problem 2.2

A continuous random variable W has the following probability density function

$$g(w) = \begin{cases} \dfrac{2w}{\beta^2} & \text{if } 0 \le w \le \beta \\ 0 & \text{otherwise.} \end{cases}$$

where β is a parameter that is to be estimated. Show that $\frac{3W}{2}$ is an unbiased estimator for β.

Solution to Problem 2.2

We have

$$E\left(\frac{3W}{2}\right) = \int_0^\beta \frac{3w}{2} \times \frac{2w}{\beta^2}\, dw = \frac{1}{2\beta^2}\int_0^\beta 6w^2\, dw = \frac{1}{2\beta^2}\left[\frac{6w^3}{3}\right]_0^\beta = \beta,$$

as required.

The Limitations of Point Estimates on their Own

The value $\bar{x}$ is termed a point estimate as it gives us an estimate of the population mean μ in the form of a single value or 'point' on the number line. In the scenario we are considering here, this estimate might be particularly useful if we wanted, for example, to make a rough comparison of the mean length of time inpatients spend in UK hospitals to the corresponding quantity in hospitals in another country. It does not, however, give any indication as to how close to μ it is actually likely to be. In order for the hospital administrator to be able to say whether or not she can predict the future availability of beds with any degree of confidence, it is important that she has an idea of the likely accuracy of her point estimates.

3. Confidence Intervals

The Problem of Assessing the Accuracy of Our Point Estimates

The point estimates for the population mean μ that we have calculated thus far are certainly very useful, but, since $\bar{X}$ is a random variable, the value $\bar{x}$ it takes will vary from sample to sample. It is not entirely clear from this how close $\bar{x}$ is likely to be to μ on an average. How might we then develop a statistical method that will allow the hospital administrator to obtain an idea of the likely accuracy of her point estimates? Well, we can obtain an estimate of a population parameter by specifying two numbers between which the parameter may be considered to lie (in a way that we shall make precise). This is called an *interval estimate* of the parameter, and will require some explanation. Incidentally, a statement concerning the precision of an estimate is often termed its *reliability* (Warner 2008).

Discussion

In this section we will assume first of all that X follows a normal distribution with known variance, since this allows us easily to construct so-called *confidence intervals* for the population mean for a sample of any size. (Note, incidentally, that the estimation of the population mean in Section 2 was, by way of contrast, independent of the underlying distribution.) This requires students to be reasonably confident about the notion of a distribution (Wild 2006) and the potential variability of samples taken from this distribution (Reading and Reid 2006). The concept of a confidence interval is not an easy one for students to grasp initially, and there is the potential for the misinterpretation of the information it actually conveys (Gilliland and Melfi 2010). It is in a very real sense superior to a point estimate since it conveys more information. In fact, if the confidence interval is symmetric then we can immediately calculate the mean from it.

We consider here the notion of confidence intervals from the classical point of view. This can lead to confusion, and is thus something that we make a point of explaining. Let us assume first of all that the length of stay in hospital X is normally distributed. In other words,

$$X \sim N(\mu,\sigma^2)$$

for population parameters μ and σ. This will mean that we will not have to resort to the central limit theorem initially, and our confidence intervals will be 'exact'.

Furthermore, we assume for the time being that the value of the population variance is known; indeed, let us suppose that $\sigma^2=49$ (where the units are in days squared). This assumption is of course rather unrealistic, but we aim for simplicity initially; this restriction will be lifted in due course. We then have

$$X \sim N(\mu,7^2).$$

Note now that, using the transformation

$$Z = \frac{X-\mu}{\sigma} = \frac{X-\mu}{7},$$

and noting that $P(Z<1.96)=0.975$, we may obtain the following

$$P(\mu-7\times 1.96 < X < \mu+7\times 1.96)=0.95.$$

In words this tells us that a randomly chosen length of stay has a 95% chance of being within $7\times 1.96=13.72$ days of the population mean μ.

A graphical illustration of this is given in Fig. 3, for the case in which $\mu=26$. The shaded area is 0.95. Then, noting the symmetry of the situation, we see that the area of each of the 'tails' of the distribution is equal to 0.025.

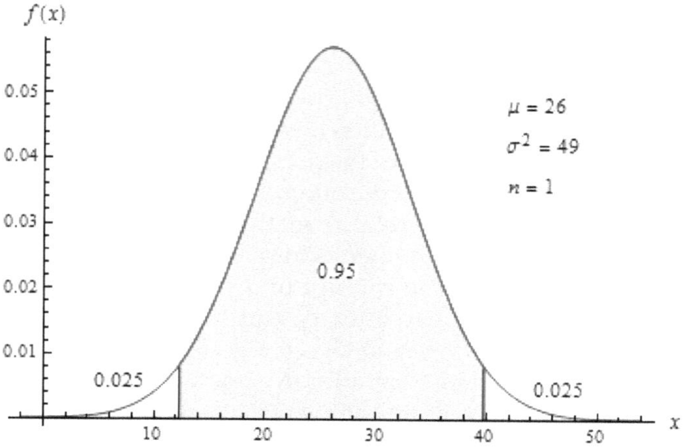

Figure 3. The probability density function of $X \sim N(26,7^2)$, showing both its 2.5% 'tails'.

From an intuitive point of view, we might say that if a large number of random observations are taken from this distribution then approximately 95% of them will lie between 26–13.72=12.28 and 26+13.72=39.72 (this is admittedly a somewhat loose statement, but this idea can be made rigorous). Furthermore, only 2.5% of the observations will exceed 39.72, and so on.

Similarly, if we now take a sample of size n, the sample mean is distributed as

$$\bar{X} \sim N\left(\mu, \frac{7^2}{n}\right),$$

and we have

$$P\left(\mu - \frac{7 \times 1.96}{\sqrt{n}} < X < \mu + \frac{7 \times 1.96}{\sqrt{n}}\right) = 0.95.$$

The mean of a randomly chosen sample of size n has a 95% chance of being within $\frac{13.72}{\sqrt{n}}$ days of the population mean μ. In Fig. 4 we see a graphical illustration of the situation in which μ=26 and n=5. Notice how, even for such a small increase in the sample size (from 1 to 5), the distribution of the sample mean narrows quite noticeably (noting further that the vertical scales in Figs. 3 and 4 are not equal).

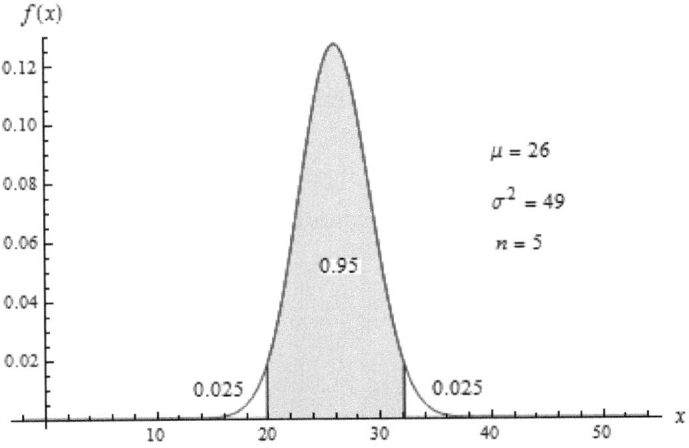

Figure 4. The probability density function of $\bar{X} \sim N(26, \frac{7^2}{5})$, along with its 2.5% 'tails'.

Computer Comment

The following R code and subsequent output tells us that if X follows a normal distribution with mean 26 and population variance 49 and a sample of size 30 is drawn from this distribution, then the probability that the sample mean lies between 23.494 and 28.504 is 95%.

```
> popmean=26; popsd=7; n=30
> error=qnorm(0.975)*popsd/sqrt(n)
> left=popmean-error; right=popmean+error
> left;right
[1] 23.49513
[1] 28.50487
```

Introducing Confidence Intervals

So far we have been considering the *fixed* intervals

$$(\mu - 13.72, \mu + 13.72) \text{ and } \left(\mu - \frac{13.72}{\sqrt{n}}, \mu + \frac{13.72}{\sqrt{n}}\right),$$

the centres of which are located at μ (a parameter with an unknown value). Let us now obtain a sample of size n from the population, calculate the sample mean $\bar{x}$ and then construct the corresponding interval for which the centres lie at $\bar{x}$. We then have the following:

$$\left(\bar{x} - \frac{13.72}{\sqrt{n}}, \bar{x} + \frac{13.72}{\sqrt{n}}\right),$$

A key point to note here is that this interval has exactly the same width as

$$\left(\mu - \frac{13.72}{\sqrt{n}}, \mu + \frac{13.72}{\sqrt{n}}\right),$$

but will not (necessarily) be in the same position. The equality of the interval widths implies that μ lies in

$$\left(\bar{x} - \frac{13.72}{\sqrt{n}}, \bar{x} + \frac{13.72}{\sqrt{n}}\right),$$

if, and only if, $\bar{x}$ lies in

$$\left(\mu - \frac{13.72}{\sqrt{n}}, \mu + \frac{13.72}{\sqrt{n}}\right).$$

This idea is illustrated in Figs. 5 and 6 (where n=5 in both cases). First, in Fig. 5 we see that $\bar{x}$=30 lies in the 95% interval for μ, and that the interval constructed about $\bar{x}$ in this case does indeed trap μ (noting that it has been drawn above the x-axis for the sake of clarity). On the other hand, from Fig. 6 it is clear that $\bar{x}$=35 does not lie in the 95% interval for μ, and the interval constructed around it does not trap μ.

We may now imagine taking a whole series of samples of size n and obtaining the mean $\bar{x}$ each time. By way of the definition of X, we would expect that 95% of these realisations of the sample mean to lie between

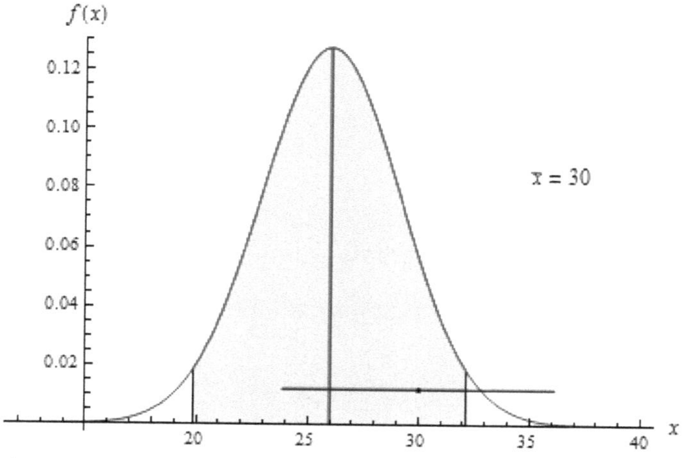

Figure 5. A normal distribution with a superimposed 95% confidence interval trapping μ.

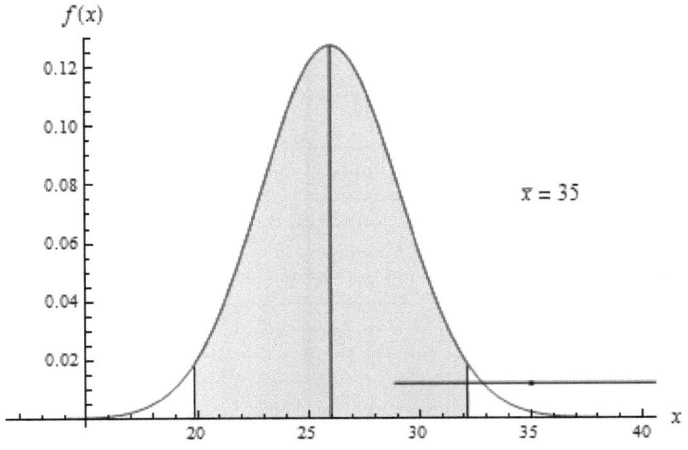

Figure 6. A normal distribution with a superimposed 95% confidence interval that does not trap μ.

$\mu - \frac{13.72}{\sqrt{n}}$ and $\mu + \frac{13.72}{\sqrt{n}}$. From these observations it follows that 95% of the constructed intervals

$$\left(\bar{x} - \frac{13.72}{\sqrt{n}}, \bar{x} + \frac{13.72}{\sqrt{n}}\right),$$

would contain or 'trap' the population mean μ. This is an example of a 95% *confidence interval*. (Incidentally, the above are termed *symmetric* confidence intervals, as the sample mean lies exactly at the centre.)

Figure 7 helps further to understand what is going on here. The population mean is some fixed value μ that we do not know. Imagine obtaining a random sample of observations from X and subsequently constructing the interval

$$\left(\bar{x} - \frac{13.72}{\sqrt{n}}, \bar{x} + \frac{13.72}{\sqrt{n}}\right).$$

This is depicted as the highest interval in the simple schematic diagram shown in Fig. 7 (noting that the intervals are at different heights simply to distinguish one from another). You can see that it does indeed trap μ (in other words, μ lies within the interval). A second interval is then obtained in a similar manner, and so on. In this example, only the seventh interval does not trap μ. From the comments above, this should not be too surprising since we would expect roughly 95% of the intervals constructed in this way to contain the population mean.

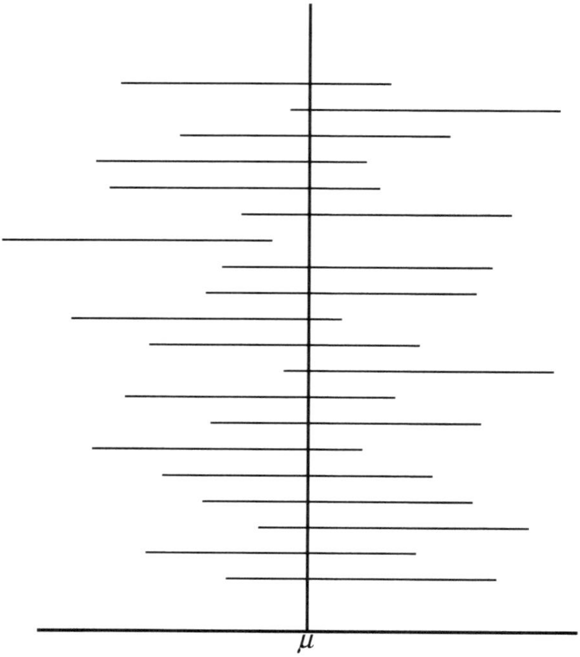

Figure 7. A series of confidence intervals, all but one of which trap the mean.

This visual approach to the teaching and learning of such a difficult notion are important in order to establish an intuitive understanding that will in turn lead on to 'inference by eye' (Cumming 2007). Care is needed with respect to how these intervals are to be interpreted; this is indeed a major area of misconception amongst students. The classical viewpoint is that 95% of the confidence intervals constructed in the manner referred to in the previous paragraph will actually contain μ. Students tend to calculate an interval and then state that it has a 95% chance of containing μ, which is something slightly different.

As the name implies, confidence intervals provide us with a degree of confidence with respect to the location of μ in the sense that although we cannot know for certain that a particular interval traps μ, it is to be expected that 95 out of 100 intervals constructed in this way will do so. There is scope here for illustrating these ideas by carrying out simulations using R (Hagtvedt, Jones and Jones 2008). This would allow students to see how the width of the confidence intervals changes as we vary the parameters.

How do we deal with the situation in which the variance is not known? If the population is normally distributed then we can carry on in a similar manner as before, but this time using the unbiased estimate for the variance; otherwise, we use the central limit theorem and the unbiased estimate for the variance once more (assuming the sample size is large enough for this to be valid). For small samples we can use the t distribution, under assumptions of normality.

Use of the Central Limit Theorem

Let us now deal with the situation for which X is not normally distributed. We will need to invoke the central limit theorem, which tells us that for a sufficiently large sample size n, $\overline{X}$ is approximately normal (in a sense that may be made rigorous). We assume from now on, therefore, that the sample size is large enough for this to apply ($n \geq 30$ suffices in general). Then $\overline{X}$ may well approximated by the random variable

$$Y \sim \mathrm{N}\left(\mu, \frac{\sigma^2}{n}\right),$$

where, as before, $\mathrm{E}(X) = \mu$ and $\mathrm{Var}(X) = \sigma^2$. Assuming $\sigma = 7$ once again, we may now construct the symmetric confidence intervals

$$\left(\bar{x} - \frac{13.72}{\sqrt{n}}, \bar{x} + \frac{13.72}{\sqrt{n}}\right).$$

Although things look much the same as they did when X was normal, it is important to appreciate a major difference; the confidence interval above is no longer exact. However, it will still give us good information about the likely precision of our estimate so long as n is large.

Next, we consider the situation in which the random variable is not necessarily normally distributed and the population variance is unknown. Note that the latter of these scenarios will almost always be true. We utilize both the central limit theorem and the unbiased estimate for the population variance. Again here, we require the sample size to be sufficiently large.

Example 3.1

A total of 120 new light bulbs of a certain type were fitted in the rooms of a hotel as part of a major refurbishment. At the end of the year all of them had burnt out, and the hotel owner had kept a record of exactly when each of them needed to be replaced. With B representing the lifetime of a randomly chosen light bulb in days, the summary data for the lifetimes is as follows:

$\sum b = 13200$ and $\sum b^2 = 1542210$.

We will obtain here a symmetric 99% confidence interval for the mean lifetime of this particular type of light bulb.

The size of the sample means that the mean life of the bulbs is approximately normal (by the central limit theorem) and also that we use the unbiased estimate of the variance for the population variance. We have

$$\bar{b} = \frac{\sum b}{n} = \frac{13200}{120} = 110$$

and

$$s^2 = \frac{1}{n-1}\left(\sum b^2 - n\bar{b}^2\right) = \frac{1}{119}(1542210 - 120 \times 110^2) = 758.067.$$

Since P(Z<2.576) = 0.995 we may obtain our 99% confidence interval as

$$\left(\bar{b} - 2.576\frac{27.533}{\sqrt{120}}, \bar{b} + 2.576\frac{27.533}{\sqrt{120}}\right) = (103.525, 116.475)\cdot$$

Note that, as usual, we need to assume that this set of 100 light bulbs did in fact constitute a random sample. Of course, in a case like this, such an assumption might not necessarily be valid; the bulbs may have all been taken from the same box and manufactured on the same day on the same machine by the same operator!

Self-evaluation Problem 3.1

The hospital administrator wishes to calculate a symmetric 90% confidence interval for the mean length of stay in hospital. She wants to be sure that her estimate is likely to be very accurate, and so wants the interval to be no greater than 3 days. What is the smallest sample size n that would give rise to a confidence interval with this width?

Self-evaluation Problem Solution 3.1

Assuming $\sigma = 7$ once more, we have

$$\mathrm{P}\left(\mu - \frac{7 \times 1.645}{\sqrt{n}} < X < \mu + \frac{7 \times 1.645}{\sqrt{n}}\right) = 0.9,$$

so a 90% confidence interval is given by

$$\left(\mu - \frac{7 \times 1.645}{\sqrt{n}}, \mu + \frac{7 \times 1.645}{\sqrt{n}}\right).$$

We thus need to solve the following equation for n:

$$\left(\mu + \frac{7 \times 1.645}{\sqrt{n}}\right) - \left(\mu - \frac{7 \times 1.645}{\sqrt{n}}\right) = 3.$$

From this we obtain $n \approx 58.931$, and thus the smallest possible sample size is in fact 59.

It is also possible to obtain confidence intervals for population proportions (Richardson and Haller 2003), as we demonstrate in the following example:

Example 3.2

Returning to Example 2.6, suppose that, as well as obtaining a point estimate for the proportion of components that fail within 10 years, our engineer wishes to gauge the likely accuracy of this estimate. He therefore obtains a confidence interval for the population proportion. Recall that 200 components were put through a rigorous testing regime, equivalent to 10 years normal usage, and 30 of them had failed by that point. We adopt the same notation, with $C \sim B(200,p)$ denoting the number in the sample that fail within 10 years, and $P_s = \frac{C}{200}$ the random variable representing the proportion of components in the sample that fail within 10 years.

In order to calculate the confidence interval, we need to obtain a normal approximation to P_s. Note first that

$$\mathrm{Var}(P_s) = \mathrm{Var}\left(\frac{C}{200}\right) = \frac{1}{200^2}\mathrm{Var}(C) = \frac{200p(1-p)}{200^2} = \frac{p(1-p)}{200}.$$

Thus P_s is distributed approximately as

$$\mathrm{N}\left(p, \frac{p(1-p)}{200}\right).$$

From the given sample, therefore, a 95% confidence interval is given by

$$\left(p_s - 1.96 \times \sqrt{\frac{p_s(1-p_s)}{n}}, p_s + 1.96 \times \sqrt{\frac{p_s(1-p_s)}{n}}\right)$$

$$= \left(0.15 - 1.96 \times \sqrt{\frac{0.15 \times 0.85}{200}}, 0.15 + 1.96 \times \sqrt{\frac{0.15 \times 0.85}{200}}\right)$$

$$= (0.1005, 0.1995),$$

where p_s denotes a realisation of P_s (in this case 0.3).

The t Distribution

Let X be a normally distributed random variable. Then we can write $X \sim N(\mu, \sigma^2)$ for some mean μ and variance σ^2. If σ^2 is known then the distribution of the sample mean is given by

$$\bar{X} \sim \mathrm{N}\left(\mu, \frac{\sigma^2}{n}\right),$$

where n is the sample size. On the other hand, if the variance σ^2 is not known then we will have to estimate it from the sample, using the unbiased estimator S^2 for the population variance, where

$$S^2 = \frac{1}{n-1}\left(\sum_{i=1}^{n} X_i^2 - n\bar{X}^2\right).$$

The additional variability this introduces implies that the sample mean is no longer exactly normal. Instead we consider the following random variable

$$T = \frac{\bar{X} - \mu}{S},$$

which follows what is known as a t distribution with n–1 degrees of freedom; it is also known as *Student's t distribution* (Spiegel and Stephens 1999; Warner 2008). In fact, this gives rise to an infinite family of distributions; one distribution corresponding to each degree of freedom. In other words, there is a separate distribution for each possible sample size. As with the normal distribution, critical values of the t distribution can be looked up in statistical tables. However, there is now a separate set of critical values for each degree of freedom. Going from bottom to top in Fig. 8, we see the probability density functions of the t distributions having degrees of freedom 1, 2 and 40, respectively.

Note that when n is large, the variability in S is rather small, and T has a distribution that is, to all intents and purposes, standard normal. In such cases, we tend in practice to assume that s^2 is a good approximation to σ^2 and that $\bar{X}$ is normally distributed (or at least extremely close to being so). Remember though, that so long as the random variable X is normally distributed and the population variance is not known, then $(\bar{X}-\mu)/S$ is distributed exactly as a t distribution with n–1 degrees of freedom.

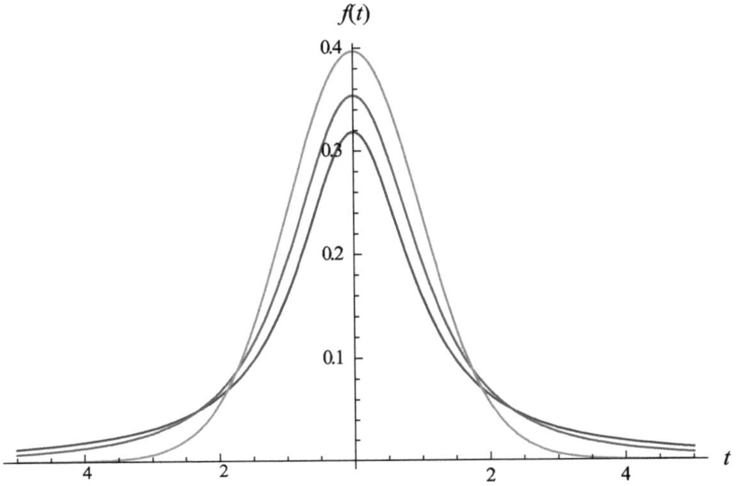

Figure 8. From bottom to top, t distributions having degrees of freedom 1, 2 and 40, respectively.

Color image of this figure appears in the color plate section at the end of the book.

Problem 3.1

A company manufactures batteries for use in graphical calculators. The engineer in charge of the manufacturing process wishes to obtain a 95% confidence interval for the mean lifetime of these batteries. In order to do so, she tests four of them and finds that they have the following lifetimes, in hours: 47.2, 51.3, 48.0, 46.3. Construct the confidence interval, stating any assumptions that you make.

Solution to Problem 3.1

Two points to note initially are, that the population variance does not appear to be known (which of course is virtually always the case in practice) and the sample size is very small. We may therefore consider using the t distribution. The only additional assumption that needs to be made here in order for it to be valid to use this distribution is that the batteries' lifetimes follow a normal distribution; this would certainly seem to be plausible here. Let us first calculate an unbiased estimate of the sample variance. We have

$$s^2 = \frac{1}{n-1}\left(\sum_{i=1}^{n} x_i^2 - n\bar{x}^2\right) = \frac{1}{3}(9307.22 - 4 \times 48.2^2) = 4.7533.$$

Then, noting that we are interested in t_3, the t distribution with three degrees of freedom, and $P(t_3 \leq 3.182)=0.975$, the confidence interval is given by

$$\left(\bar{x} - 3.182 \times \sqrt{\frac{s^2}{n}}, \bar{x} + 3.182 \times \sqrt{\frac{s^2}{n}}\right)$$

$$= \left(48.2 - 3.182 \times \sqrt{\frac{4.7533}{4}}, 48.2 + 3.182 \times \sqrt{\frac{4.7533}{4}}\right)$$

$$= (44.7312, 51.6687).$$

Incidentally, we may wonder why a small sample would ever be used in estimating a population mean or in constructing a confidence interval. One reason is that some forms of testing actually destroy the product; for example, testing the lifetimes of batteries renders them useless. Consider also the manufacture of car windscreens; in testing their strength, the samples may end up getting broken. Furthermore, it is the case that obtaining large samples may be very costly in terms of both time and money.

Computer Comment

The following R code carries out the above calculation.

```
> x=c(47.2,51.3,48.0,46.3)
> samplemean=mean(x); samplesd=sqrt(var(x))
> error=qt(0.975,length(x)-1)*samplesd/sqrt(length(x))
> left=samplemean-error; right=samplemean+error
> left;right
[1] 44.73079
[1] 51.66921
```

Note that the confidence interval also appears in the output of the following simple piece of code; this illustrates a clear connection with the confidence intervals in this chapter and hypothesis testing in the next.

```
> t.test(x,mu=0,conf.level=0.95)
        One Sample t-test
data: x
t = 44.2158, df = 3, p-value = 2.546e-05
alternative hypothesis: true mean is not equal to 0
```

```
95 percent confidence interval:
 44.73079 51.66921
sample estimates:
mean of x
 48.2
```

Self-evaluation Problem 3.2

i) Construct a 90% confidence interval for the mean length of stay in the hospital from the following data given at the beginning of this chapter (assuming the underlying distribution is normal):
25, 28, 1, 3, 32, 24, 27, 3, 137, 29.

ii) Comment on the validity of your calculation.

Self-evaluation Problem Solution 3.2

i) Since the length of stay is assumed to follow a normal distribution, the population variance is unknown and the sample size is relatively small, we shall employ the *t* distribution here. We have n=10, $\bar{x}$=30.9 and s^2=1535.433. We thus consider the *t* distribution with 9 degrees of freedom (noting that $P(t_9 \le 1.883)=0.95$) to give

$$\left(\bar{x} - 1.833 \times \sqrt{\frac{s^2}{n}}, \bar{x} + 1.833 \times \sqrt{\frac{s^2}{n}}\right)$$
$$= \left(30.9 - 1.883 \times \sqrt{\frac{1535.433}{10}}, 30.9 + 1.883 \times \sqrt{\frac{1535.433}{10}}\right)$$
$$= (7.5673, 54.2327).$$

ii) There are several points worth noting here. First, in the above calculation we have assumed that the distribution of the length of stay is normal. This is unlikely to be the case here; indeed the data in the sample would appear to suggest that the distribution is not normal. Second, referring back to the beginning of this chapter, we will recall that the data comprised the lengths of stay of the last 10 people to leave the sample. This is hardly a random sample, and does invalidate the above calculation to a certain extent. Finally, the fact that there is a high degree of variability in the lengths of stay means that the confidence interval is really rather wide for a sample of size 10, and not particularly useful.

4. Further Topics of Interest

Confidence Interval for the Difference Between Two Means

We consider here another situation in which it is possible to obtain confidence intervals. Suppose that sunflowers were grown in two different types of soil, A and B say. Let μ_A and μ_B denote the mean height of the sunflowers grown in soil types A and B, respectively. The heights of a sample of 40 sunflowers from A and 50 from B were measured. The mean heights of these samples were 1.7 m and 1.65 m respectively. Furthermore, unbiased estimates of the population variances were obtained from this sample, giving $s_A^2 = 10$ and $s_B^2 = 14$. We will calculate a symmetric 90% confidence interval for the difference between the two population means μ_A and μ_B.

Since the sample sizes are relatively large, we know, without having to make assumptions about normality, that the difference between two means is, to a very good approximation, distributed as

$$\bar{A} - \bar{B} \sim \mathrm{N}\left(\mu_A - \mu_B, \frac{s_A^2}{n_A} + \frac{s_B^2}{n_B}\right) = \mathrm{N}\left(\mu_A - \mu_B, \frac{10}{40} + \frac{14}{50}\right).$$

A 98% confidence interval is given by

$$\left(\bar{a} - \bar{b} - 2.326\sqrt{\frac{10}{40} + \frac{14}{50}}, \bar{a} - \bar{b} + 2.326\sqrt{\frac{10}{40} + \frac{14}{50}}\right) = (-1.6434, 1{,}7434).$$

Pre-empting some of the work on hypothesis testing in Chapter 6, we see that since 0 is comfortably contained in this confidence interval, there is no reason to believe (on the basis of the above calculation at least) that there is a difference in the mean heights of the sunflowers grown in the two different types of soil.

Confidence Intervals for the Variance

We are also able to obtain confidence intervals for the variance of a normal population. In order to do so we introduce the chi-squared distribution (which is also utilized in the following chapter). Let the random variables $Z_1, Z_2, \dots, Z_n$ each be distributed as the standard normal distribution, so that $Z_1 \sim \mathrm{N}(0,1)$, and so on. Then the sum of their squares,

$$Z_1^2 + Z_2^2 + \cdots + Z_n^2,$$

follows a chi-squared distribution with $n-1$ degrees of freedom, which is denoted by χ_{n-1}^2. Note that this sum may be rearranged to give

$$Z_1^2 + Z_2^2 + \cdots + Z_n^2 = \frac{(n-1)S^2}{\sigma^2},$$

where S^2 is the unbiased estimator or the population variance (encountered several times earlier in the chapter) and σ^2 is the true population variance.

We may use our knowledge of the chi-squared distribution, which is tabulated, to obtain a 95% confidence interval for σ^2, as follows. With $\chi^2_{n-1}(0.025)$ and $\chi^2_{n-1}(0.975)$ denoting the critical values of the chi-squared distribution for which 2.5% of the area lies in the lower and upper tail, respectively, we have

$$\chi^2_{n-1}(0.025) < \frac{(n-1)S^2}{\sigma^2} < \chi^2_{n-1}(0.975),$$

which may alternatively be written as

$$\frac{(n-1)S^2}{\chi^2_{n-1}(0.975)} < \sigma^2 < \frac{(n-1)S^2}{\chi^2_{n-1}(0.025)},$$

leading to our 95% confidence interval:

$$\left(\frac{(n-1)S^2}{\chi^2_{n-1}(0.975)}, \frac{(n-1)S^2}{\chi^2_{n-1}(0.025)}\right).$$

As with the t distribution, when we say "the" chi-squared distribution we actually mean an infinite family of distributions, indexed by their degrees of freedom. In Fig. 9 we see graphs of the probability density functions of

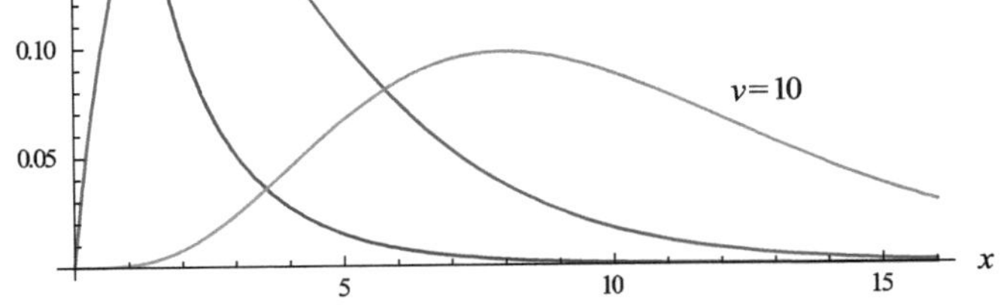

Figure 9. Chi-squared distributions having 1, 4 and 10 degrees of freedom.
Color image of this figure appears in the color plate section at the end of the book.

chi-squared distributions having 1, 4 and 10 degrees of freedom. Note that these functions are non zero only when x is positive; this should indeed be apparent on considering the definition of the chi-squared distribution.

Self-evaluation Problem 4.1

i) A random sample of just four 18-year-old females was selected from a school in London. Their heights, in centimeters, were 167, 159, 177 and 168. Assuming normality, obtain a 90% confidence interval for the variance of the heights of 18-year-old females in the United Kingdom.

ii) Comment on the validity of this statistical experiment.

Self-evaluation Problem Solution 4.1

i) First,

$$s^2 = \frac{1}{n-1}\left(\sum h^2 - n\bar{h}^2\right) = \frac{1}{3}(112723 - 4 \times 167.75^2) = 166.$$

We require

$$\left(\frac{(n-1)S^2}{\chi^2_{n-1}(0.95)}, \frac{(n-1)S^2}{\chi^2_{n-1}(0.05)}\right) = \left(\frac{166(4-1)}{7.815}, \frac{166(4-1)}{0.3518}\right) = (63.7236{,}1415.5770),$$

where the tabulated values of the chi-squared distribution (3 degrees of freedom) were used.

ii) This statistical experiment is not particularly satisfactory. First, the fact that all the students came from the same school and thus from the same geographical location, means that the sample is hardly random. Second, in circumstances such as this, in which it would be relatively easy and inexpensive to obtain a much larger sample, it would be worthwhile doing so. After all, the confidence interval obtained above is not really of much use in gaining an accurate picture of the variation amongst the heights of 18-year-old females in the United Kingdom. A considerably larger sample size would lead to a 90% confidence interval with a much narrower width.

Example of a Simple Maximum-likelihood Estimator

Although we always make the assumption, when tossing a coin, that heads and tails are equally likely to occur, the fact is that coins often exhibit minor irregularities that may cause them to be slightly biased towards either heads or tails. Suppose that for a particular coin, P(H)=p and P(T) = 1–p (ignoring the fact that there is a non-zero probability that the coin will land on its edge). One way of estimating p is to use the proportion of heads in a sample of n tosses as an estimator. For example, if the coin was tossed 100 times, 53 of which resulted in heads, then our estimate for p would be 0.53.

There is, however, another way in which this estimation may be carried out. It requires a subtle change in viewpoint. We now ask ourselves the question: What is the most likely value of p that would give rise to 53 heads resulting from 100 tosses of the coin?

In this case the likelihood function $L(p)$ is given by

$$L(p) = \binom{100}{53} p^{53}(1-p)^{47}.$$

We differentiate this in order to maximize its value with respect to p:

$$\frac{dL}{dp} = \binom{100}{53}\left(53p^{52}(1-p)^{47} - 47p^{53}(1-p)^{46}\right)$$

$$= \binom{100}{53} p^{52}(1-p)^{46}(53(1-p) - 47p)$$

$$= \binom{100}{53} p^{52}(1-p)^{46}(53 - 100p).$$

On setting this to zero, we obtain our estimate of p=0.53 (disregarding the solutions 0 and 1), which gives the same result as the estimate obtained by using the sample proportion. This process is termed 'maximum-likelihood estimation'. Note that in general it is not possible to perform such calculations by hand. Indeed, plenty of computing power is often needed to obtain the estimate by using this method. Furthermore, maximum likelihood estimates are not necessarily unbiased, and in some complicated problems they might be unsuitable or even fail to exist.

Asymmetric Confidence Intervals

Thus far we have only considered symmetric confidence intervals, in which the sample mean is located at the centre. As we now describe, it is also possible to construct asymmetric confidence intervals; indeed, as we shall see, this is of particular relevance to the following chapter.

To take an example, we previously considered the calculation of a confidence interval for the mean life of a particular type of light bulb. The sample size was large, so the central limit theorem could be invoked. We performed the following calculations, noting that $P(Z<2.576)=0.995$:

$$\bar{b} = \frac{\sum b}{n} = \frac{13200}{120} = 110$$

and

$$s^2 = \frac{1}{n-1}\left(\sum b^2 - n\bar{b}^2\right) = \frac{1}{119}(1542210 - 120 \times 110^2) = 758.067.$$

From this we obtained a symmetric 99% confidence interval as

$$\left(\bar{b} - 2.576\frac{27.533}{\sqrt{120}}, \bar{b} + 2.576\frac{27.533}{\sqrt{120}}\right) = (103.525, 116.475).$$

However, noting that P(Z<2.326)=0.99, we see that the following

$$\left(-\infty, \bar{b} + 2.326\frac{27.533}{\sqrt{120}}\right) = (-\infty, 115.846).$$

is a 99% confidence interval for the population mean in the sense that 99% of intervals constructed in this manner will in fact contain the true mean. It is an asymmetric confidence interval; the sample mean no longer occupies the central position. Likewise, (104.154,∞) is an 99% asymmetric confidence interval for the population mean.

5. Closing Comments

From the work carried out in this chapter we can see that the larger the confidence level the larger the confidence interval. In other words, we need larger and larger confidence intervals if we want to increase the percentage of them that trap the mean. However, the fact is that the larger the confidence interval, the less certain we are about the location of the parameter. It may be seen, therefore, that the choice of confidence level is something of a compromise between accuracy and certainty and there is a trade off between the percentage of intervals that will actually contain μ and the precision the administrator is hoping for in her estimate.

There is of course also the issue of the effect of sample size on the width of confidence intervals to take into account (Kozack 2009, Mulekar and Siegel 2009), and indeed this is something to be discussed in greater detail in the following chapter, in which the relationship between confidence intervals and hypothesis tests will become clear.

In order not to overcomplicate things initially, one of our assumptions was that we had prior knowledge of the population variance; this allowed us to model the sample mean exactly as a normal random variable (so long as the underlying distribution was itself normal). This is not at all realistic, and we thus went on to utilize the *t distribution*, which takes account of the fact that we do not know the population variance (although the underlying random variable is still required to be normal). In this case the sample mean follows exactly the t distribution with n–1 degrees of freedom. This necessitates the consideration of *degrees of freedom* (Eisenhauer 2008), an important mathematical concept which appears in statistics and in other areas of mathematics. Indeed, students might be encouraged to do some independent reading around this notion. Taking things further still, for large samples we do not even need to assume normality, and may employ the central limit theorem. Finally, able and interested students may be introduced to the idea of Bayesian estimation (Elfessi and Reineke 2001).

Mixed Problems

1. The probability that a particular coin turns up heads when it is tossed is p.
   i) The random variable H represents the number of times the coin is tossed up and including, the occurrence of the first head. It is well-known that H follows a geometric distribution with parameter p, with $E(H)=\frac{1}{p}$. Show, however, that $\frac{1}{H}$ is in fact a biased estimator for p.
   ii) Devise a piece of R code to provide some numerical evidence for the result in (i).
   iii) The random variable L represents the number of times the coin is tossed up and including, the occurrence of the second head. Show that the random variable
   $$\frac{1}{L-1}$$
   is an unbiased estimator for p.
2. A scientist devises an experiment to determine g, the acceleration due to gravity. He carries out the experiment five times, and obtains the results 9.84, 9.80, 9.82, 9.78, 9.79, each of which has the units ms^{-2}.
   i) Calculate a point estimate for g using the above data.
   ii) Obtain a symmetric 99% confidence for g, stating clearly any assumptions that you make in your calculation.
   iii) Is this confidence interval exact? Justify your answer.
3. A continuous random variable Y has the following probability density function:
   $$f(y) = \begin{cases} \frac{1}{2}q^3y^2e^{-qy} & \text{if } y \geq 0 \\ 0 & \text{otherwise.} \end{cases}$$
   where q is a parameter. Decide whether or not $\frac{2}{Y}$ is an unbiased estimator for q.
4. A total of 80 London residents were chosen at random and asked whether or not they had attended the 2012 Olympic Games. Out of these 80 people, 9 said they had attended.

i) Use this data to construct a symmetric 90% confidence interval for the proportion of Londoners attending the games.
ii) Is the interval exact? Explain your answer.
iii) The City of London Council claimed that 15% of all London residents had attended the games. Does the calculated confidence interval support their claim?

5. A random variable V has mean μ and variance σ^2. Two observations, V_1 and V_2, are made of V.
i) Let a be a real number. Show that the random variable $\alpha V_1 + (1-\alpha) V_2$ is an unbiased estimator for μ.
ii) Find the value of a for which this estimator is the most efficient.

References

Batanero, C. 2004. Statistics education as a field for research and practice, 1–15. *In:* M. Niss [ed.] Proceedings of ICME 10, Regular Lecture, Copenhagen, Denmark, Roskilde University.

Crawshaw, J. and J. Chambers. 1994. A Concise Course in A-Level Statistics. Stanley Thornes, Cheltenham, UK.

delMas, R. and Y. Liu. 2005. Exploring students' conceptions of the standard deviation. Statistics Education Research Journal 4: 55–82.

Elfessi, A. and D. Reineke. 2001. A Bayesian look at classical estimation: the exponential distribution. Journal of Statistics Education 9: 1.

Eisenhauer, J. 2008. Degrees of freedom. Teaching Statistics 30: 75–78.

Gilliland, D. and V. Melfi. 2010. A note on confidence interval estimation and margin of error. Journal of Statistics Education 18: 1.

Grimmett, G. and D. Stirzaker. 2001. Probability and Random Processes (Third Edition). Oxford University Press, Oxford, UK.

Hagtvedt, R., G. Jones and K. Jones. 2008. Teaching confidence intervals using simulation. Teaching Statistics 30: 53–56.

Kaplan, J., D. Fisher and N. Rogness. 2009. Lexical ambiguity in statistics: what do students know about the words association, average, confidence, random and spread? Journal of Statistics Education 17: 3.

Kozack, M. 2009. How to show that sample size matters. Teaching Statistics 31: 52.

Lavy, I. and M. Mashiach-Eizenberg. 2009. The interplay between spoken language and informal definitions of statistical concepts. Journal of Statistics Education 17: 1.

Lecoutre, M.-P., K. Rovira, B. Lecoutre and J. Poitevineau. 2006. People's intuitions about randomness and probability: an empirical study. Statistics Education Research Journal 5: 20–35.

Mulekar, M. and M. Siegel. 2009. How sample size affects a sampling distribution. Mathematics Teacher 103: 34–42.

Mvududu, N. and G. Kanyongo. 2011. Using real-life examples to teach abstract statistical concepts. Teaching Statistics 33: 12–16.

O'Brien, T. 2008. Teaching statistical concepts, fundamentals and modeling. *Teaching Statistics* 30: 81–85.

Peters, S. 2011. Robust understanding of statistical variation. Statistics Education Research Journal 10: 52–88.

R Development Core Team. R: A language and environment for statistical computing. Vienna: R Foundation for Statistical Computing (see http://www.r-project.org)

Reading, C. and J. Reid. 2006. An emerging hierarchy of reasoning about distribution: from a variation perspective. Statistics Education Research Journal 5: 42–68.
Richardson, M. and S. Haller. 2003. Confident in a Kiss. Teaching Statistics 25: 6– 11.
Sanqui, S. and A. Arnholt. 2011. Why is *S* a biased estimate of *s*? Teaching Statistics 33: 23–28.
Schau, C. and N. Mattern. 1997. Assessing students' connected understanding of statistical relationships, 91–104. *In:* I. Gal and J. Garfield [eds.] . The Assessment Challenge in Statistics Education. IOS Press, Amsterdam, Netherlands.
Spiegel, M. and L. Stephens. 1999. Theory and Problems of Statistics (Third Edition). McGraw-Hill, New York, USA.
Warner, R. 2008. Applied Statistics: From Bivariate Through Multivariate Techniques. Sage Publications, Los Angeles, USA.
Watier, N., C. Lamontagne and S. Chartier. 2011. What does the mean mean? Journal of Statistics Education 19: 2.
Wild, C. 2006. The concept of distribution. Statistics Education Research Journal 5: 10–26.

CHAPTER 6

Tests of Hypotheses

Martin Griffiths

1. Introduction to Hypothesis Testing

One of the purposes of statistics is to enable us to make informed decisions based on some set of data. We will look here at a particular decision making process called *hypothesis testing*, an area of inferential statistics (in which we are trying to 'infer' something from the data). This aspect of statistics is often misunderstood, not only by students (Sotos, Vanhoof, Van den Noortgate and Onghena 2009) but also by scientists or other professionals who use statistical inference as part of their everyday work though are non-specialists in this area. The aim of this chapter is to introduce the ideas that lie behind hypothesis testing and to dispel the many misconceptions that arise commonly. We will keep this first section fairly non-technical for the basic ideas behind hypothesis testing to be absorbed. It is important to have the opportunity to see the 'big picture' before being concerned with the minutiae.

Introductory Problem

Suppose that a factory produces a particular electronic component that is used in mobile phones. In an efficiency drive, the engineer in charge of the production line is considering switching over to a new assembly procedure that he believes will reduce the time it takes to produce each component. He already knows, from measurements taken over several years, that when using the old procedure the mean production time for the component is 57 s and the standard deviation is 8 s. In order to test his hypothesis, he

Mathematical Institute, University of Oxford (United Kingdom).

measured the time it took for each of 20 randomly-chosen components to be produced using the new procedure. The mean of this sample was 52. On the basis of this evidence, should the engineer change to the new production procedure?

Some Initial Points

The sample mean of the production time when using the new procedure was 5 s lower than the population mean for the old one. May we infer from this that there has been an improvement in performance, and hence the engineer should adopt the new procedure? Some students may argue that, because the sample mean is below 57, it follows that the new procedure does indeed reduce the mean production time for the components. There is of course an inherent danger in making inferences about a population based on a small sample of individuals from that population. A key point to highlight here is that this difference of 5 s could simply be down to the presence of natural variation in the sample data (noting, in particular, that the variance is relatively large). It would undoubtedly be very expensive to equip the whole factory with new production lines, and so, before doing so, the engineer would want to be reasonably confident that there would be a significant decrease in the mean production time.

Relating this new scenario to the work covered in Chapter 5, the population of interest comprises the set of times required to produce the components using the new assembly procedure. This population will have some (unknown) mean μ. We already know from Chapter 5 that in estimating a population mean μ by way of a sample mean $\overline{x}$, we can never be totally certain as to how close the estimate actually is, unless the sample comprises the whole population (which, for reasons previously discussed, will virtually never be the case). The best we can hope to do in this situation is to construct confidence intervals in order to gain an idea of the likely accuracy of our estimate. This is a particularly important point; we will generally not be in a position to be able to make claims with absolute certainty and any statement that is made about the estimate will be necessarily probabilistic.

Discussion

Via our introductory problem, let us start by considering the notion of hypothesis testing in a fairly informal sense. We eventually want to make a decision concerning whether or not the new assembly procedure has actually reduced the mean production time of the component. In making our decision, we will adopt an 'innocent before proven guilty' approach; that is, we will have no reason to doubt the statement

> "The new assembly procedure has not significantly reduced the mean production time"

unless there is strong evidence to suggest otherwise. We will clarify below, within the current context, exactly what is meant by the phrase "strong evidence to suggest otherwise".

For reasons that will become apparent, we initially consider a subtle variation on the introductory problem. Instead of devising a test to see whether or not the mean production time has decreased, we want to formulate one that indicates whether or not the mean has changed in either direction (in other words, we want to see whether it has decreased or increased). Consider the following method of coming to a decision as to whether or not the mean production time μ has changed, where X is the time taken to produce a component using the new procedure:

1) Obtain a random sample of n production times, and find the sample mean $\bar{x}$. We already know that these are n=20 and $\bar{x}$=52.
2) Assume that the new procedure does not alter the variance of the production times. The population variance is thus given by σ^2=64 . Now assume (for the time being at least) that X is normally distributed. We thus have

$$\bar{X} \sim \mathrm{N}\left(\mu, \frac{\sigma^2}{n}\right) = \mathrm{N}\left(\mu, \frac{64}{20}\right).$$

3) Construct a symmetric 95% confidence interval for μ. This is given by

$$\left(\bar{x} - 1.96\frac{\sigma}{\sqrt{n}}, \bar{x} + 1.96\frac{\sigma}{\sqrt{n}}\right) = \left(52 - 1.96\frac{8}{\sqrt{20}}, 52 + 1.96\frac{8}{\sqrt{20}}\right)$$

$$= (48.494, 55.506).$$

4) Notice that 57 is not contained this interval. Then, since 95% of all intervals constructed in this manner will contain the true mean, we might come to the conclusion that μ is unlikely to be equal to 57. In other words, on the balance of probability, there would appear to be strong evidence that the mean is not equal to 57.

Just to emphasize previous comments, however, it is the case that although the above calculation would appear to indicate that the mean has changed, we are not in a position to state that this is definitely the case.

To gain further insight and to link this in with the notion of confidence intervals discussed in Chapter 5, we may consider Fig. 1. The probability density function shown is that of a normal random variable with mean 57

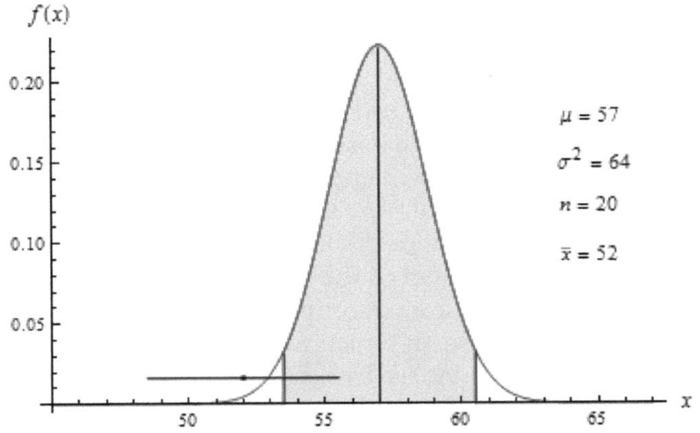

Figure 1. A normal distribution with a superimposed symmetric 95% confidence interval that does not trap μ.

and variance 64/20. Assuming that the production times do in fact follow this distribution, 95% of all intervals constructed as above will contain 57. Notice that the value of $\bar{x}$ falls into the left-most tail of the 2.5% tails of the distribution, and, as a consequence, the corresponding confidence interval does not trap 57. This gives us a visual interpretation of the fact that we might have cause to doubt, on the basis of our sample, the correctness of the assumed distribution of production times.

Let us now return to the original formulation of the problem. In reality of course, we would only be interested in adopting a new procedure if it reduces the mean production time (clearly, if the mean production time either remained unchanged or increased then it would be pointless to adopt it). This leads us to consider, a 95% asymmetric one instead of a 95% symmetric confidence interval; recall that we briefly discussed such an interval in Chapter 5. In the above series of four steps, the first two remain the same. However, in the third step we now construct the asymmetric 95% confidence interval for μ given by

$$\left(-\infty, \bar{x} + 1.645\frac{\sigma}{\sqrt{n}}\right) = \left(-\infty, 52 + 1.645\frac{8}{\sqrt{20}}\right)$$

$$= (-\infty, 54.943).$$

In the fourth step we notice, as before, that 95% of intervals constructed in this manner will 'trap' the true mean μ. However, from the nature of these asymmetric confidence intervals, we see that in this case they can lie entirely below 57 but not entirely above it. In other words, if the upper limit of the interval is at least 57 then the interval will contain 57. We may therefore conclude that, on the balance of probability, there would appear to be strong evidence that the mean is less than 57 (as opposed to "not equal to 57").

This argument is summarized diagrammatically in Fig. 2. Since we are now only interested in the possibility of a decrease in the production time, we consider only the lower end of the assumed distribution to see if the random sample can give us any useful information in this regard. We are still looking at a 95% confidence interval though, but this time it is an asymmetric one. The lower tail of the distribution thus has an area of 5% and the upper tail has zero area (as opposed to each tail having an area of 2.5% for the symmetric case, in which they 'share' the 5% between them). Notice two things about the asymmetric confidence interval shown in Fig. 2. First, it does not trap the assumed population mean, leading us to doubt our assumption. Second, the confidence interval does in fact stretch leftwards to $-\infty$.

This is essentially a simple example of a hypothesis test, which might be thought of as a method of deciding between two opposing theories. In our example, the opposing theories might be "The new assembly procedure reduces the mean production time" and "The new assembly procedure does not reduce the mean production time". Key concepts to introduce at this early stage are the *null* and *alternative* hypotheses (Crawshaw and Chambers 1994, Warner 2008). The students might think of a hypothesis as a theory which is assumed to be true unless they can find evidence to indicate otherwise. In particular, 'null' means 'nothing' (or empty) in mathematics, so we may think of the null hypothesis as a theory of no change from what we would expect based on past experience. In the above problem this would correspond to the new procedure being no better than the old one, where 'better' is synonymous with 'quicker' in our particular scenario.

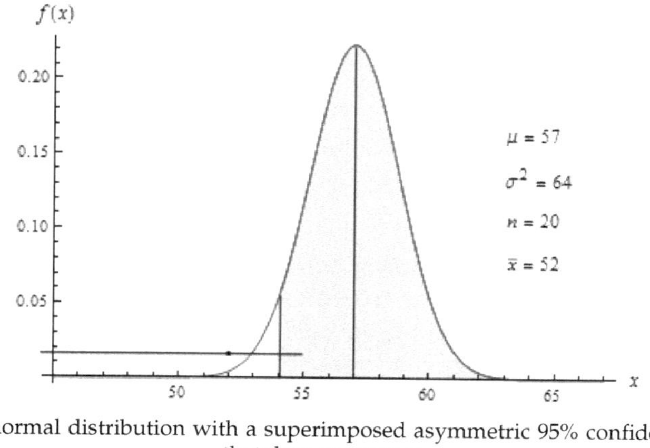

Figure 2. A normal distribution with a superimposed asymmetric 95% confidence interval that does not trap μ.

The alternative hypothesis is, on the other hand, a theory of change. In our example it is the theory that the new procedure reduces the time it takes to produce each component.

Another point that may be discussed and clarified here is the notion of *one-* and *two-tail tests*. In the example currently being considered, the alternative hypothesis proposes that the change will be in one direction only, namely a reduction in time; refer back to the asymmetric confidence interval constructed earlier. There are also scenarios when an alternative hypothesis will propose an increase in the variable being tested; for example, the yield of wheat in a particular field after some fertilizer had been used. These two situations are called one-tailed tests since they posit changes in only one direction (either up or down). There are of course many other situations in which we are merely interested to know whether there is any change at all, irrespective of whether it is up or down. An example of this would be a test to determine whether the percentage salinity of the water in a particular stretch of sea had changed. This would give rise to a two-tailed test (this idea is further clarified in the following section). It is worth pointing out that whether a test is one- or two-tailed sometimes depends on the context, and some interpretation may be required. For example, from the point of view of a consumer buying a battery, there would only be cause for concern if the mean life time of the battery was below that of the claimed mean, thereby resulting in a one-tailed test as opposed to two-tailed. We would not complain if the mean lifetime of the batteries was longer than claimed by the manufacturer! The two contrasting theories might thus be in this case "The mean lifetime of the batteries is as claimed by the manufacturer" and "The mean lifetime of the batteries is lower than that claimed by the manufacturer". These statements give the null and alternative hypothesis, respectively.

It is important that students and teachers are both comfortable with these concepts when conducting tests. Before going any further, we ask the reader to carry out the following self-evaluation problems:

Self-evaluation Problems 1.1

For each of the scenarios below, state both the null and alternative hypotheses, focussing particularly on whether or not they will result in one- or two-tailed tests. Justification must be provided for each choice, and, in the case of one-tailed tests, the direction of change being proposed should be clearly indicated by way of the alternative hypothesis.

1. It was stated in a report that the current mean age of death of a UK resident is 81 years. A hypothesis test is carried out in order to test this claim.

2. London is known to receive, on average, 4.3 hrs of sunshine per day in June. A holiday maker claims that Malaga enjoys more hours of sunshine per day on an average than does London. A test is conducted to check the plausibility of his claim.
3. A baker claims that her chocolate muffins have a mean mass of 200 g. A test is carried out to check her claim.

Self-evaluation Problem Solutions 1.1

1. The null hypothesis here is "The mean age of death of a UK resident is in fact 81 yrs, as claimed by the report" while a suitable alternative hypothesis is given by the statement "The mean age of death of a UK resident is different from 81 yrs of age". This is a two-tailed test since there is nothing in the scenario to imply that we suspect the mean age of death to be lower than 81 years or, alternatively, to be greater than 81.
2. The null and alternative hypotheses are given by "Malaga receives, on average, 4.3 hrs of sunshine per day in June" and "Malaga receives in excess of 4.3 hrs of sunshine per day on average in June", respectively. There is a clear direction in the holiday maker's claim here; he suspects that Malaga receives strictly more hours of sunshine in June than London. This is thus a one-tailed test.
3. Although, on the face of it, this scenario might appear to be a candidate for a two-tailed test, the fact is that if your muffin was heavier than claimed, then this would be of no great concern. On the other hand, if the muffin was underweight then there would be cause for complaint. Our null and alternative hypotheses might therefore be given as "The chocolate muffins have a mean mass of 200 g, as claimed by the baker" and "The mean mass of the chocolate muffins is below 200 g", respectively. This is a one-tailed test. Admittedly, however, there is possibly an element of ambiguity here; indeed, this might make for an interesting discussion point.

2. The Steps of a Hypothesis Test

In the previous section we considered the notion of hypothesis testing by way of the construction of confidence intervals; something with which we were familiar. The steps outlined there may be formalized somewhat to result in what is known as a *significance test* or *hypothesis test*. The logic behind these tests corresponds precisely to the logic behind the confidence-interval approach.

Returning now to our introductory example, let μ denote the mean time to produce a component under the new manufacturing procedure. Then, bearing in mind the discussion the previous section, our null and alternative hypotheses are denoted by

H_0: $\mu = 57$

and

H_1: $\mu < 57$,

respectively. This is an example of a one-tailed test since we are only interested here in potential change in one direction (i.e., a possible reduction in production time). The interpretation of these hypotheses can sometimes cause confusion, and it is worth spending some time examining precisely what they represent.

It is easiest first to consider H_1: $\mu < 57$. This represents the statement "The new assembly procedure reduces the mean production time". On the other hand, although we may write H_0: $\mu = 57$ and state the null hypothesis as "The mean production time is 57 seconds", it best to interpret it as "The mean production time has not been reduced". This then means that not only is there no 'overlap' between the two hypotheses, but between them they actually cover all possibilities.

The key idea now is that we carry out the calculations in the test under the assumption that the null hypothesis is true; the phrase "under H_0" is often used to signify this. We would like to have an indication, under this assumption, of how likely it is for our sample mean $\bar{x}$ to be 52 minutes or below. In very rough terms, if this probability is very small then it is unlikely that our sample came from a population with mean 57 seconds and standard deviation 8 seconds.

As mentioned in the introductory section, the whole process is a little like an 'innocent until proven guilty' scenario. In other words, we assume that $\mu = 57$ unless there is strong evidence to suggest otherwise. Any such evidence would come in the form of a probability. We now need a way of determining the outcome of the test using the known parameter values, the data that has been collected and our knowledge of the underlying distribution.

To keep things relatively simple to start with, let us suppose once again that the engineer assumes the new production time X follows a normal distribution and that the standard deviation remains the same when using the new procedure. Thus we may write $X \sim N(\mu, 8^2)$. In particular, under the null hypothesis H_0, it is the case that $X \sim N(57, 8^2)$. Under this restriction the *sampling distribution* of the mean (when the sample size is 20) is then given by

$\bar{X} \sim \mathrm{N}\left(57, \frac{8^2}{20}\right) = \mathrm{N}\left(57, \frac{16}{5}\right).$

It is at this point that we may introduce the notion of a *significance level*. Let us consider the possible values that the sample mean $\bar{X}$ may take. We can split up these possible outcomes into two regions. One region corresponds to the set of values of $\bar{X}$ that would cause us to accept the null hypothesis and the other corresponds to those values of $\bar{X}$ that would result in us rejecting the null hypothesis in favour of the alternative hypothesis. Needless to say, these are termed the *acceptance* and *rejection regions*, respectively. A key point is that the rejection region is chosen so that it is unlikely for the sample to fall in it if H_0 is true. In our case, we might look for some point below which there is only a probability of 5% of $\bar{X}$ falling there under H_0. This is known as a 5% significance level (this might admittedly seem a somewhat arbitrary figure, and this is discussed further in forthcoming sections). Note how a 5% significance level essentially corresponds to a 95% confidence interval.

In Fig. 3 we see the distribution of the sample mean $\bar{X}$ under the null hypothesis, and a visual interpretation of the idea of a significance level. We can clearly see that since the alternative hypothesis proposes a decrease, we are only interested in the lower end of this distribution, leading to a one-tail test. Figure 4 highlights both the rejection and the acceptance regions for this test; the critical value separating these regions will be calculated shortly. There are a number of misconceptions associated with the significance level of a test (Batanero 2004), and these will be discussed in due course.

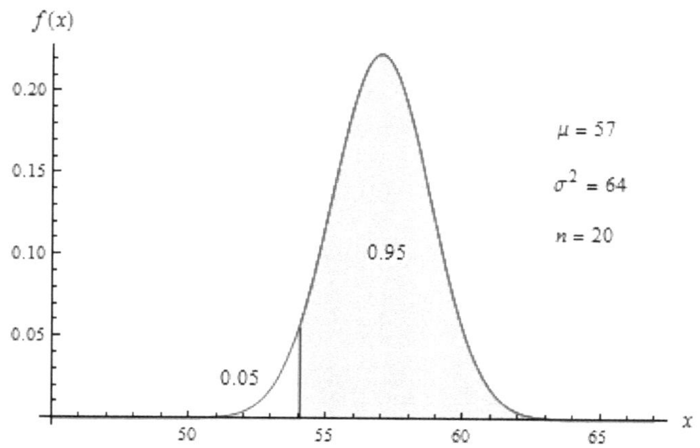

Figure 3. The distribution of the sample mean $\bar{X}$ under the null hypothesis, including a visual interpretation of a one-tailed test at the 5% significance level.

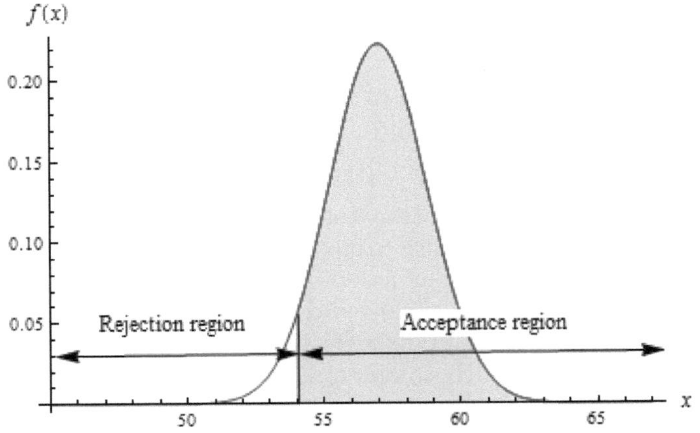

Figure 4. A depiction of the acceptance and rejection regions at the 5% significance level.

Putting it All Together

At this point we will slowly work through the above example, showing how each of the points discussed above leads to a calculation that allows us to make some inferences concerning out initial question. Some familiarity with respect to the handling of calculations involving the normal distribution is assumed here. Furthermore, we need to be aware of the fact that we are essentially calculating a *conditional probability* (this links in with concepts introduced in earlier chapters).

There are essentially three standard approaches to carrying out the calculations in the hypothesis test, although they are all equivalent. Each of these will be considered in the current section. Possibly the most transparent and intuitive is the notion of a *p*-value.

Using the p-value

In our case this would be the probability that the sample mean is 52 or below, given that H_0 is true; recall that this is a one-tailed test. To recapitulate, the distribution of $\bar{X}$ under H_0 is given by

$$\bar{X} \sim \mathrm{N}\left(57, \frac{8^2}{20}\right) = \mathrm{N}(57, 3.2).$$

Therefore,

$$P(\bar{X} \leq 52) = P\left(Z \leq \frac{52-57}{\sqrt{3.2}}\right)$$

$$= P(Z \leq -2.795)$$
$$= 0.0026,$$

which is termed the *p*-value of the test outcome. Since $0.0026<0.05$, we may conclude that there is evidence to suggest that the null hypothesis is false and that the mean of the new assembly procedure is indeed lower than that of the previous one (in other words, we reject H_0 in favour of H_1). We term such an outcome "significant". This is because, assuming the null hypothesis to be true, such an outcome is very unlikely to occur; for this reason we call into question the veracity of the null hypothesis.

Rejection Regions

We may also approach the test by calculating the rejection and acceptance regions for $\bar{X}$. In order to find the rejection region in this particular example we want to find the value c of $\bar{X}$ such that $P(\bar{X} \leq c)=0.05$. We need to solve

$$P\left(Z \leq \frac{c-57}{\sqrt{3.2}}\right) = 0.05,$$

which gives

$$\frac{c-57}{\sqrt{3.2}} = -1.645,$$

and hence $c=54.057$. Thus our rejection region is $\bar{X} \leq 54.057$ (this is illustrated in Fig. 4). Then, since 52 is in the rejection region, we reject H_0 in favour of H_1 (which of course is what we would expect, having just carried out exactly the same test via a slightly different method above).

The z Value

We may also carry out the test by calculating the z value. In this case, since we are performing the test at the 5% significance level, we reject H_0 if $Z \leq -1.645$. Now,

$$z = \frac{52-57}{\sqrt{3.2}} = -2.795,$$

which causes us to reject H_0 in favour of H_1 once more. Notice that this number also appeared in our calculations to obtain the *p*-value.

The reader should convince themselves that the three approaches to carrying out a hypothesis test given above are in fact equivalent. This might be aided by sketching appropriate normal curves, one of which has been transformed to the standard normal distribution.

Computer Comment

As will be seen at various points throughout this chapter, R has many built-in routines to carry out hypothesis tests. However, we first show how to use R to carry out the above test 'by hand'. Given the sophistication of R, this is a somewhat primitive way of doing things, but it is always a good idea to start from the basics. Note that the code below corresponds essentially to the z value approach.

```
> mu0=57; xbar=52; sigma=8; n=20
> z=(xbar-mu0)/(sigma/sqrt(n)); z
[1] -2.795085
> z.alpha=-qnorm(0.95); z.alpha
[1] -1.644854
```

Example 2.1

A company manufactures resistors and claims that the mean resistance of these resistors is 10 ohms. A random sample of eight resistors was taken, with the following resistances:

9.8, 10.5, 10.2, 10.5, 9.7, 10.4, 9.8, 10.6.

It is known, from past experience, that the variance of the resistance of these resistors is 0.5. It may be assumed that the resistance is normally distributed. Using the p-value approach, carry out a hypothesis test, at the 5% significance level, to determine whether or not the company's claim is valid. Then obtain the rejection region for this test.

First, let R be the random variable representing the resistance of a randomly-chosen resistor. We know that

$$\bar{R} \sim \mathrm{N}\left(\mu, \frac{0.5}{8}\right) = \mathrm{N}(\mu, 0.0625),$$

where μ is the true population mean. Note that this is an example of a two-tailed test since we are looking for any deviation in the resistance from 10 ohms rather in just one direction. In other words, we simply want to find out whether the manufacturer's claim is correct; we are not interested here in whether or not the mean is specifically lower than 10 ohms or specifically higher.

The null and alternative hypotheses are given by:

H_0: $\mu = 10$

and

H_1: $\mu \neq 10$.

From the sample, we may obtain the sample mean $\bar{r} = 10.1875$. Under H_0 we have $\bar{R} \sim N(10, 0.0625)$. Thus, noting that we require $P(\bar{R} \geq 10.1875)$ rather than $P(\bar{R} \leq 10.1875)$ because 10.1875 falls into the upper half of the distribution (so we are checking the upper tail here):

$$P(\bar{R} \geq 10.1875) = P\left(Z \geq \frac{10.1875 - 10}{\sqrt{0.0625}}\right)$$

$$= P(Z \geq 0.75)$$

$$= 0.2266.$$

Since this is a two-tailed test at the 5% significance level, we apportion a probability of 0.025 to each of the tails. Finally, as 0.2266>0.025, we see that the outcome of this test is not significant. There would appear to be insufficient evidence to dispute the claim.

In order to find the rejection regions we need to find c_1 and c_2 such that

$$P\left(Z \leq \frac{c_1 - 10}{\sqrt{0.0625}}\right) = 0.025 \quad \text{and} \quad P\left(Z \geq \frac{c_2 - 10}{\sqrt{0.0625}}\right) = 0.025.$$

These lead to

$$\frac{c_1 - 10}{\sqrt{0.0625}} = -1.96 \quad \text{and} \quad \frac{c_2 - 10}{\sqrt{0.0625}} = 1.96,$$

from which we obtain $c_1 = 9.51$ and $c_2 = 10.49$. The rejection regions are thus $\bar{R} \leq 9.51$ and $\bar{R} \geq 10.49$. Note, incidentally, that 10.875 lies in the acceptance region, which, from the previous calculation, is what we would expect. Figure 5 provides us with a visual summary; the distribution is of the sample mean $\bar{R}$ under H_0 We can see the two tails of the distribution, with an area of 2.5 % apportioned to each of them.

Self-evaluation Problems 2.1

1. Averaged over many years, the mean daily number of hrs of sunshine in New York during the month of August is reported to be 5.3 hrs. Last year the daily amounts of sunshine y, in hrs, were recorded. The results are summarized as follows:

   $\Sigma y = 168$ and $\Sigma y^2 = 960$.

   Test, at the 10% significance level, whether there was an unusually high amount of sunshine in New York during August last year.

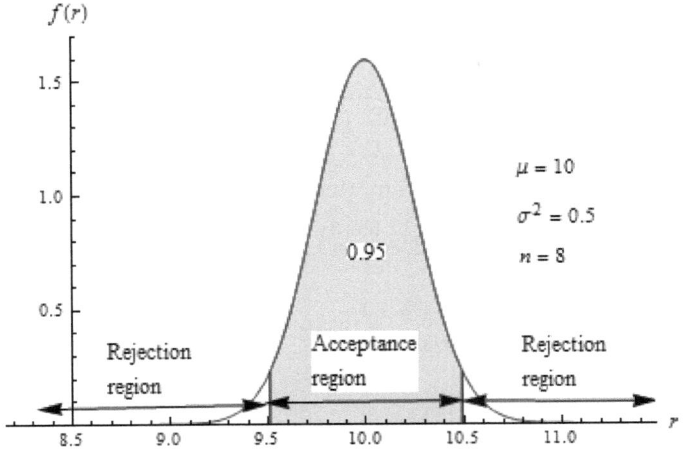

Figure 5. An illustration of the acceptance and rejection regions of the test.

2. A particular type of battery has a lifetime that is normally distributed. The claimed mean lifetime is 1,200 hrs, and a test is carried out to check the validity of this claim. These batteries are extremely costly to produce, so a small sample of size 5 is taken. The lifetimes were as follows:

   1,198; 1,214; 1,173; 1,159; 1,178.

   Test at the 5% significant level, the manufacturer's claim regarding the mean lifetime of the batteries.

Self-evaluation Problem Solutions 2.1

1. Let Y be the amount of sunshine during a randomly-chosen day in August. First, although the distribution of Y has not been given, it is not necessary to assume that it follows a normal distribution since the sample size of 31 is sufficiently large to utilize the central limit theorem (Grimmett and Stirzaker 2001). Furthermore, we can use the sample variance as an unbiased estimate for the population variance.

   The null and alternative hypotheses are as follows:

   H_0: $\mu = 5.3$

   and

   H_1: $\mu > 5.3$.

Next, an unbiased estimate for the variance of the daily amount of sunshine in August is given by

$$s^2 = \frac{1}{n-1}\left(\sum_{i=1}^{n} x_i^2 - n\bar{x}^2\right) = \frac{1}{30}(960 - 31 \times 5.4194^2) = 1.6516$$

noting that the sample mean $\bar{y}$ is equal to 5.4194. Under H_0 we have, via the central limit theorem, that $\bar{Y}$ may be well approximated by

$$A \sim \mathrm{N}\left(5.3, \frac{1.6516}{31}\right),$$

so that

$$\begin{aligned} \mathrm{P}(\bar{Y} \geq 5.3) &\approx \mathrm{P}(A \geq 5.3) \\ &= \mathrm{P}\left(Z \geq \frac{5.4194 - 5.3}{\sqrt{0.0521}}\right) \\ &= \mathrm{P}(Z \geq 0.5231) \\ &= 0.3005. \end{aligned}$$

From this we may conclude that there is not sufficient evidence to reject H_0 in favour of H_1; it would seem that there was not an unusually high amount of sunshine in New York last year.

2. Although the sample size is relatively small, the fact that the battery lifetimes are normally distributed means that we may use the *t* distribution to carry out a *t* test. Furthermore, note that this requires a one-tailed test since we will only have cause for complaint if the batteries' mean lifetime is less than that claimed by the manufacturer. Let L denote the random variable representing the lifetime of a randomly-chosen battery. The sample mean $\bar{l}$ is equal to 1184.4. Next, we obtain an unbiased estimate for the population variance:

We have $s^2 = \frac{1}{n-1}\left(\sum_{i=1}^{n} x_i^2 - n\bar{x}^2\right) = \frac{1}{4}(7015894 - 5 \times 1184.4^2) = 469.3.$

$H_0: \mu = 1200$

and

$H_1: \mu < 1200.$

Under H_0,

$$\begin{aligned} \mathrm{P}(\bar{L} < 1184.4) &= \mathrm{P}\left(T < \frac{1184.4 - 1200}{\sqrt{469.3/5}}\right) \\ &= \mathrm{P}(T < -1.602). \end{aligned}$$

By consulting a table of critical values for the *t* distribution, and noting that the number of degrees of freedom in this instance is 5–1=4, we see that the rejection region is given by $T<-2.132$. We may therefore conclude that, as with the test carried out in problem 1, there is not sufficient evidence to reject H_0 in favour of H_1. We thus have no reason to doubt the manufacturer's claim regarding the mean life time of their batteries. There are two further points that are worth making here:

i) The sample size is extremely small here, and ideally a larger sample should be taken.
ii) If, instead of 5%, the significance level had been set at 10%, then H_0 would have been rejected, and the conclusion of the test would thus have been different. Indeed, the outcome would have been significant, with the evidence pointing towards a false claim by the manufacturer concerning the lifetime of the batteries. This is because the rejection region in this case would have been $T<-1.533$. This goes to show that choosing the significance level does seem introduce an element of arbitrariness into the process. It turns out that a balance has to be struck between the significance level and another factor that has to be considered when carrying out hypothesis tests; more on this in due course.

Computer Comment

The R code below shows how to carry out the test considered in problem 2 above. Notice that this also provides us with an asymmetric 95% confidence interval (as the test is one-sided).

```
> x=c(1198,1214,1173,1159,1178)
> t.test(x,mu=1200,alternative="less",conf.level=0.95)

        One Sample t-test

data: x
t = -1.6102, df = 4, p-value = 0.09132
alternative hypothesis: true mean is less than 1200
95 percent confidence interval:
      -Inf 1205.054
sample estimates:
mean of x
  1184.4
```

We emphasize once more here that since we are talking about probabilistic processes, we cannot say for certain that our conclusions for the above tests are correct; hence the note of caution in the wording of all our test conclusions. Students tend have misconceptions in this area, and often make statements implying that the mean has either been reduced definitely or is definitely the mean as given by the null hypothesis. When interpreting the results of our calculations, therefore, the wording does need to be extremely careful.

Before moving on, students might concentrate on trying to see the 'big picture' with regard to the logic of hypothesis testing (Garfield 2002, Lawton 2009). It would be advisable to reread the current section, making notes on the key points of the procedures we have encountered and the role each step plays in helping us to come to a reasoned conclusion.

3. Hypothesis Tests Involving Discrete Random Variables

Thus far, all our hypothesis tests have concerned continuous random variables. Let us now consider the corresponding situation for discrete random variables.

Example 3.1

A Poisson distribution is used to model the number of accidents that occur along a particular stretch of road each month. A warning sign is erected, and a test carried out to see if the presence of the sign has a significant impact with respect to reducing the accident rate. Suppose then that the random variable A represents the number of accidents occurring on this stretch of road during the course of a month. We may write $A \sim \text{Po}(\lambda)$ for some positive real number λ representing the average number of accidents per week.

Before the warning sign was erected, it is known that there were, on an average, 19 accidents per month; in other words, $\lambda = 19$. During the month after the warning sign had been erected there were 11 accidents. Let us carry out a hypothesis test at the 5% significance level to ascertain whether or not the sign has had the desired effect. Some might argue that of course the sign has had the desired effect, since 11 accidents is an improvement on 19. However, this reduction could have happened by chance, and, through our test, we want to establish whether or not the outcome of 11 accidents is actually a significant event.

Our null and alternative hypotheses are

H_0: $\lambda = 19$

and

H_1: $\lambda < 19$,

respectively, noting that this is indeed a one-tailed test. Under H_0 we have $A \sim \text{Po}(19)$. Using a table of cumulative Poisson probabilities, we obtain

$P(A \leq 11)=0.0347$.

Then, since 0.345<0.05, the outcome of this hypothesis test is regarded as significant, and we may conclude that there is evidence that the warning sign has had an impact in reducing the number of accidents per month on this particular stretch of road.

A Point to be Noted

Although the procedures for carrying out hypothesis tests are certainly similar for continuous and discrete random variables, there is one subtle difference that ought to be highlighted here. When we state a significance level for a test involving a continuous random variable, we are able to find a rejection region which corresponds exactly to this probability. For example, in section 2, we found that the rejection region corresponding to the 5% significant level was, with suitable rounding, $c=54.057$. However, this is not generally possible in the discrete case. In the example we have just considered, we find under H_0 that

$P(A \leq 11)=0.0347$ and $P(A \leq 12)=0.0606$.

From this it may be seen that a rejection region of $A \leq 11$ gives rise to a significance level 3.47% while a rejection region of $A \leq 12$ leads to a significance level of 6.06%. For this reason, we say that this test has a *nominal* significance level of 5%, implying that in practice it generally cannot be achieved exactly. The general convention in such a situation is to take the rejection region corresponding to the largest probability not exceeding the nominal significance level.

Self-evaluation Problems 3.1

1. A local authority claims that more than three quarters of the people in its jurisdiction are happy with the services it provides (transport, libraries, garbage collection, etc.). From a random survey of 30 people, 26 said that they were happy with the services. Carry out a test at the 10% significance level to examine the local authority's claim.
2. In the morning, telephone calls usually come into a call centre at a rate of 14 calls per minute. A one-minute slot at lunch time was chosen at random to see whether the rate of calls at this time of the day was significantly different to that in the morning, and a test was carried out. With C representing the number of calls per minute at lunchtime, the rejection regions of this two-tailed test were deemed to be $C \leq 7$ and $C \geq 21$. Calculate the true (i.e., not nominal) significance level for this test.

Self-evaluation Problem Solutions 3.1

1. Let X denote the random variable representing the number of people in a sample of 30 that state they are happy with the services. The first thing to recognize here is that X follows a binomial distribution. In fact, we have $X \sim B(30,p)$, where p the proportion of people receiving these services who are in fact happy with them.

   Our null and alternative hypotheses are

   $H_0: p=0.75$

   and

   $H_1: p > 0.75$

   respectively, noting that this is a one-tailed test. Then, under H_0, we have $X \sim B(30,0.75)$. Using a table of cumulative binomial probabilities, we obtain

   $P(X \geq 26)=0.0979.$

   Since $0.0979 < 0.1$, we may conclude that there is evidence that more than three quarters of the people are happy with the services. Note, however, that the outcome of the test is marginal at the 10% significance level.
2. The description of C would indicate that it is a Poisson random variable. So the number of calls per minute at lunchtime is given by $C \sim Po(\lambda)$. Our null and alternative hypotheses are

   $H_0: \lambda = 14$

   and

   $H_1: \lambda \neq 14$

   respectively, noting that this is a two-tailed test. Then, under H_0, we have $C \sim Po(14)$. The significance level for this test is

   $P(C \leq 7)+P(C \geq 21) = 0.0316+0.0479=0.0795.$

As well as carrying out such hypothesis tests using the exact discrete distributions, we can do so using approximations (Lesch and Jeske 2009), depending on both the practicality of the situation and the validity of the approximations. It needs to be borne in mind of course that continuity corrections are required when using the normal distribution to approximate a discrete one. This may create a discussion point, introducing informal ideas about what is meant by 'an approximation' in this scenario.

Problem 3.1

The mayor of a town claimed that less than 40% of the population were in favour of a ring road being built around the town. A sample of 80 of the town's inhabitants were surveyed and 28 of them said they were in favour of the ring road. Carry out a test at the 5% significant level to test the mayor's claim.

Solution to Problem 3.1

Let R represent the number of people in a random sample of 80 people from the town that are in favour of the ring road. Then $R \sim \mathrm{B}(80,p)$, where p is the proportion of the town's inhabitants that are in favour. The null and alternative hypotheses are given by:

$H_0: p = 0.4$

and

$H_1: p < 0.4$,

respectively. Then, under H_0, we have $R \sim \mathrm{B}(80,0.4)$. Since $np = 80 \times 0.4 = 32 > 5$ and $n(1-p)=80 \times 0.6=48>5$ then R may be well approximated by

$$Y \sim \mathrm{N}(np, np(1-p)) = \mathrm{N}(32,19.2).$$

We then have

$$\begin{aligned} \mathrm{P}(R \le 28) &\approx \mathrm{P}(Y \le 28.5) \\ &= \mathrm{P}\left(Z \le \frac{28.5-32}{\sqrt{19.2}}\right) \\ &= \mathrm{P}(Z \le -0.7988) \\ &= 0.2122, \end{aligned}$$

from which it may be seen that there does not appear to be sufficient evidence to corroborate the Mayor's claim. In other words, there is no reason to believe that less than 40% of the population were in favour of a ring road being built around the town.

Computer Comment

As an alternative to carrying out the above hypothesis test by hand, we may easily do so using R, as follows. Both the input and output are shown.

```
> prop.test(28,80,p=0.4,alternative="less",conf.
level=0.95, correct=TRUE)

1-sample proportions test with continuity correction

data: 28 out of 80, null probability 0.4
X-squared = 0.638, df = 1, p-value = 0.2122
alternative hypothesis: true p is less than 0.4
95 percent confidence interval:
 0.0000000 0.4476956
sample estimates:
  p
0.35
```

It is worth noting that:

i) The procedure `prop.test` takes several arguments, some of which are optional. In particular, `p=0.4` states the probability under the null hypothesis and `correct=TRUE` means that a continuity correction has been applied (as indeed it had been in our calculation).
ii) The outcome of the test may be ascertained by the looking the statement `p-value=0.2122` given in the output (notice that this is exactly the same as the *p*-value we had calculated by hand).
iii) Notice that the lower limit of the 95% confidence interval for the population proportion is given as 0. This is because the confidence-interval calculation results in a negative lower limit in this example.

4. Type I and Type II Errors

When carrying out a hypothesis test, there are essentially four possible things that can happen:

i) H_0 is true and H_0 is accepted;
ii) H_0 is true but H_0 is rejected;
iii) H_0 is false but H_0 is accepted;
iv) H_0 is false and H_0 is rejected.

Note that in cases (i) and (iv) the decision is correct. Throughout the current section, however, we are interested only in cases (ii) and (iii), in which the decisions are not correct. These situations are so important with respect to hypothesis testing that they actually have their own names; a Type I error corresponds to (ii) and a Type II error corresponds to (iii). Thus, a Type

I error occurs when a true null hypothesis is rejected and a Type II error happens when a false null hypothesis is accepted. It is clear that

$$P(\text{Type I error}) = P(\text{reject } H_0 \mid H_0 \text{ true})$$

and

$$P(\text{Type II error}) = P(\text{accept } H_0 \mid H_0 \text{ false}).$$

Making errors in hypothesis tests can clearly have consequences, and it is in everyone's interest to avoid them occurring. For example, if we were developing a particular type of drug to reduce blood pressure then we might only wish to manufacture it if the prevalence of certain unpleasant side effects amongst the population is below a certain level; indeed, a decision as to whether or not to proceed with production may be based on the outcome of a hypothesis test; for example, it might be required that less than 5% of the population suffer side effects.

If a Type I error was made in this case, we may end up making a decision not to produce the drug, despite that fact that it does actually meet the requirement concerning the side effect for the population taken as a whole. The consequences here are that the company might lose out financially through not being able to manufacture and sell the drug, and the potential recipients of the drug will clearly also lose out on any benefit it may have had for their health.

On the other hand, if a Type II error occurred and the decision to manufacture the drug was made as a consequence, there could be a major problem in that we have a drug on the market that could actually be doing more harm than good.

From the definition of a Type I error, it is clear that for tests involving continuous random variables, P(Type I error) is equal to the significance level. On the other hand, for tests involving discrete random variables, there will generally not be equality between P(Type I error) and the nominal significance level.

For a fixed sample size n, there will always be a compromise between P(Type I error) and P(Type II error). As one decreases, the other generally increases. The significance level chosen may depend on the consequences of a Type I error occurring. If the consequences were not serious, you might use a significance level of 10%, while serious consequences might lead one to adopt a significance of 1% or even lower.

Note that in order to calculate P(Type I error) or P(Type II error) we need to assume a value for the mean of the population. For P(Type I error) things are simple; calculations are carried out assuming that H_0 is true. But what about P(Type II error)? Unlike the null hypothesis, the alternative hypothesis specifies a range of values of the mean rather than a specific value. What value should we assume here? Well, the answer to this is that, in order to calculate P(Type II error), we assign some nominal value to the mean, where the selection of this nominal value may have come about through prior experience or even informed guesswork.

As a further point, it is worth mentioning that when tackling questions involving hypothesis tests, and Type I and Type II errors, it is most convenient to calculate the rejection regions. This is because they will generally be required somewhere in the calculation.

Problem 4.1

In a brewery, bottles of beer are filled by a machine. If the machine is functioning correctly, the volume of beer in a bottle is normally distributed with mean 1 litre and standard deviation 0.002. The performance of the machine is tested by taking a sample of 9 bottles and calculating their mean content.

i) A test is to be carried to see whether the machine is in fact working as it should be. With the significance level set at 5%, obtain the rejection region for the sample mean.
ii) State the value of P(Type I error) for this test.
iii) Calculate P(Type II error), given that the mean content has actually fallen to 0.998 litres.

Solution to Problem 4.1

i) Let V denote the volume of beer in a randomly selected bottle. Then $V \sim \mathrm{N}(\mu, 0.002^2)$, where μ is the mean content of a bottle. The null and alternative hypotheses are given by

H_0: $\mu = 1$

and

H_1: $\mu < 1$,

respectively. Then, under H_0, we have $V \sim \mathrm{N}(1, 0.002^2)$, and hence

$$\bar{V} \sim \mathrm{N}\left(1, \frac{0.002^2}{9}\right).$$

For a one-tailed test at the 5% significance level, the rejection region for the test statistic Z (which follows the standard normal distribution) is given by $Z \leq -1.645$. It is thus the case that

$$\frac{\bar{V} - 1}{\sqrt{0.002^2/9}} \leq -1.645.$$

From this we may obtain the rejection region as $\bar{V} \leq 0.9989$.

ii) As previously mentioned, P(Type I error) is equal to the significance level of the test, which in this case is 0.05.

iii) We know that

$$\text{P(Type II error)} = \text{P(accept } H_0 | H_0 \text{ false)}$$

$$= \text{P}(\bar{V} > 0.9989 | \mu = 0.998)$$

$$= \text{P}\left(Z > \frac{0.9989 - 0.998}{\sqrt{0.002^2/9}}\right)$$

$$= 0.0885.$$

In Fig. 6 we see a diagram which includes the distributions under both H_0 and H_1 (given that the mean content has actually fallen to 0.998 litres in the latter case). This serves to provide a clear picture of the acceptance and rejection regions for two distributions. Then, in Fig. 7, the relationship between P(Type I error) and P(Type II error) is depicted for this particular example. It is straightforward to visualize from this that a decrease in significance level will result in an increase in P(Type II error), assuming all other quantities remain the same. Also note that if the actual value is very close to the hypothesized value then P(Type II error) will actually be quite large (see Fig. 8). This problem can be alleviated somewhat by taking a larger sample.

The notion of errors in hypothesis testing can cause difficulty for students. Furthermore, some find it unsatisfactory that the result of a hypothesis test should depend on the significance level chosen (this aspect

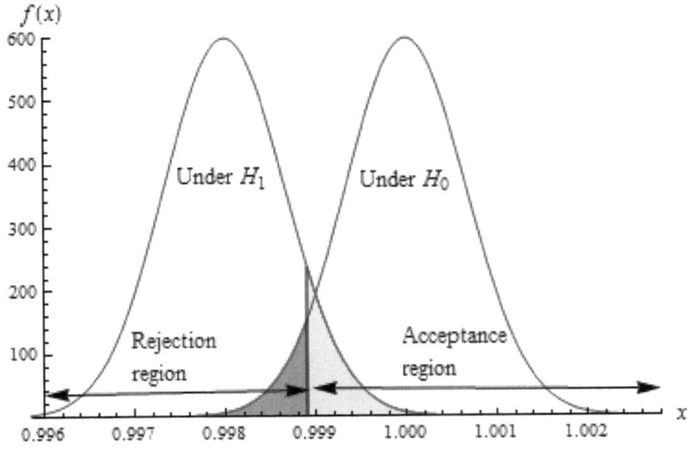

Figure 6. The distribution of the volume under both H_0 and H_1.

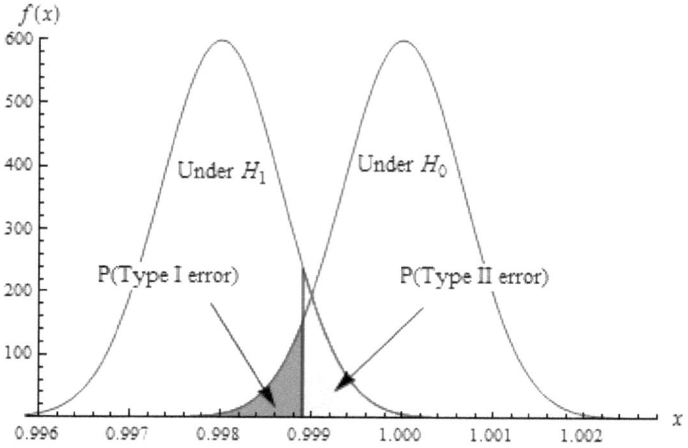

Figure 7. The relationship between P(Type I error) and P(Type II error).

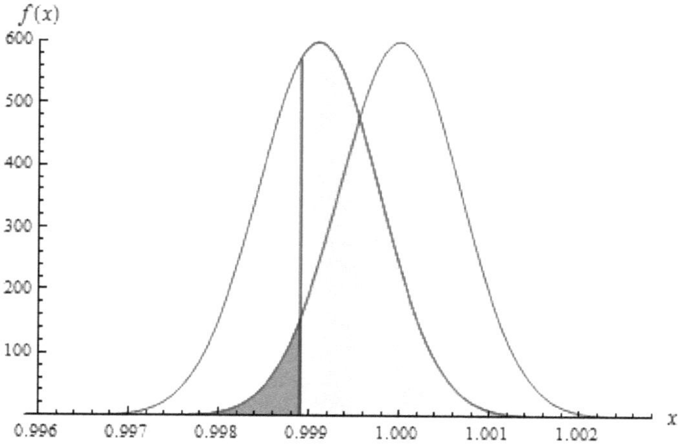

Figure 8. A demonstration showing that when the effect size is small, P(Type II error) may be large.

of the test might seem to them somewhat arbitrary). However, there is always a compromise that has to be made between the two types of error. In the first instance, the choice of significance level might depend on how serious the consequences of making a type I error are. The more serious the consequences of wrongly rejecting H_0 are, the lower the value of the significance level.

This leads naturally on to the *power* of a test. In order to gain an informal notion of this concept, we might start by, for example, considering the effect the sample size has on the underlying sampling distribution (Kozack

2009, Mulekar and Siegel 2009). This can then be related pictorially to type II errors, and aided by way of appropriate software such as Java applets (Wiseman 2004, Anderson-Cook and Dorai-Raj 2003).

5. The Statistical Power of a Test

The power of a hypothesis test is defined to be the probability that the test rejects the null hypothesis when the null hypothesis is indeed false. In other words, the power of a test is the probability of not making a Type II error, and is equal to 1–P (Type II error). By way of an example, let us consider Problem 4.1 involving bottles of beer. We might want to carry out a test to see if there was any evidence that the machine is under filling the bottles. The power is the probability that the test will detect a statistically significant difference between the volume that should be dispensed and that which is actually being dispensed. From an intuitive point of view, it is clearly a function of the size of the difference between these two quantities. This is because we would expect to be able to detect differences more easily when those differences are large.

There are actually three factors that influence the power:

a) The significance level of the test.
b) The effect size (which is the difference between the null hypothesis and the true value of the mean, discussed above briefly).
c) The sample size used to detect the effect.

In order to illustrate this, recall that the rejection region for Problem 4.1 was shown to be $\bar{V} \le 0.9989$. More generally, this rejection region is a function of the significance level α, the sample size n and the value of the mean under the null hypothesis, μ_0 say. Indeed, it may be seen that

$$\bar{V} \le \mu_0 - z_\alpha \sqrt{0.002^2 / n}$$

where μ_0 is the value of the mean under the null hypothesis and z_α is the z value corresponding to the significance level α. Thus, since the power of the test is given by

$$1 - \mathrm{P(Type\ II\ error)} = 1 - \mathrm{P}(\bar{V} > \mu_0 - z_\alpha\sqrt{0.002^2/n} \mid \mu = \mu_1),$$

where

$$\bar{V} \sim N\left(\mu, \frac{0.002^2}{9}\right)$$

and μ is the true population mean, it may be seen that each of the factors (a), (b) and (c) above do indeed exert an influence on the power of a hypothesis test.

Power tables can be used to get a rough idea of the sample size to use, given the significance level and an educated guess (based in prior experience) concerning the effect size (Warner 2008). Statistical power of 0.8 is often cited as a reasonable goal in any hypothesis test. Note that 'raw' effect sizes are somewhat arbitrary in nature since they depend on the units of measurement that are used. In power tables, therefore, we generally utilize a standardized effect size:

$$d = \frac{\mu - \mu_0}{\sigma},$$

sometimes referred to as Cohen's little *d* (Warner 2008).

It is not a good idea to increase the significance level as this leads to a corresponding increase in the probability of making a Type I error. Furthermore, the power of a test can be extremely low if the hypothesized value of the mean is in fact very close to the true population mean. A demonstration of this phenomenon is depicted in Fig. 8; note that P(Type II error)>0.5 so that the power of the test is less than one half. In such cases the test might not be at all effective in distinguishing between the hypothesized and the actual mean unless the sample size was increased significantly. Of course, increasing the sample size generally means an increase in time and expense. It can thus be seen, therefore, that the whole business of hypothesis testing is largely based on compromise, with both practical and financial constraints to be taken account of.

Self-evaluation Problem 5.1

Describe what happens to the power of a test when:

i) The effect size and the sample size remain constant but the significance level increases.
ii) The effect size and the significance level remain constant and the sample size increases.
iii) The significance level and the sample size remain constant but the effect size increases.

Self-evaluation Problem Solution 5.1

The power (i) increases; (ii) increases; (iii) increases.

Computer Comment

The following two pieces of R code calculate for us the sample size required to achieve a power of 0.8, given the effect size (denoted by `delta`), population standard deviation and significance level. Notice in this case the dramatic increase in sample size required when going from an effect size of 3 to one of 0.5. Indeed, that for the former is approximately 24, while for the latter it is almost 800.

```
> power.t.test(delta=3,sd=5,sig.level=.05,power=.8,
type="one.sample",alternative="two.sided")

 One-sample t test power calculation

   n          = 23.79457
   delta      = 3
   sd         = 5
   sig.level  = 0.05
   power      = 0.8
   alternative= two.sided

> power.t.test(delta=0.5,sd=5,sig.level=.05,power=.8,
type="one.sample",alternative="two.sided")

 One-sample t test power calculation

   n          = 786.8109
   delta      = 0.5
   sd         = 5
   sig.level  = 0.05
   power      = 0.8
   alternative= two.sided
```

6. Testing for Differences in Means Between Two Populations

Suppose that a particular mathematical problem is given to a group of 40 men and 40 women and each of these 80 subjects is timed to see how long, in minutes, it takes them to complete the problem. We want to carry out a hypothesis test at the 10% significance level to ascertain whether or not there is any difference between the genders in their problem-solving performances. Let M and W denote the random variables representing the time taken for a randomly selected man and a randomly selected woman, respectively, to solve the problem. Suppose that the mean times for the men and women are given by 17 and 16, respectively, and unbiased estimates for the variance are 7 and 8. Let the population means be μ_m and μ_w.

Our null and alternative hypotheses are

H_0: $\mu_m - \mu_w = 0$

and

H_1: $\mu_m - \mu_w \neq 0$,

respectively, noting that this is indeed a two-tailed test. Under H_0 we have, since the sample size is relatively large,

$$\bar{M} - \bar{W} \sim N\left(0, \frac{7}{40} + \frac{8}{40}\right),$$

from which we see that the value of the test statistic Z is given by

$$z = \frac{17 - 16}{\sqrt{15/40}} = 1.6330.$$

When carrying out a two-tailed test with a 10% significance level, we reject the null hypothesis when $Z \leq -1.645$ or $Z \geq 1.645$. In this case we see, therefore, that there is not sufficient evidence to reject H_0 in favour of H_1 (although the outcome is, admittedly, somewhat marginal). It would appear then that there is no significant difference in the problem-solving abilities of men and women (at least with regard to this one particular problem).

Note, however, that if we had replaced the alternative hypothesis above with the one given by H_1: $\mu_m - \mu_w > 0$, then we would obtain a one-tailed test to see whether women performed better than men (i.e., to see if women took less time on average to solve the problem). In this instance we would we reject the null hypothesis when $Z \geq 1.282$. Thus, since $z = 1.6330$, it may be seen that on changing the alternative hypothesis, our test now gives a significant result. It seems that on the basis of this new test, women do perform better with respect to this mathematical problem. This would appear to demonstrate the apparently arbitrary nature sometimes of the choice of significance level (recall, however, the comments made earlier regarding the compromises that have to be made when carrying out hypothesis tests).

Example 6.1

For smaller samples we may carry out a two-sample t test. This, however, does require some slightly restrictive assumptions, namely that the two populations are normally distributed and the population variances are equal (though unknown, as above). Suppose now that only four randomly chosen men and six randomly chosen women tackled the problem, with the following times:

Men: 20, 25, 18, 17; Women: 18, 18, 21, 21, 16, 17.

From this we obtain $\bar{m} = 20$ and $\bar{w} = 18.5$. We then obtain an unbiased estimate of the common population variance s_c^2 by way of the following:

$$s_c^2 = \frac{s_m^2(4-1) + s_w^2(6-1)}{(4-1)+(6-1)} = 7.4375.$$

Incidentally, we sometimes refer to this as the 'pooled' estimate of the population variance. Using the same null and alternative hypotheses as in the previous example, and testing again at the 10% level, we see that the value of the T statistic under H_0 is

$$t = \frac{20 - 18.5}{\sqrt{s_c^2\left(\frac{1}{4} + \frac{1}{6}\right)}} = 0.8521,$$

with $\bar{m} = (4–1) + (6–1) = 8$ degrees of freedom. It is easily checked that this is not a significant outcome.

Computer Comment

In R we can carry out both two-sample and paired t tests. Below is the code and the output to perform the two-sample test above. Notice the argument specifying the assumption that the two populations have equal variances.

```
> maledata=c(20,25,18,17);
femaledata=c(18,18,21,21,16,17)
>t.test(maledata,femaledata,alternative="two.
sided",conf.level=0.9,var.equal=TRUE)

     Two Sample t-test

data: male data and female data
t = 0.8521, df = 8, p-value = 0.4189
alternative hypothesis: true difference in means is not
equal to 0
90 percent confidence interval:
 -1.773522 4.773522
sample estimates:
mean of x mean of y
       20.0      18.5
```

Example 6.2

The government of a particular country was concerned at the number of road accidents caused by drivers who had been drinking alcohol just before they set off. It wanted to carry out a test to see if the presence of alcohol in the blood increased the reaction time taken by motorists to apply their brakes, and set up a committee to investigate this.

A government official in this committee suggested the following way in which the test could be carried out. A sample of eight motorists is chosen at random. Then, without having had consumed any alcohol, the reaction time for each of them is measured. Another random sample of eight motorists is then obtained, each of whom is given a specified volume of alcohol to drink. Their reaction times are measured, and a two-sample t test is carried out in order ascertain whether or not the intake of alcohol does indeed increase reaction time.

One of the other members of the committee was a statistician, and he suggested that there might be an alternative, possibly more effective, way of conducting the test, as follows. Instead of obtaining two separate (independent) samples, each consisting of eight motorists, just obtain one random sample of eight motorists. Then record their reaction times both before taking alcohol and after. Thus, associated with each person is a pair of reaction times. A paired t test could then be carried out, in which we consider the random variable, D say, representing the difference between an individual's reaction time before and after consuming alcohol. If the mean difference of the sample was above a certain value (to be determined whilst carrying out the test) then we might argue that there is significant evidence to indicate that the presence of alcohol does indeed increase the reaction time when applying the brakes.

Self-evaluation Problem 6.1

Explain why, under certain conditions the paired t test might be regarded as superior to the two-sample t test.

Self-evaluation Problem Solution 6.1

If we examined the reaction times of the people in the sample before they had consumed any alcohol, there would, in practice, be plenty of variation amongst them. However, we are not particularly concerned here in the natural variation between individuals' reaction times. Rather, we are interested in the mean effect the alcohol has on the times. The pairing of individuals' times essentially removes a source of variation in the hypothesis test (that created by natural differences amongst the individuals in the sample). One slight constraint with this test is that we do need to assume that the difference between each pair is normally distributed. However, this is a fair assumption to make in many instances.

There is in fact another reason why in this case the paired t test might be a better option than a two-sample t test. In the latter it would be necessary to assume that the variance of the reaction times of the group not taking alcohol is the same as that of the group after they had taken a specified quantity of alcohol. This may not in fact be a realistic assumption since individuals have widely differing tolerance to alcohol, and it is indeed possible that the intake of alcohol does lead to an increase in the variance of reaction times.

Example 6.2 (continued)

Table 1 shows the reaction times, in hundredths of a second, before and after the alcohol was consumed, and the increase in the times. We will carry out a paired t test at the 1% significance level to see if the intake of alcohol significantly increases reaction time.

Table 1

Motorist	A	B	C	D	E
Before	34	42	50	37	48
After	41	45	58	45	53
Difference	7	3	8	8	5

Let D be the random variable representing the difference in reaction times for a randomly-chosen individual, and let μ_d denote the population mean. We will assume that D is distributed normally, and in fact here,

$$T = \frac{\bar{D} - \mu_d}{S}.$$

Our null and alternative hypotheses are given by

H_0: $\mu_d = 0$

and

H_1: $\mu_d > 0$.

The sample mean $\bar{d}$ is equal to 6.2, and an unbiased estimate for the variance is given by

$$s^2 = \frac{1}{n-1}\left(\sum_{i=1}^{n} x_i^2 - n\bar{x}^2\right) = \frac{1}{4}(211 - 5 \times 6.2^2) = 4.7.$$

Under H_0 the value of the test statistic T is thus

$$t = \frac{6.2}{\sqrt{4.7/5}} = 6.3948.$$

The rejection region, at four degrees of freedom, is $T \geq 3.747$. We may conclude from this that there is strong evidence to reject the null hypothesis. It appears that the intake of the specified amount of alcohol reduces reaction time significantly when operating the brakes.

Computer Comment

This is the R code to carry out the above paired t test.

```
> before=c(34,42,50,37,48);after=c(41,45,58,45,53)
> t.test(after,before,alternative="greater",conf.
level=0.99,paired=TRUE)

      Paired t-test

data: after and before
t = 6.3948, df = 4, p-value = 0.001535
alternative hypothesis: true difference in means is
greater than 0
99 percent confidence interval:
 2.5672 Inf
sample estimates:
mean of the differences
            6.2
```

Of course, it is not always possible to arrange a paired t test, and indeed it may not lend itself to some situations in which means are being compared. For example, suppose a gardener grew two batches of sunflowers in her garden. The first batch was grown in ordinary soil while the second batch was grown in soil that had had compost added to it. She wanted to carry out a test to see whether or not the sunflowers growing in the compost attained a significantly greater height than those growing in the ordinary soil. It is clearly not possible to collect data so that they are paired in this instance; the same sunflower could not be grown in both types of soil (at least in such a way as to obtain any meaningful data). This actually raises an interesting point in that when using a hypothesis test to test for the difference in means, students sometimes have difficulty in deciding whether or not the data are paired when the samples are of the same size.

Problem 6.1

Devise a procedure for testing not just for a difference in means, but say for a difference that is greater than some positive number b.

7. Goodness-of-fit Tests

In Chapter 5 we obtained confidence intervals for the population variance of a normally distributed random variable by way of the chi-squared distribution. In that instance χ^2_{n-1} was being used as an exact distribution. Here, however, we use it for a slightly different purpose. Indeed, in this chapter we carry out tests on tabulated data in order to ascertain whether or not that set of data was likely to have come from some specified distribution (such as binomial, Poisson, geometric, and so on); hence the name 'goodness-of-fit' tests. In this context, the chi-squared distribution is used only as an approximating one.

Example 7.1

A manufacturer produces wine glasses, and they are packed in boxes with 6 glasses in each. A random sample of 80 boxes is chosen, and the number of imperfect glasses in each is counted. Table 2 below summarizes the information.

Table 2

Number of imperfect glasses in a box	0	1	2	3	4	5	6
Number of boxes	18	30	19	10	2	1	0

The manager believes that approximately 30% of all glasses manufactured at the factory are imperfect, and furthermore that distribution of such glasses is binomial. Carry out a test, at the 10% significance level, to see whether a B(6,0.3) model is indeed appropriate here for the number of imperfect glasses in a box of six.

First, our null and alternative hypotheses are as follows:

H_0: The data can be modelled by B(6,0.3)

and

H_1: The model B(6,0.3) is not appropriate for this data.

Let X~B(6,0.3). We will need to compare the observed frequencies in the table above with the expected frequencies under H_0, and so next need to calculate the expected frequencies for the B(6,0.3) model, bearing in mind that there are 80 boxes in all. The expected frequencies are given by

$$80 \times P(X = 0) = 80 \times \binom{6}{0} \times \left(\frac{3}{10}\right)^0 \times \left(\frac{7}{10}\right)^6 = 9.4119,$$

$$80 \times P(X = 1) = 80 \times \binom{6}{1} \times \left(\frac{3}{10}\right)^1 \times \left(\frac{7}{10}\right)^5 = 24.2021,$$

and so on. This is summarized in Table 3:

Table 3

Number of imperfect glasses in a box	0	1	2	3	4	5	6
Observed Number of boxes O_i	18	30	19	10	2	1	0
Expected number of boxes E_i	9.4119	24.2021	25.9308	14.8176	4.7628	0.8165	0.0583

When the expected frequencies are very small, there can be problems with the reliability of the calculations. A generally-accepted rule to take account of this is to avoid having classes for which the expected frequency is less than five. We thus combine the last three classes in Table 3 to give Table 4, noting that we have also added a fourth row to include the quantities $(O_i - E_i)^2/E_i$, which are required in order to carry out the test:

Table 4

Number of imperfect glasses in a box	0	1	2	3	≥ 4
Observed Number of boxes O_i	18	30	19	10	3
Expected number of boxes E_i	9.4119	24.2021	25.9308	14.8176	5.6376
$(O_i - E_i)^2/E_i$	7.8364	1.3980	1.8525	1.5663	1.2340

Note that the random variable X^2 defined by

$$X^2 = \sum \frac{(O_i - E_i)^2}{E_i}$$

provides us with a measure of how much the observed data deviates from the expected frequencies under H_0. It can be shown that this is very well approximated by χ^2_{n-1}, where n is the number of classes used in the final calculation. From the above table we see that there are 5 classes and hence 4 degrees freedom.

We have

$X^2 = 7.8364 + 1.3980 + 1.8525 + 1.5663 + 1.2340 = 13.8872.$

At the 10% significance level, the critical value of the chi-squared distribution (at 4 degrees of freedom) is given by $\chi^2_4 \geq 7.779$. From this we conclude that there is sufficient evidence to reject H_0. It would seem that the model B(6,0.3) is not appropriate for this data.

Computer Comment

Here is some R code to perform this goodness-of-fit test. It is worth noting that there is a slight discrepancy between the value of X^2 calculated by hand and the one obtained via the R routine. This is more than likely caused by R treating the classes on the right-hand side of the tables (with very small expected frequencies) differently to the way in which we treated them. Notice in fact that a warning message is given; this is no doubt in connection with the presence of the aforementioned classes with small expected frequencies.

```
> x=c(0:6);obs=c(18,30,19,10,2,1,0)
> chisq.test(obs,y=NULL,p=dbinom(x,6,0.3))

        Chi-squared test for given probabilities

data: obs
X-squared = 14.3463, df = 6, p-value = 0.026

Warning message:
In chisq.test(obs, y = NULL, p = dbinom(x, 6, 0.3)) :
  Chi-squared approximation may be incorrect
```

It is important to mention, however, that the outcome of this hypothesis test could be due to any of the following:

a) The proportion of imperfect glasses is not 30%;
b) The number of imperfect glasses in a box does not follow a binomial distribution;
c) The proportion of imperfect glasses is not 30%, and the number of imperfect glasses in a box does not follow a binomial distribution.

Let us suppose that we suspect the distribution is binomial, but believe that the manager has not estimated the proportion p of imperfect glasses correctly. We can actually estimate p from the data, and then carry out another goodness-of-fit test based on this updated proportion. The only extra thing we need to bear in mind is that in estimating the proportion from the data, we have one additional constraint, and will therefore lose one degree of freedom through this.

On using the data in Table 2, an estimate for p is given by

$$\frac{18 \times 0 + 30 \times 1 + 19 \times 2 + 10 \times 3 + 2 \times 4 + 1 \times 5 + 0 \times 6}{6 \times 80} = 0.23125.$$

So we have

H_0: The data can be modelled by B(6,0.23125)

and

H_1: The model B(6,0.23125) is not appropriate for this data

In Table 5 we see updated expected frequencies based on a proportion of 0.23125:

Table 5

Number of imperfect glasses in a box	0	1	2	3	4	5	6
Observed Number of boxes O_i	18	30	19	10	2	1	0
Expected number of boxes E_i	16.5120	29.8022	22.4122	8.9892	2.0280	0.2440	0.0122

This time we need to combine the final three classes, to give:

Table 6

Number of imperfect glasses in a box	0	1	2	≥ 3
Observed Number of boxes O_i	18	30	19	13
Expected number of boxes E_i	16.5120	29.8022	22.4122	11.2734
$(O_i - E_i)^2/E_i$	0.1371	0.0013	0.5195	0.2644

In this case

$X^2 = 0.1371 + 0.0013 + 10.5195 + 0.2644 = 0.9223.$

There are four classes in our final table, and there are two further constraints. As before, the first of these is imposed by the fact that the total frequency must sum to 80. As already mentioned, however, we used the data to estimate the population proportion, so another degree of freedom is lost, making total of 4–2 = 2. At the 10% significance level, the critical value of the chi-squared distribution (at two degrees of freedom) is given by $\chi_2^2 \geq 4.605$. From this we conclude that there is no reason to reject H_0. The rather small value of X^2 (in comparison to the critical value of the chi-squared distribution) does lend strong weight to the hypothesis that the data fits the model B(6,0.23125) well.

8. Closing Comments

Notice first the many links between Chapters 5 and 6; in particular, the relationship between hypothesis testing and confidence intervals. They may in fact be regarded as different sides of the same coin, and it is important to emphasize this connection. As an introduction to hypothesis testing, students might like to carry out some informal 'intuitive' hypothesis testing using a probability simulation on a piece of software such as R (Lee et al. 2010). It is also possible to develop applets for teaching hypothesis testing (Schneiter 2008). Following this informal introduction, students must gain an appreciation for the need to conduct tests in a more rigorous manner, paying careful attention to the conditions under which the various distributions are valid, and demonstrating an awareness as to whether or not the sample mean follows them exactly or approximately.

Taking things further, tests of independence may be carried out using the chi-squared distribution in conjunction with contingency tables. It is also possible to conduct hypothesis tests involving regression coefficients (Eisenhauer 2009). Students may suggest as to why it could be useful to compare the variances of two normal distributions (Gallagher 2006) and then carry out such a test using the F distribution (Spiegel and Stephens 1999). It is also possible to take things further by considering non-parametric tests such as the Wilcoxon rank-sum test (Spiegel and Stephens 1999). As a further point of interest, readers might like to consider, in the context of the work in this chapter, a psychological phenomenon known as 'belief bias' (Kaplan 2009), in which subjects rate the strength of arguments based on their own personal view regarding the believability of the conclusions. In particular, the subjects in this study were more likely to question the experimental design when they did not believe the conclusions reached by the hypothesis test.

Hypothesis Testing is an aspect of classical statistics with a rich history. Many of the ideas were developed in the first half of the twentieth century by Ronald Fisher and Jerzy Neyman (Lehmann 2011). Familiarity with the historical aspects of mathematics and statistics allows students to gain a broad overview of these disciplines.

Mixed Problems

1. A parcel delivery company claims that at least 75% of the parcels they deal with reach their destination within 48 hrs of posting. A survey was carried out to test this claim, and out of a random selection of 150 parcels, 104 were found to have arrived within 48 hrs of posting. Test the company's claim at the 5% significance level, stating any assumptions and approximations that you make in your calculations.

HINT: The claim involves the phrase "at least". The alternative hypothesis will thus involve the phrase "less than".

2. A survey was carried out to compare the sales of *The Hobbit* in bookshops situated across the Atlantic from one another. A random selection of 50 bookshops in Chicago was made, and similarly of London. The sales of *The Hobbit* are summarized as follows, with x and y referring to Chicago and London, respectively:

$$\sum x = 3{,}991, \quad \sum x^2 = 325{,}269, \quad \sum y = 3{,}745, \quad \sum y^2 = 299{,}803.$$

i) Test, at the 10% significance level, for a difference in the mean sales of *The Hobbit* between bookshops in Chicago and London.
ii) Obtain a 90% confidence interval for the difference in means, and relate this to your answer to part (i).
iii) Explain what further information would be required in order to ascertain whether or not the total sales of *The Hobbit* were actually greater in Chicago than they were in London.

3. Table 7 gives both the arm span and height (in cm) of six randomly-chosen adults from a particular town.

Table 7

Person	John	Mary	Rafi	Alex	Raj	Sundeep
Arm span (cm)	178	167	180	172	172	159
Height (cm)	181	166	182	177	172	165

i) Carry out an appropriate test at the 1% significance level to determine whether the mean height of this town's population is greater than the mean arm span, stating any assumptions you make in order to perform your calculations.
ii) Now suppose that, instead of selecting just one group of individuals in which each have both their arm span and their height measured, two randomly-chosen groups of six people are obtained. The first group have their arm spans recorded and the second their heights. The arm spans in cm of the first group are 177, 181, 159, 168, 173, 176 and the heights are 182, 165, 184, 179, 161, 172. Using this data, carry out an alternative test at the 1% significance level to determine whether the mean height of this town's population is greater than the mean arm span. Once more, you should state any assumptions you make in order to perform your calculations.
iii) State, with justification, which of the two tests carried out in (i) and (ii) might be deemed preferable.

4. During the course of an hour, the number of telephone calls arriving each minute at a busy switchboard were recorded. Table 8 summarizes the data:

Table 8

Number of calls (in one minute)	0	1	2	3	4	≥5
Frequency	26	14	9	7	4	0

Carry out a goodness-of-fit test, at the 5% significance level, to determine whether or not this data fits a Poisson distribution. Note that since a nominal rate per minute has not been given, this will need to be estimated from the data.

5. During a quality control procedure at a bakery, 5 randomly chosen cakes were weighed. You may assume that the masses of the cakes are normally distributed with variance 7.29. A test of the null hypothesis $\mu = 247$ against the alternative hypothesis $\mu \neq 247$ is carried out at the α% significance level. A decision is made to accept the null hypothesis if the sample mean lies somewhere between 245 and 249 g.

i) Calculate the value of α.
ii) Given that the true value of μ is 250 g, find P(Type II error).
iii) What can you say about P(Type II error) when the true value of μ is greater than 250 g?

6. Provide a heuristic argument as to why combining classes in a goodness-of-fit test unnecessarily can lead to an increase in P(Type II error).

References

Anderson-Cook, C. and S. Dorai-Raj. 2003. Making the concepts of power and sample size relevant and accessible to students in introductory statistics courses using applets. Journal of Statistics Education 11: 3.

Batanero, C. 2004. Statistics education as a field for research and practice, 1–15. *In:* M. Niss [ed.] Proceedings of ICME 10, Regular Lecture, Copenhagen, Denmark, Roskilde University.

Crawshaw, J. and J. Chambers. 1994. A Concise Course in A-Level Statistics. Stanley Thornes, Cheltenham, UK.

Eisenhauer, J. 2009. Explanatory power and statistical significance. Teaching Statistics 31: 42–46.

Gallagher, J. 2006. The F test for comparing two normal variances: correct and incorrect calculation of the two-sided p-value? Teaching Statistics 28: 58.

Garfield, J. 2002. The Challenge of Developing Statistical Reasoning. Journal of Statistics Education 10: 3.

Grimmett, G. and D. Stirzaker. 2001. Probability and Random Processes (Third Edition). Oxford University Press, Oxford, UK.

Kaplan, J. 2009. Effect of belief bias on the development of undergraduate students' reasoning about inference. Journal of Statistics Education 17: 1.

Kozack, M. 2009. How to show that sample size matters. Teaching Statistics 31: 52.

Lawton, L. 2009. An exercise for illustrating the logic of hypothesis testing. Journal of Statistics Education 17: 2.

Lee, H., R. Angotti and J. Tarr. 2010. Making comparisons between observed data and expected outcomes: students' informal hypothesis testing with probability simulation tools. Statistics Education Research Journal 9: 68–96.

Lehmann, E. 2009. Fisher, Neyman, and the Creation of Classical Statistics. Springer, New York, USA.

Lesch, S. and D. Jeske. 2009. Some suggestions for teaching about normal approximations to Poisson and Binomial distribution functions. The American Statistician 63: 274–277.

Mulekar, M. and M. Siegel. 2009. How sample size affects a sampling distribution. Mathematics Teacher 103: 34–42.

Schneiter, K. 2008. Two applets for teaching hypothesis testing. Journal of Statistics Education 16: 3.

Sotos, A., S. Vanhoof, W. Van den Noortgate and P. Onghena. 2009. How confident are students in their misconceptions about hypothesis tests? Journal of Statistics Education 17: 2.

Spiegel, M. and L. Stephens. 1999. Theory and Problems of Statistics (Third Edition). McGraw-Hill, New York, USA.

Warner, R. 2008. Applied Statistics: From Bivariate Through Multivariate Techniques. Sage Publications, Los Angeles, USA.

Wiseman, F. 2004. A useful example when teaching hypothesis testing. Teaching Statistics 26: 59–61.

CHAPTER 7

Analysis of Variance

*David L. Trumpower** and *Sait Atas*

1. Introduction to Analysis of Variance—A Two Group Example

Suppose that two scientists develop a baseball bat made of genetically-engineered wood. The scientists think that their bats will hit baseballs further than bats made of normal wood. To test their hypothesis they devise an experiment in which they use a robotic arm to hit baseballs with exactly the same force. In their experiment they use the robotic arm to hit several randomly selected baseballs with baseball bats made with normal wood, and to hit several other randomly selected baseballs with their baseball bats made with genetically engineered wood in random order. Now, suppose that the distances (in meters) of the balls hit with the two different types of baseball bats were as follows:

Table 1. Distances of baseballs by type of bat—hypothetical outcome #1.

Type of bat	
Normal wood	**Genetically engineered wood**
230	420
370	280
300	350

Based on these results, do you think that the type of baseball bat really does have an effect on the distances that the baseballs travel? Before answering, consider each of these additional questions:

1. On an average, how much further did the balls hit by bats made with genetically engineered wood travel than the balls hit by bats made with normal wood?

University of Ottawa (Canada).
*Corresponding author

2. Why do you think the three baseballs hit by the bats made with genetically engineered wood did not all travel the same distance even though they were each hit with the same force using the same type of bat? Likewise, why do you think the three baseballs hit by the bats made with normal wood did not all travel the same distance?
3. Is it possible that the balls hit by bats made with genetically engineered wood went further, on an average, than the balls hit by bats made with normal wood because of such factors as you listed in response to question #2, rather than because of the type of bat with which they were hit? Why or why not?
4. How can we decide if the observed difference between the average distances of the balls hit by the bats made with normal and genetically engineered wood was more likely due to the type of bat or to those other factors that you mentioned in response to question #2?

In answering the first question, you should have noticed that, on an average, the balls hit by bats made with normal wood travelled 300 m, whereas those hit by bats made with genetically engineered wood travelled 350 m. This 50 m difference *may* have been due to the type of bat with which the balls were hit. That is, the difference might have resulted because the bats made with genetically engineered wood are actually better.

However, before making this conclusion, we should also consider that baseballs hit with the same force and the *same* type of baseball bats did not all travel the same distance. Why not? In response to question #2, you may have identified many uncontrolled factors, including fluctuations in wind, humidity, air temperature, elasticity of the balls, etc. These are all plausible suggestions. They all represent random factors that could have affected the distance travelled by baseballs hit with the same type of bat and the same force. We refer to these as *random factors* because they vary randomly across the experiment. For example, a strong gust of wind in the same direction as the flight of the balls could occur at any time during the experiment. It is as likely to occur when a ball is being hit with a bat made with genetically engineered wood as when a ball is being hit with a bat made with normal wood. Note that errors in the experimental procedure (e.g., in measurement of the distances, in the precise functioning of the equipment, etc.) could also be considered random factors.

How large an influence do you think wind speed and other random factors have on the distance that balls travel? Our best estimate can be obtained by considering the distance travelled by balls hit with the same type of bat. The three balls hit with the bats made of normal wood differed only with respect to random factors—they were all hit with the same force and with the same type of bat. Thus, any observed variation within this group of balls must have been due to random factors. The average amount of variation in the distances of those three balls can be quantified by the standard deviation. Likewise, all the balls hit with bats made with genetically engineered wood must have differed only with respect to random factors. Here, too, we can quantify the amount of variation in the distances of this group of three balls using the standard deviation. Overall, the influence of random factors, then, can be estimated by the average standard deviation within these two groups. Looking at the distances of balls in the normal column, we can see that they deviate by 70 m from ball to ball (this can be determined informally by just looking at the distances, or more formally by calculating the standard deviation of these three values). Likewise, the distances of balls in the genetically engineered column deviate by 70 m. Thus, we see that random factors are responsible for about a 70 m difference from ball to ball, on an average.

Recall that we earlier determined that the balls hit with bats made of genetically engineered wood travelled 50 m further, on an average, than the balls hit with bats made of normal wood, and noted that this difference *may* have been due to type of bat. However, question #3 asks if it is possible that the difference was not due to type of bat but was, rather, entirely the result of random factors. The answer is, yes, this too is a possibility. How do we know? Well, suppose for example that a strong tailwind happened to gust every time a ball was being hit with a bat made of genetically engineered wood, but that no such random gusts occurred when the balls were being hit with the bats made with normal wood. Even if the type of bat makes no difference, the balls hit by the bats made of genetically engineered wood would have, therefore, travelled further because of the random gusts of wind. Thus, it should be evident that whereas the variation observed *within* each group of balls must only be due to random factors (it cannot be due to type of bat since all balls in a group were hit with the *same* type of bat), the observed difference *between* groups could be due to type of bat, to random factors, or to a combination of both.

Now, the billion dollar question: How do we determine how much of the difference between groups is likely to be due to type of bat versus other random factors? Given that the variation between groups could be due to both type of bat *and* random factors, and that the within group variation must be due to *only* random factors, we can use the ratio of the former to the latter to determine the influence of the type of bat relative to the influence of random factors:

$$\frac{between\ group\ variation}{within\ group\ variation} = \frac{effect\ of\ random\ factors + effect\ of\ type\ of\ bat}{effect\ of\ random\ factors}$$

Another way of thinking about this ratio is that it indicates proportionally how much more variability there is between group means than can be expected due to random factors alone. If in reality, the type of bat has no effect, then the between group difference is completely due to random factors, and the between group variation and within group variation are both independent estimates of the same thing (i.e., the effect of random factors). In such a situation, the ratio should be close to one (i.e., the size of the observed difference between groups is the same as what we estimate to be due to random factors). Alternatively, if the type of bat does have an effect, then the between group variation should be larger than the within-group variation (i.e., there is a larger difference between groups than can be expected based on random factors alone). All else being equal, the larger the ratio, the stronger the evidence of an effect of the type of bat (and the smaller the ratio, the weaker the evidence). In our example, the between group variation (i.e., the 50 m difference between the means of the two groups) is actually a bit smaller than the within group variation (i.e., the 70 m average difference from ball to ball within either group), so the evidence for an effect of type of bat is weak.

At this point, it should be realized that the magnitude of the observed between-group difference is only meaningful in relation to the magnitude of within-group differences. In and of itself, the absolute magnitude of an observed between-group difference is not interpretable. To fully understand this point, consider a few other outcomes that could have occurred in our example. Suppose that instead of the results described earlier, the distances of the balls had looked like this:

Table 2. Distances of baseballs by type of bat—hypothetical outcome #2.

Type of bat	
Normal wood	**Genetically engineered wood**
290	350
310	340
300	360

Here, the difference between the means of the balls hit with bats made of normal and genetically engineered wood is the same as in our previous example (50 m), but there is much less variation within either group of balls (on an average, balls within either group differ by 10 m). This suggests that the difference between groups of balls is less likely to have been completely due to random factors. Whereas random factors seem to account for 10 m differences from ball to ball, this is five times less than the observed 50 m difference between the means. Even though it is still *possible* that the observed difference between groups could have occurred if type of bat has no real effect, it is *less likely* than in the previous outcome.

Finally, consider this alternative outcome:

Table 3. Distances of baseballs by type of bat—hypothetical outcome #3.

Type of bat	
Normal wood	**Genetically engineered wood**
299	305
301	304
300	306

Here, the observed difference between the means of the two groups of balls is much smaller than in the previous outcomes, but so too is the variation within either group. In this outcome, there is a 5 m difference observed between the means of balls hit by bats made with normal and genetically engineered wood, but random factors seem to account for differences of only 1 m from ball to ball (i.e., the average standard deviation within the two groups of balls is 1 m). So, in the first outcome, there was very little evidence that the observed difference between means was due to the type of bat since it was no larger than what was estimated to be due to random factors. In the two latter outcomes, there was stronger evidence since the observed differences between means were each five times larger than what was estimated to be due to random factors. However, in the last outcome, a smaller observed difference between means generated evidence that was similar in strength as in the second outcome. This was because the estimated effect of random factors was much smaller in the last than in the second outcome (Perhaps the experiment that generated the last outcome was conducted in a climate-controlled dome?).

At this point, we should summarize the informal analyses that we have been conducting and put things in a more general perspective. Also, a few important points should be highlighted before turning to a more formal discussion. Thus far, we have been estimating the variation between means of two groups of data, as well as the variation within those groups of data, and then comparing the between-group variation to the within-group variation. Formally, this procedure is (aptly) referred to as analysis of variance, or ANOVA for short. Applied to our example, ANOVA is based on the logic that the observed difference *between* the two groups of balls is due to the effects of random factors and type of bat (if it actually has an effect), whereas the observed differences *within* groups must be due only to the effects of random factors. The ratio of between-group to within-group variation, therefore, indicates the effect of type of bat relative to the effect of random factors. In formal ANOVA terminology, this ratio is referred to as the *F*-ratio.

2. Points to Ponder

2.1 General Logic of ANOVA

More broadly, ANOVA is the name given to a set of inferential statistical procedures for testing hypotheses about differences in means across a variety of conditions. Whenever we measure a dependent variable in an experiment (distance travelled by the balls, in our example), the scores on the dependent variable are not all the same—they vary. Otherwise, we would not call them dependent *variables*. ANOVA partitions this variability in scores on the dependent variable into the variability observed *within* conditions formed by the independent variable (type of bat, in our example) and the variability observed *between* those conditions. ANOVA then analyzes these two variance estimates based on the logic described earlier that within-group variation must be due to random, uncontrolled factors present in the experiment, whereas between-group variation may be due to the combination of random factors and the effect of the independent variable. ANOVA culminates in calculation of the ratio of the between- to within-group variance estimates, or *F*-ratio. As we have seen in our example, the larger the ratio, the stronger is the evidence for rejecting the null hypothesis that the independent variable has no effect on the dependent variable. This comparison of between to within-group variation can be likened to finding a signal (the effect of an independent variable) in noise (the effect of random factors). The signal must be larger than the noise in order to be detected. If there is very little noise, then a fainter signal may be detected; if there is much noise, then the signal must be stronger in order to be detected.

Although we have now seen that bigger *F*-ratios provide stronger evidence against the null hypothesis, how big is big enough to actually allow us to conclude that an independent variable has a real effect on a dependent variable? Later in the chapter we will show you the equations and corresponding procedures in a computer software package, SPSS, that can be used to calculate the *F*-ratio in order to answer this question in a formal way. But before doing so, we would like to highlight a few more important conceptual points about ANOVA.

2.2 From Samples to Populations

The first point that we must always keep in mind is that in experiments being analyzed with ANOVA, we are observing samples in order to make inferences about populations. In our example, we want to know if there is a difference in how far bats made of genetically engineered wood and bats made of normal wood, *in general*, can hit baseballs. But, it would be impossible for us to hit balls with every bat made of genetically engineered and normal wood. So instead, we hit some balls with a *sample* of bats made of genetically engineered wood and normal wood to make an informed conclusion. And anytime we measure anything about two or more samples, it is extremely rare that the means of the samples are identical. So, to find out that the mean distance of our sample of balls hit by bats made of genetically engineered wood is larger than the mean distance of our sample of balls hit with bats made of normal wood is not too surprising or informative. If the objective of our experiment was to determine if there was a difference in how far these three particular bats made of genetically engineered wood and these three particular bats made of normal wood hit these particular baseballs on this particular day, then we could simply calculate the sample means of the two groups of balls and conclude that, yes, there is a difference without the need for ANOVA. However, what we really want to know is something about the reliability of the results—if I use a bat made of genetically engineered wood in the future, how convinced am I that it will hit the ball farther than if I use a normal wood bat? This is the type of question for which ANOVA, and the *F*-ratio in particular, provides evidence. If the evidence shows that random effects alone could likely produce differences as extreme, or more so, than observed in the experiment, then I am not very convinced that a bat made of genetically engineered wood will reliably hit balls farther than normal wood bats in the future. In other words, I conclude that the observed differences very well may have been due to random factors and not due to type of bat and I am, therefore, not convinced that the type of bat matters. On the other hand, if the evidence shows that random effects alone would be very *unlikely* to produce differences as extreme, or more so, than observed in the experiment, then I am more confident that a bat made of genetically engineered wood will hit balls farther than normal wood bats in the future; that is, I conclude that the type of bat *does* make a difference. The likelihood that random effects alone, without any influence of the independent variable, could generate results as extreme as, or more so than, observed in an experiment, is referred to as the *probability value* (or, *p*-value) and is determined by three things—the magnitude of the observed difference between sample means, the amount of variation within samples and the size of the samples—which we discuss below.

2.3 Sample Size and Representativeness

Larger samples are more representative of the populations from which they are drawn than are smaller samples. This point is intuitively obvious to most people. If you want to know how good a movie is, you would probably be more confident after reading the reviews of many people that had seen the movie, rather than after reading the review of just one or two people. With respect to ANOVA, results obtained from larger samples provide stronger evidence than those obtained from smaller samples, all else being equal. To understand this point more fully, reconsider the third hypothetical outcome in our example, in which three balls hit with bats made of genetically engineered wood went, on average, 5 m farther than three balls hit with bats made of normal wood. The results in this example indicated that random factors were responsible for only a 1 m difference from ball to ball, on an average. Because the difference between groups is larger than what we might expect based on random factors alone, we can say that such an outcome would be unlikely if the type of bat has no effect. However, we should acknowledge that it *is* possible. Perhaps several random events all happened to occur together (e.g., a gust of wind, drop in air pressure, etc.) when each of the three balls were being hit with bats made of genetically engineered wood, but not when the three balls were hit with the bats made of normal wood. Now, imagine the same outcome (a 5 m difference between the means of the two groups coupled with an average 1 m difference from ball to ball within each group) being obtained in an experiment in which three hundred balls were hit with each type of bat. If type of bat has no effect, how likely would it be that several random events happened to occur together when each of the *three hundred* balls were being hit with bats made of genetically engineered wood, but not when any of the three hundred balls were hit with bats made of normal wood? It should be clear that it is much less likely for such random events to occur when three hundred particular balls are being hit than when only three particular balls are being hit. Thus, if such an outcome occurred with a much larger sample, it would provide even stronger evidence against the null hypothesis and, conversely, increase our confidence that type of bat really does have an effect.

2.4 Absolute Magnitude of Between-group Differences versus Ratio of Between- to Within-group Variation

As we indicated earlier, the difference between group means in an experiment is not very informative in and of itself. Also, we have been stressing throughout the chapter that it is the ratio of between-group differences relative to within-group variation, rather than just the absolute magnitude of between-group differences, that provides evidence about the effect of the independent variable on the dependent variable. It is extremely important that you understand why this is so. Research has shown that individuals often fail to consider the amount of within and between-group variation together in a relative sense (Obrecht et al. 2007, Trumpower and Fellus 2008, Trumpower 2013). Rather, we tend to think of bigger differences between groups as better. We also tend to think of within- and between-group variation as being independent. But they are not, since both can arise from random factors. Thus, using words like "large" and "small" to describe the absolute magnitude of a between-group difference is somewhat meaningless when we are trying to determine evidence for an effect of an independent variable. As we saw in the three hypothetical outcomes discussed earlier, a mean difference that is "big" in an absolute sense may or may not provide strong evidence for a real effect, depending on how much random factors influence the dependent variable. Also, a mean difference that is "small" in an absolute sense may provide stronger evidence for a real effect than a larger mean difference, depending on how much random variation exists. It all boils down to how likely the difference observed in the sample means would be if there is really no effect. That likelihood depends on the magnitude of the difference between group means *relative to* the amount of random variation (i.e., the ratio of between- to within-group variation). If the independent variable has no effect, then a smaller ratio is more likely to be obtained than a larger ratio. Thus, larger ratios indicate stronger evidence against the null hypothesis, whereas smaller ratios indicate weaker evidence.

2.5 Practical Importance versus Statistical Significance

When students describe the observed difference between group means in an experiment as "large", what they usually mean is that this large would be very meaningful *if it were real*. For example, if I had a new bat made of genetically engineered wood that could hit baseballs 50 m farther than normal bats, then that would be practically important to me as a baseball player. However, the problem is that we cannot be confident that bats made of genetically engineered wood really can reliably hit balls 50 m farther than normal bats if there were other random factors present in the experiment that also cause some balls to travel at least 50 m farther or shorter than other balls. So, to say that an observed 50 m difference between sample means is "large" presupposes that this difference was really caused by the type of bat and not by other random factors. Whereas our subjective impression of the practical importance of the effect of type of bat may be based on the size of the difference between means (bigger is better), our confidence that type of bat really does have an effect is based on the ratio of between- to within-group variation as we've been discussing. When the ratio is large enough, we conclude that the results would not be likely if type of bat does not make a difference and, consequently, we say that the results are reliable —we are confident that if we again hit balls with bats made of genetically engineered wood and normal wood as in the experiment, the group of balls hit by genetically engineered wood bats would go farther. Technically, we say that there is a *statistically significant* difference in the distance of balls hit by bats made of genetically engineered and normal wood. Failure to realize this distinction between practical importance and statistical significance may lead one to believe incorrectly that large absolute differences in means are indicative of a real effect. But, as we saw in the preceding section, this is not true.

Just as a large absolute difference between means that would seemingly be of practical importance if it were real may not be statistically significant, the converse may also be true. That is, a statistically significant difference may not be large enough to be considered practically important. Statistical significance is determined by the difference between group means relative to the differences in scores within groups, as well as the size of the samples. All else equal, larger between-group differences (i.e., larger effect sizes), smaller within-group differences (i.e., less random error) and larger samples (i.e., more observations per group), all increase the likelihood of a statistically significant result. Thus, a very small absolute difference between group means that would not be considered practically important could, nonetheless, be statistically significant if the random error is also very small and/or the sample sizes are really big. For example, suppose that 3000 baseballs are hit with bats made of genetically engineered wood and that the distance they travel differs by *less than 1 m* from ball to ball. Also, another 3000 balls are hit with bats made of normal wood and they, too, differ by less than 1 m from ball to ball. But, the average distance travelled by the balls hit with the genetically-engineered wood bats is *exactly 1 m* farther than the average distance of the balls hit with normal wood bats. The 1 m difference between the group means is larger than the amount of variation observed within either group and these observations are based on quite large samples of balls. In this example, the difference will be statistically significant. However, suppose the bats made of genetically-engineered wood cost $100 more than normal wood bats. It is unlikely that many recreational baseball players would consider a 1 m difference practically important, especially if they have to spend $100 to achieve it! So, whereas we can be confident that bats made with genetically engineered wood will hit balls farther than bats made of normal wood in this example, the difference is not large enough to justify the cost—it is not practically important. As you can see, practical importance is determined more subjectively, and is based in part on the absolute magnitude of the difference between group means, whereas statistical significance is determined more objectively and is based on the magnitude of the difference between group means relative to the amount of random variation. Therefore, statistical significance and practical importance are not always consistent with one another. Both should be considered when interpreting the outcome of experiments.

Self-Check #1. Before reading further, make sure that you can explain, in your own words, why we cannot automatically assume that a difference between sample group means that appears large in an absolute sense is indicative of a real effect of the independent variable. Suppose that a running magazine reports that it tested the cushioning of three different brands of new running shoes. The magazine obtained eight pairs of each brand of shoes and sent one pair to 24 different runners. The runners wore their new pair of shoes for a long run and then rated the level of cushioning on a 10-point scale. The results showed that the mean rating for brand X, brand Y, and brand Z was 5.50, 6.00, and 9.25, respectively. Based on these means, the report concludes that you should buy brand Z if you want the most cushioned shoes. Explain why the magazine should conduct a formal statistical test, such as ANOVA, before making such a conclusion. Try to use language that even a non-statistician could understand, and be sure to consider: a) samples vs. populations, b) between-group variability vs. the ratio of between- to within-group variability, c) and statistical significance vs. practical importance, and in your explanation.

3. Determining the *F*-ratio

We will now present and examine the formal equations for determining the obtained *F*-ratio in an ANOVA. In reality, you should never have to actually use these equations. Computer software packages, such as SPSS, can compute *F*-ratios and determine the associated probability values for you. We will take a look at how SPSS can be used to do so later in the chapter. However, we believe that recognition of the basic components of the equations, as well as seeing how they apply to a small example, can help in understanding the logic of ANOVA. As a result, after pointing out the components of the equations, we will apply them to the first hypothetical outcome in our baseball bat example.

3.1 Equations for Determining the F-ratio

Below, five versions of the *F*-ratio are presented, in order of increasing generality.

$$F = \frac{variability\ due\ to\ independent\ variable\ and/or\ random\ factors}{variability\ due\ to\ random\ factors}$$

$$F = \frac{variability\ between\ groups}{variability\ within\ groups}$$

$$F = \frac{MS_B}{MS_W}$$

$$F = \frac{SS_B / df_B}{SS_W / df_W}$$

$$F = \frac{\sum n_k(\bar{x}_k - \bar{\bar{x}})^2 / (a-1)}{\sum (n_k - 1)\, s_k^2 / (N-a)}$$

where

$$k = specific\ group\ (i.e., specific\ level\ of\ the\ independent\ variable)$$

$$n_k = number\ of\ observations\ or\ scores\ in\ the\ kth\ group$$

$$N = total\ number\ of\ scores\ across\ all\ groups$$

$$\bar{x}_k = sample\ mean\ of\ the\ kth\ group$$

$$\bar{\bar{x}} = overall\ mean\ (i.e., mean\ of\ the\ group\ means)$$

$$a = total\ number\ of\ groups$$

$$s_k = sample\ standard\ deviation\ of\ the\ kth\ group$$

$$df_B = between\ groups\ degrees\ of\ freedom\ = a - 1$$

$$df_W = within\ groups\ degrees\ of\ freedom\ = N - a$$

$$SS_B = between\ groups\ sum\ of\ squares = \sum_k n_k(\bar{x}_k - \bar{\bar{x}})^2$$

$$SS_W = within\ groups\ sum\ of\ squares = \sum_k (n_k - 1)s_k^2$$

$$MS_B = between\ groups\ mean\ square = {}^{SS_b}/_{df_b}$$

$$MS_W = within\ groups\ mean\ square = {}^{SS_w}/_{df_w}$$

The first two versions of the equation are intended to display the logic of ANOVA that we've been discussing throughout this chapter, namely that ANOVA is an analysis of the variability that exists between groups relative to the variability that exists within groups (equation 1) and that the variability that exists between groups could be due both to the independent variable and random factors whereas the variability that exists within groups must be due to random factors alone (equation 2). The third version of the equation shows some of the formal terminology used in ANOVA to describe our estimates of the variability between groups (between-groups Mean Square or MS_B) and of the variability within groups (within-groups Mean Square or MS_W) that exist in the population (equation 3). The fourth version of the equation provides some additional ANOVA terminology. More specifically, it indicates that the mean squares, *MS*, are determined by dividing sums of squares, *SS*, by degrees of freedom, *df* (equation 4). If you are not familiar with these terms, don't worry—the fifth version of the equation shows how sums of squares and degrees of freedom are determined (equation 5). Briefly, the between groups sum of squares, SS_B, refers to the sum of the squared differences of the group means and the grand mean. The between groups degrees of freedom is one less than the number of groups. By dividing the *SS* by the *df*, we get an estimate of how much the group means differ from the grand mean, on an average, across all groups (i.e., the between groups mean square, MS_B). Similarly, the within groups sum of squares, SS_W, refers to the sum of the squared differences of the scores within groups and the mean of the group to which the scores belong. Here, by dividing the *SS* by the *df*, we get an estimate of how much the individual scores within groups differ from their associated group means, on average (i.e., the within groups mean square, MS_W). Although it is not essential for you to understand exactly how these terms are calculated, you should realize that all the concepts we've been discussing thus far are present in this final version of the equation: differences between group means, variability within groups, and sample size.

Now, let's apply the equations to the data from the first hypothetical outcome in our example. Below we have redisplayed the data along with some descriptive statistics (means and standard deviations) that we will use in our calculations.

Table 4. Hypothetical outcome #1 with descriptive statistics.

	Type of bat		
	Normal wood	**Genetically engineered wood**	
	230	420	
	370	280	
	300	350	
Mean:	300	350	Overall Mean=325
Standard deviation:	70	70	

Now, as we calculate each component of the *F*-ratio, we will provide a description of what, conceptually, is being determined. First, we will calculate the numerator, the between-groups mean square:

$$MS_B = {SS_B}/{df_B}$$

$$= \sum_k n_k(\bar{x}_k - \bar{\bar{x}})^2 / (a-1)$$

$$= n_{Normal}(\bar{x}_{Normal} - \bar{\bar{x}})^2 + n_{Genetically\ Engineered}\left(\bar{x}_{Genetically\ Engineered} - \bar{\bar{x}}\right)^2 / (a-1)$$

$$= 3(300-325)^2 + 3(350-325)^2 / (2-1)$$

$$= 3750/1$$

$$= 3750$$

Notice that this provides, essentially, the average amount that the group means differ (in squared units) from the overall mean. It is not exactly the average amount, because we have divided by *a*-1 rather than *a*. In reality, this is only an *estimate* of how much the *population* group means differ based on the observed sample means, so we divide by the number of groups minus one to obtain an unbiased estimate. Don't worry too much about following this logic—the main thing to realize is that if the sample group means are identical, then the between-group mean square, MS_B, will be zero. But, if there is any difference between group means, then the larger the difference, the larger the MS_B will be. And, as mentioned earlier, any observed differences between groups could be due to the effect of the independent variable, if it exists, but also could be due to random factors. Thus, the numerator of the *F*-ratio provides an indication of the combined effects of the independent variable (i.e., type of bat) and any random factors present in the experiment.

By contrast, the denominator of the *F*-ratio, the within-groups mean square, provides an indication of the effect of only the random factors and is calculated as follows:

$$MS_W = SS_W / dfw$$

$$= \sum_k (n_k - 1)s_k^2 \Big/ (N - a)$$

$$= (n_{Normal} - 1)s_{Normal}^2 + (n_{Genetically\ Engineered} - 1)s_{Genetically\ Engineered}^2 \Big/ (N - a)$$

$$= (3 - 1)70^2 + (3 - 1)70^2 \Big/ (6 - 2)$$

$$= 19600 / 4$$

$$= 4900$$

Here, we have essentially calculated the average of the group standard deviations, which indicates the amount (in squared units) that scores within groups differ. Again, this is not exactly the case, but the important thing to realize is that the more scores within groups differ from one another (i.e., the larger the standard deviations within the groups), the larger the within-groups mean square, MS_W, will be. If the group standard deviations, s_k, are zero, then the MS_W will be zero. But, the larger the standard deviations, the larger will be the MS_W.

At this point, we have an estimate of the magnitude of the combined effect of the independent variable and random factors (MS_B) and an estimate of the effect of random factors alone (MS_W). Together, the ratio of the former to the latter indicates how much greater the combined effect is than the effect of random factors alone. If the independent variable has no real effect, then the combined effect and the effect of random factors alone should be the same. Alternatively, if the independent variable does have a real effect, then the combined effect will be larger than the effect of random factors alone. Thus, for a given study, a larger *F*-ratio would indicate stronger evidence against the null hypothesis than a smaller *F*-ratio would. In the present example, the MS_B is actually a bit *smaller* than the MS_W, such that the ratio of between- to within-group variation is less than one, indicating that the difference between groups is not larger (and, in fact, smaller) than we might expect due to random factors alone:

$$F = \frac{MS_B}{MS_W} = \frac{3750}{4900} = 0.765$$

Before going further, notice the similarity between our informal analysis conducted at the beginning of this chapter and the formal analysis conducted here. In our informal analysis, we noted that, on an average, balls hit by bats made of genetically engineered wood traveled 50 m farther than balls hit by bats made of normal wood, but that balls hit with the *same* type of bat varied in distance by 70 m on average. Thus, the difference between groups was less than the variation within groups. This informal analysis simply looked at the between-group and within-group variation present in the *sample* data. In our formal analysis, we estimated the amount of between-group and within-group variation in the *population* based on what we observed in the samples. Our formal calculations revealed, again, that the variation between groups (MS_B=3750) is less than the variation within groups (MS_W=4900). We therefore came to the same conclusion that there is not sufficient evidence to say that the type of bat has an effect on distance traveled by balls—random factors alone have a large enough effect to produce a difference between the mean distances traveled by two samples of balls that is as large or larger than the difference observed in this experiment. That is, assuming that type of bat has no influence on how far balls travel, a difference between groups as large as, or larger than, observed in this experiment would still be fairly likely.

In this example, the conclusion may have seemed obvious, even without a formal analysis. But what if the obtained evidence had been stronger, leading to a larger ratio of between-group to within-group variation? All else equal, larger F-ratios provide stronger evidence against the null hypothesis of no effect. But how large is large enough to conclude that the independent variable does have a reliable effect on the dependent variable? If the data obtained in this experiment had shown that the between group variability was two times larger than the within group variability (i.e., an F=2), would that have been large enough? What if it had been 10 times larger (i.e., F=10)? The answer depends on how likely it would be to conduct a study like you did and obtain a similarly large, or larger, F-ratio *if in reality* there were no effect of the independent variable on the dependent variable. This likelihood, or *probability value*, in turn, depends on both the size of the F-ratio and the size of your study (recall that larger samples provide better estimates of populations). In the old days, researchers would use tables of critical values to determine if the obtained F-ratio could be considered statistically significant given the size of the experiment as indicated by df_B and df_W. Today, however, statistical software packages such as SPSS will automatically determine the probability value associated with the obtained F-ratio. In general, the probability value is the likelihood of results as or more extreme than those obtained in your experiment, assuming that the null hypothesis is true. As applied to ANOVA, then, the probability value indicates the likelihood of obtaining data in a replication of your experiment (i.e., with the same number of observations and the same levels of the independent variable) that would result in an *F*-ratio as large or larger than you actually obtained in your experiment, assuming that the independent variable has no real effect on the dependent variable. In order to determine the probability value of our present example, let's now conduct our ANOVA via SPSS.

3.2 Determining the F-ratio via SPSS

First, we will need to enter the data into the SPSS data editor (see Fig. 1). We will need one column for our independent variable (type of bat) and one column for our dependent variable (distance the balls were hit). The *type of bat* variable is coded as "1" for balls hit by bats made of normal wood and "2" for balls hit by bats made of genetically engineered wood. Because this variable is nominal, the values "1" and "2" are purely arbitrary, intended only to distinguish the two groups. The *distance* variable indicates how far each ball was hit in meters. Now, to get SPSS to perform the ANOVA, we click on "Analyze", followed by "Compare Means", and then "One-Way ANOVA..." (see Fig. 2). We then move our dependent variable named *distance* into the box labelled "Dependent List:" and move our independent variable named *type of bat* into the box labelled "Factor:". We can also click on "Options..." and the checkbox labelled "Descriptives" if we want SPSS to display the means, standard deviations, and other descriptive statistics for each group. Finally, we click "OK" (see Fig. 3).

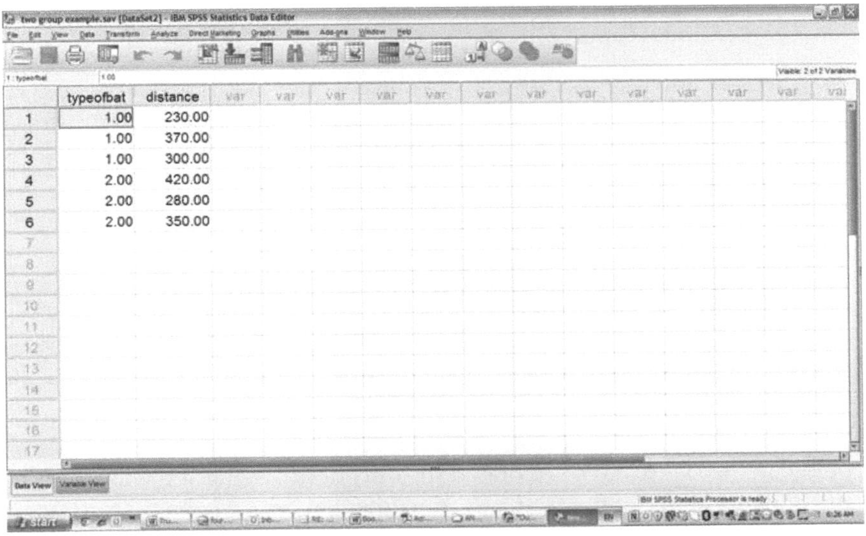

Figure 1. Setting up data for a one-way, between-subjects ANOVA in SPSS for two-group example.

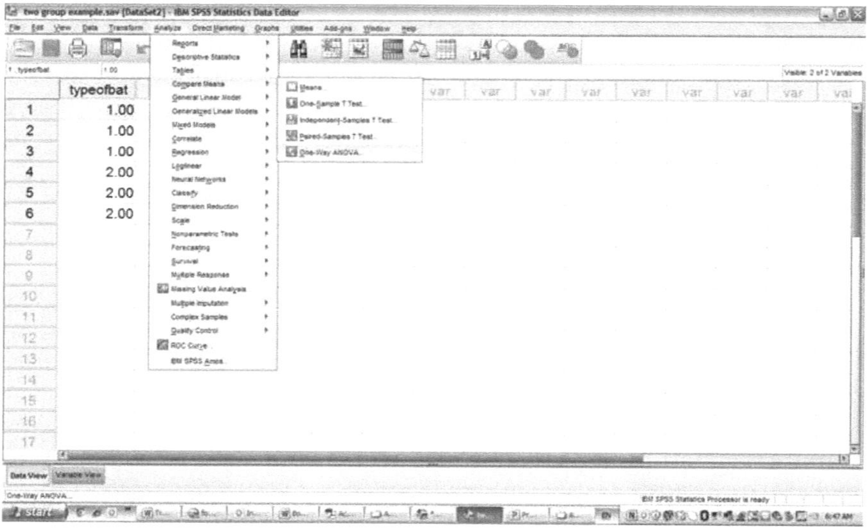

Figure 2. Choosing the one-way, between-subjects ANOVA procedure in SPSS for two-group example.

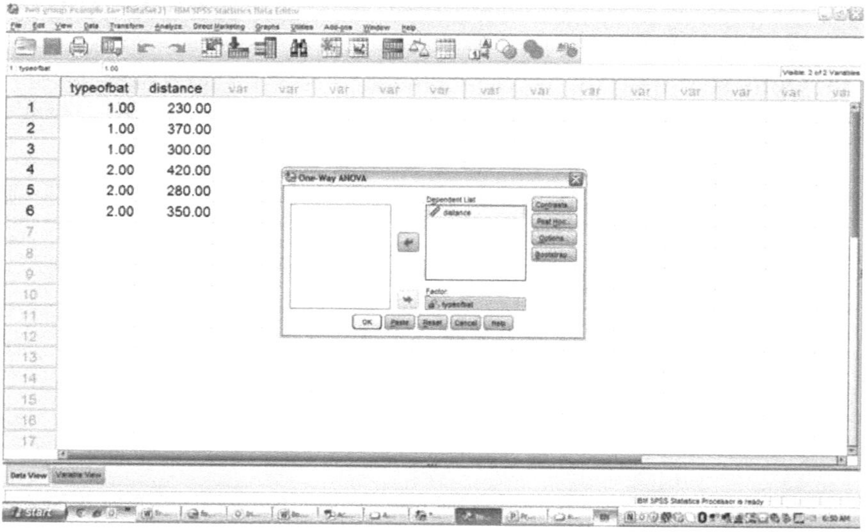

Figure 3. Specifying the independent and dependent variables in SPSS for two-group example.

The output generated by SPSS in response to these steps is shown below in Fig. 4. We have annotated the output to highlight the location and meaning of many of the ANOVA concepts that we've been discussing.

As you can see, SPSS has determined that the probability-value associated with the obtained *F*-ratio of .765 in our example (in which there

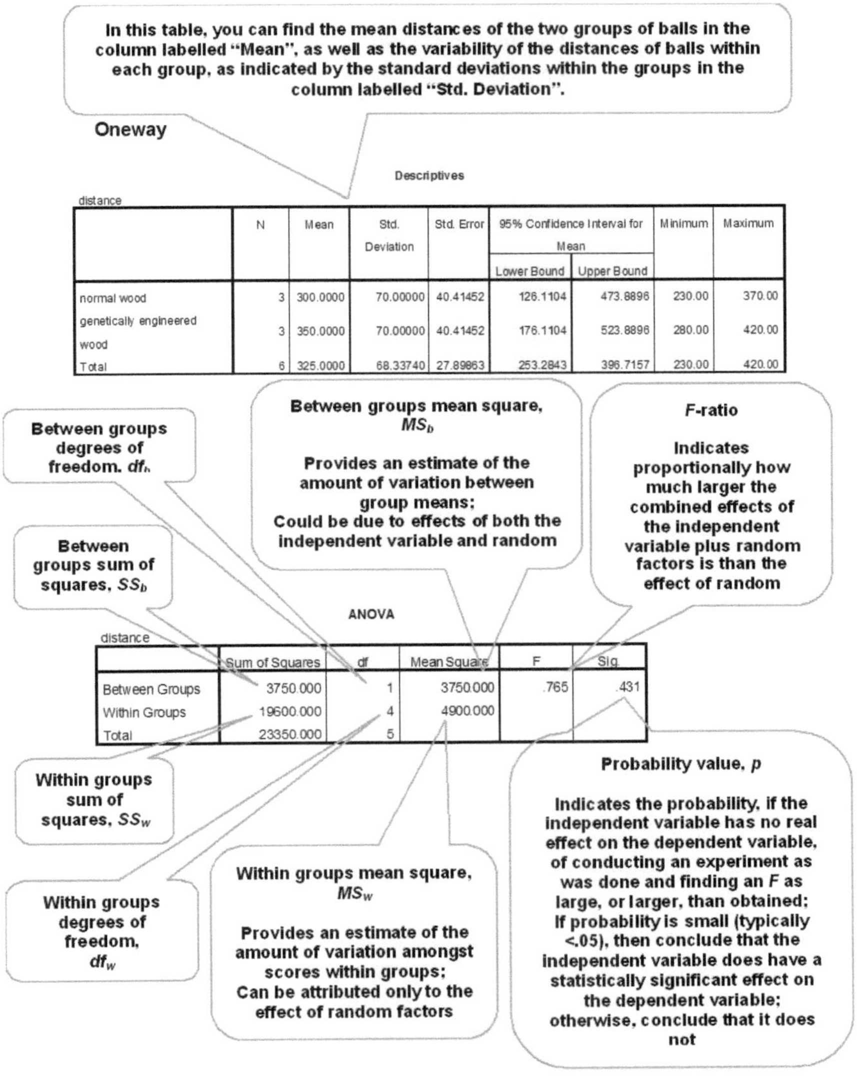

Descriptives

distance

	N	Mean	Std. Deviation	Std. Error	95% Confidence Interval for Mean		Minimum	Maximum
					Lower Bound	Upper Bound		
normal wood	3	300.0000	70.00000	40.41452	126.1104	473.8896	230.00	370.00
genetically engineered wood	3	350.0000	70.00000	40.41452	176.1104	523.8896	280.00	420.00
Total	6	325.0000	68.33740	27.89863	253.2843	396.7157	230.00	420.00

ANOVA

distance

	Sum of Squares	df	Mean Square	F	Sig.
Between Groups	3750.000	1	3750.000	.765	.431
Within Groups	19600.000	4	4900.000		
Total	23350.000	5			

Figure 4. One-way, between-subjects ANOVA output generated by SPSS for two-group example.

were 6 observations and 2 groups) is .431. This means that if in reality the type of bat has no effect on distance and we were to replicate our experiment, the probability that we would obtain data that leads to an *F*-ratio ≥ 0.765 is .431. Or, in other words, if we were feeling ambitious and were to replicate our experiment many times, we would expect group differences that result in an *F*-ratio as big or bigger than what we found in this experiment 431 out of 1000 times even if the only reason for differences between sample group means is random factors alone (i.e., if the null hypothesis is true)! To expect results as or more extreme than what we actually obtained, even if there is no effect of the independent variable, with such a high probability suggests that our evidence against the null hypothesis is not very strong. The observed difference between groups could have easily occurred, even if the independent variable has no effect. Thus, we must fail to reject the null hypothesis and conclude that the evidence is not strong enough to support the alternative—alas, we cannot say that type of bat affects the distance that balls travel when hit by them.

Self-Check #2. Calculate the F-ratio (by hand and/or computer) and p-value (by computer) for the second and third hypothetical outcomes of our example. But, before doing so, estimate whether the resultant between-group mean squares, within-group mean squares, and F-ratios will be larger, smaller, or the same as outcome #1.

4. A Four-group Example

Thus far we have been considering a relatively simple, contrived example involving just two groups. This was done so that you could easily estimate between- and within-group differences and get a feel of the logic of ANOVA. However, ANOVA is a more general procedure that allows comparison of means from two or more groups or conditions. Let's now look at a more realistic example involving four groups. Suppose that the scientists who developed the genetically engineered wooden bats had decided to compare them not only to normal wooden bats, but also to aluminum bats and composite graphite bats. Therefore, they included four groups of balls in their experiment (those hit with bats made of genetically engineered wood, normal wood, aluminum, and graphite). Data from the experiment, including the means and standard deviations for the four groups, are shown below. As can be seen, sixteen balls were hit with each type of bat. ANOVA applied to this example addresses the null hypothesis that type of bat has no effect on how far balls hit by the bats travel. But note that "type of bat" has now been expanded to include aluminum and composite bats, whereas in our earlier examples it referred only to normal wood and genetically engineered wood bats. When there are more than two groups being compared, the null hypothesis can be rejected if *any* of the groups is found to be significantly different from any of the other groups (or, if any *combination* of groups is different from any other combination). However, the same basic logic of ANOVA still applies: if the differences *between* any of the groups is no larger than we would expect based on the amount of variability observed *within* the groups (i.e., if the ratio is small enough), then we must concede that there is not strong enough evidence to conclude that type of bat has an effect on distance. With that in mind, let's now analyze the variance in this four-group example.

Table 5. Distance of baseballs by type of bat—four group example.

	Type of bat			
	Normal wood	**Genetically engineered wood**	**Aluminum**	**Graphite**
	337	352	354	373
	344	350	360	369
	366	359	352	366
	349	349	351	374
	350	351	364	364
	330	348	354	380
	337	359	346	395
	350	362	344	398
	357	346	370	351
	355	335	372	347
	360	369	374	352
	319	379	334	356
	321	335	383	358
	326	347	338	384
	328	361	376	389
	353	374	342	383
Mean:	342.62	354.75	357.12	371.19
Standard deviation:	14.62	12.46	14.75	16.01

First, how much do the four groups differ *between* one another? By looking at the group means, we see that the four samples do vary. The mean for the sample of balls hit with normal wood bats is smallest at 342.62 m, whereas the mean for the sample of balls hit with composite bats is largest at 371.19 m. The means of the samples of balls hit with the bats made of genetically engineered wood and aluminum are 354.75 m and 357.12 m, respectively. Thus, the groups differ by as much as about 30 m and as little as 2 m. More formally, we can calculate the MS_B as we did in the two-group example, which provides something akin to the average amount that group means differ from the overall mean in squared units. SPSS analysis of the four-group example data reveals this value to be **2195.557** (see Fig. 5). Be sure to recall that this between-group variation could be due to the effect of type of bat (if it really has an effect), but could also be due to the effect of random factors.

Next, how much do individual balls hit with the same type of bat (i.e., *within* the same group) vary? To begin, look at the sixteen balls hit with normal wood bats. You will notice from the reported standard deviation

Oneway

Descriptives

distance

	N	Mean	Std. Deviation	Std. Error	95% Confidence Interval for Mean		Minimum	Maximum
					Lower Bound	Upper Bound		
normal wood	16	342.6250	14.61905	3.65476	334.8351	350.4149	319.00	366.00
genetically engineered wood	16	354.7500	12.46060	3.11515	348.1102	361.3898	335.00	379.00
aluminum	16	357.1250	14.74619	3.68655	349.2673	364.9827	334.00	383.00
graphite composite	16	371.1875	16.00924	4.00231	362.6568	379.7182	347.00	398.00
Total	64	356.4219	17.46998	2.18375	352.0580	360.7857	319.00	398.00

Between groups mean square, MS_B

F-ratio

ANOVA

distance

	Sum of Squares	df	Mean Square	F	Sig.
Between Groups	6586.672	3	2195.557	10.421	.000
Within Groups	12640.938	60	210.682		
Total	19227.609	63			

Probability value, p

Within groups mean square, MS_W

Figure 5. One-way, between-subjects ANOVA output generated by SPSS for four-group example.

that they differ by 14.62 m. Likewise, the three balls hit with genetically engineered wood bats differ by 12.46 m, the three balls hit with aluminum bats differ by 14.75 m, and the three balls hit with composite bats differ by 16.01 m. Recall that within-group variation is due only to the effect of random factors. On an average, then, it appears that random factors account for approximately 14 m differences from ball to ball, on an average. Formally, this is calculated by MS_W, which is essentially the average of the variances (or squared standard deviations) of the four groups. SPSS has determined this value to be **210.682** in our four-group example.

Now that we have estimates of the between- and within-group variation, we can see proportionally how much more groups differ from one another than we might expect based on the random variation found within groups. This is indicated by the *F*-ratio, which SPSS has determined is **10.421**. Thus, the variation between groups is much larger than the variation within groups. If in reality the null hypothesis that type of bat has no effect on distance is true, then it would seem very unlikely that random factors (which we determined account for only about 14 m differences in distance from ball to ball) would generate differences between groups that are as large or larger than what we observed in this experiment (which were as large as 30 m). More formally, this probability (i.e., unlikelihood) is shown by SPSS to be **.000**. In actuality, the probability is not exactly 0. SPSS rounds to three significant digits, such that a probability value reported as .000 indicates that the true probability value is simply less than .001. Because this resultant probability value is so small (less than the typical criteria of .05), we realize that it would be very *unlikely* to get experimental results like these if there is no effect of type of bat. Therefore, we conclude that type of bat *does* have an effect—the four different types of bat do not all hit balls the same distance.

However, does this conclusion indicate that all four types of bat are different? No, not necessarily. Rejection of the null hypothesis that all population means are identical is not the same as saying that all population means are different from one another. Although this is certainly one possibility, it is not the only alternative to the null hypothesis when more than two groups are involved. In our four group example, another possibility is that bats made of normal wood, genetically engineered wood, and aluminum hit balls the same average distance, but they all hit balls significantly shorter than bats made of graphite. Another possibility is that normal wood and genetically engineered wood bats hit balls the same average distance and that aluminum and graphite bats hit balls the same average distance, but the two types of wooden bats hit balls significantly shorter than the two types of non-wood bats. Yet another possibility is that genetically engineered wood bats and aluminum bats hit balls the same average distance, but both hit balls significantly farther than normal wood bats and significantly shorter than composite graphite bats. These are just a few of the many situations that can lead to rejection of the null hypothesis (see Table 6). The point is that the test that we've conducted is a general one that shows us that the group means are not all identical. Such a test is called an *omnibus*, or overall, test. But, to determine *how* the groups differ, we need to conduct more specific comparisons.

Table 6. Some possible alternatives if the omnibus null hypothesis, $H_o: \mu_{normal\ wood} = \mu_{genetically\ engineered\ wood} = \mu_{aluminum} = \mu_{graphite}$, is rejected.

$\mu_{normal\ wood} < \mu_{genetically\ engineered\ wood} < \mu_{aluminum} < \mu_{graphite}$
$\mu_{normal\ wood} = \mu_{genetically\ engineered\ wood} = \mu_{aluminum} < \mu_{graphite}$
$\mu_{normal\ wood} = \mu_{genetically\ engineered\ wood} < \mu_{aluminum} = \mu_{graphite}$
$\mu_{normal\ wood} < \mu_{genetically\ engineered\ wood} = \mu_{aluminum} < \mu_{graphite}$

5. Multiple Comparisons

To determine more precisely how groups differ when more than two groups are included in an experiment, we must conduct tests of more specific hypotheses about differences between pairs of group means or between combinations of group means. Such tests are referred to as *comparisons* or *contrasts*. For example, in the four-group baseball bat experiment, one might test the specific null hypothesis that the mean distances travelled by balls hit with bats made of genetically engineered wood and aluminum are the same (i.e., $H_0: \mu_{genetically\ engineered} = \mu_{aluminum}$). Such a comparison is referred to as a *pair wise* comparison because it compares only two of the groups in the experiment. Or, one might test the specific null hypothesis that the average distance travelled by balls hit with bats made of normal and genetically engineered wood is the same as the average distance travelled by balls hit with bats made of aluminum and graphite (i.e., a comparison of wood versus metal bats, or

$$H_0: \left(\mu_{normal\ wood} + \mu_{genetically\ engineered\ wood}\right)/2 = \left(\mu_{aluminum} + \mu_{graphite}\right)/2).$$

This type of comparison is referred to as a *complex* comparison because it involves a more complex combination or combinations of means so that more than two of the groups in the experiment are included.

It should now be realized that a comparison is simply a test of a difference between two groups or between two combinations of groups. In fact, any comparison can be defined by the set of coefficients that specify which group means, or combinations of means, are to be compared. The coefficients can be determined by considering the specific null hypothesis to be tested. For example, the null hypothesis that tests the pairwise comparison of the bats made with genetically engineered wood versus bats made with aluminum can be written as:

$$H_0: \mu_{genetically\ engineered} = \mu_{aluminum}$$

However, we can rewrite this same null hypothesis as a linear combination of group means by rearranging the terms in the equality with a bit of algebra as such:

$$\mu_{genetically\ engineered} - \mu_{aluminum} = 0$$

$$1\mu_{genetically\ engineered} - 1\mu_{aluminum} = 0$$

$$(1)\mu_{genetically\ engineered} + (-1)\mu_{aluminum} = 0$$

$$(0)\mu_{normal\ wood} + (1)\mu_{genetically\ engineered} + (-1)\mu_{aluminum} + (0)\mu_{graphite} = 0$$

In this last version of the hypothesis, we see that the coefficients of 0, 1, –1, 0 applied to the normal wood, genetically engineered wood, aluminum and graphite bat means, respectively, define this pairwise comparison. In general, the groups to be compared receive equal and opposite coefficients and all groups not involved in the comparison receive coefficients of zero.

Similarly, the set of coefficients that can be used to test the complex comparison of the average of the normal and genetically engineered wood bats versus the average of the aluminum and graphite bats is determined as follows by rearranging terms in the corresponding null hypothesis:

$$H_0: \left(\mu_{normal\ wood} + \mu_{genetically\ engineered\ wood}\right)/_2 = \left(\mu_{aluminum} + \mu_{graphite}\right)/_2$$

$$\tfrac{1}{2}\mu_{normal\ wood} + \tfrac{1}{2}\mu_{genetically\ engineered\ wood} = \tfrac{1}{2}\mu_{aluminum} + \tfrac{1}{2}\mu_{graphite}$$

$$\tfrac{1}{2}\mu_{normal\ wood} + \tfrac{1}{2}\mu_{genetically\ engineered\ wood} - \tfrac{1}{2}\mu_{aluminum} - \tfrac{1}{2}\mu_{graphite} = 0$$

$$\tfrac{1}{2}\mu_{normal\ wood} + \tfrac{1}{2}\mu_{genetically\ engineered\ wood} - \tfrac{1}{2}\mu_{aluminum} - \tfrac{1}{2}\mu_{graphite} = 0$$

Here, we can see that the coefficients of ½, ½, -½, and - ½ applied to the normal wood, genetically engineered wood, aluminum, and graphite bat means, respectively, define this complex comparison.

Self-Check #3. Using this same procedure, determine the coefficients needed to test the following pairwise and complex comparisons. Begin by writing the null hypothesis for each comparison, and then rearrange terms to find an appropriate set of coefficients:

- Genetically engineered wood bats versus graphite bats
- Graphite bats versus aluminium bats
- Graphite bats versus the average of the other three types of bats
- Genetically engineered wood bats versus the average of the aluminum and graphite bats

5.1 Testing Comparisons via SPSS

Because any comparison is simply a test of a difference between two groups (i.e., pairwise comparison), or between two combinations of groups (i.e., complex comparison), it can be tested using the same general logic of ANOVA. That is, we can determine how much the group means, or the combinations of group means, differ, estimate the influence of random factors by determining how much variation exists within groups, and then use the ratio of the former to the latter to determine how likely it would be to obtain a difference between groups, or combinations of groups, as large or larger than observed, if the specific null hypothesis being tested were true. This is exactly what occurs when we specify a set of coefficients to test a comparison using the "Contrast…" option in SPSS. To do so, we begin by entering our data and setting up a one-way ANOVA as shown earlier. We next click on the "Contrast…" button and then enter the coefficients needed to test the comparison of interest, one at a time. We first specify the coefficient that corresponds to our first group (i.e., the group labelled as 1 in SPSS) and click "Add", then specify the coefficient that corresponds to our second group (i.e., the group labelled as 2 in SPSS) and click "Add", and so on until a coefficient is specified for each group (see Fig. 6). When finished, the Coefficient Total should sum to zero. If so, we can click on the "Next" button to specify a new set of coefficients to test another comparison of interest in the same manner. When all of the comparisons that we want to test are specified, we click on "Continue" and "OK" to have SPSS perform the comparisons.

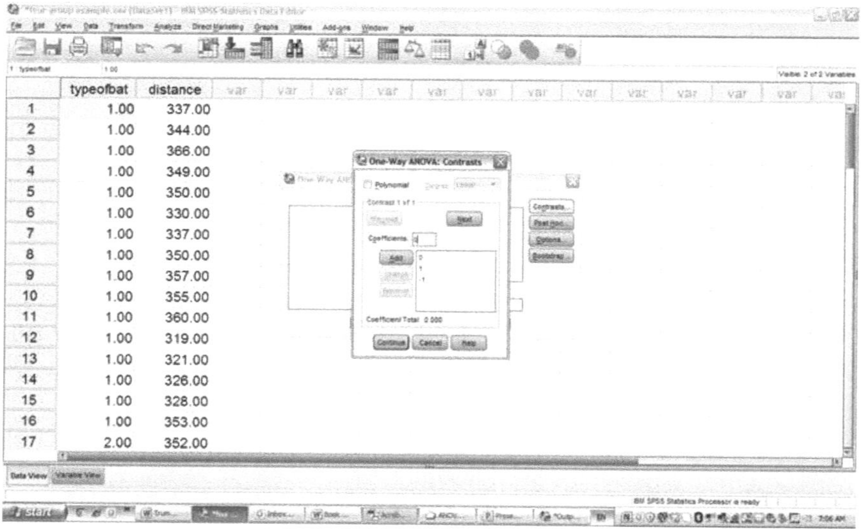

Figure 6. Specifying coefficients to test comparisons via SPSS for four-group example.

The SPSS generated output for four comparisons is shown below (see Fig. 7). By examining the "Contrast Coefficients" table, you should be able to determine that three of the comparisons were pairwise, comparing the genetically engineered wood bats to each of the other three types of bats. The fourth comparison was complex, comparing the average of the normal and genetically engineered wood bats to the average of the aluminum and graphite bats. It should also be noted by examining the "Contrast Tests" table that, rather than an *F*-ratio, SPSS generates a *t* statistic when performing comparisons. The *t* statistic, when squared, is identical to the *F*-ratio. Thus, the corresponding *p*-values are interpreted the same as for an *F*-ratio. Take, for instance, the comparison of the genetically engineered wood and aluminum bats (labelled as Contrast #2 in the output). The $t = -.463$ (which, if squared, is equivalent to an $F = .214$), and has a corresponding p-value $= .645$. This implies that if in reality there is no difference between genetically engineered wood and aluminum bats, then the likelihood of replicating this experiment and finding a difference between the sample means of the genetically engineered wood and aluminum bats as larger or larger than what we obtained in this experiment (which turned out to be a 2.375 m difference) is .645. Therefore, it would be very likely to obtain this large or larger difference even if the null hypothesis is true, so we should conclude that there is not strong enough evidence to say that aluminum and genetically engineered wooden bats hit balls different distances.

5.2 Controlling Against Type I Errors

Although the same general logic of ANOVA that we've been discussing throughout this chapter is used to determine such probability values, researchers often use different criteria against which to judge these probabilities when performing multiple comparisons. That is, instead of using the typical .05 or smaller level required for considering a probability value to be statistically significant, the criterion is adjusted, depending on several factors which we discuss next.

As we've seen, when the omnibus null hypothesis that all group means are the same can be rejected, many alternatives are possible. To determine more precisely how groups differ, one typically conducts *multiple* comparisons. In our present four group experiment, for example, we were primarily interested in how the genetically engineered wood bats compared with other types of bats. Thus, we conducted three pairwise comparisons ($\mu_{\text{genetically engineered wood}}$ vs. $\mu_{\text{normal wood}}$, $\mu_{\text{genetically engineered wood}}$ vs. μ_{aluminum}, $\mu_{\text{genetically engineered wood}}$ vs. μ_{graphite}). In addition, we threw in a fourth complex comparison for good measure! Now, suppose that we were to consider any of these comparisons to be statistically significant if it has a probability value less than the typical criterion of .05. Our probability of making a Type I error

Oneway

ANOVA

distance

	Sum of Squares	df	Mean Square	F	Sig.
Between Groups	6586.672	3	2195.557	10.421	.000
Within Groups	12640.938	60	210.682		
Total	19227.609	63			

In this table, you can find the coefficients that were used to define the four comparisons that were tested.

Contrast 1 is a pairwise comparison of the Normal wood versus Genetically engineered wood groups. Likewise, comparisons 2 and 3 are each pairwise comparisons.

Contrast Coefficients

Contrast	typeofbat			
	normal wood	genetically engineered wood	aluminum	graphite composite
1	-1	1	0	0
2	0	1	-1	0
3	0	1	0	-1
4	.5	.5	-.5	-.5

Contrast 4 is a complex comparison of the average of the Normal and Genetically engineered wood groups versus the average of the Aluminum and Graphite groups.

In this table you can find the t-statistics and corresponding p-values for each comparison. The top half of results labelled "Assume equal variances" can be used unless the equal variances assumption of ANOVA has not been met.

The t-statistics, when squared, are identical to F-ratios.

These are the probability values used to make conclusions for each comparison.

Contrast Tests

		Contrast	Value of Contrast	Std. Error	t	df	Sig. (2-tailed)
distance	Assume equal variances	1	12.1250	5.13179	2.363	60	.021
		2	-2.3750	5.13179	-.463	60	.645
		3	-16.4375	5.13179	-3.203	60	.002
		4	-15.4688	3.62872	-4.263	60	.000
	Does not assume equal variances	1	12.1250	4.80223	2.525	29.266	.017
		2	-2.3750	4.82647	-.492	29.188	.626
		3	-16.4375	5.07175	-3.241	28.295	.003
		4	-15.4688	3.62872	-4.263	58.291	.000

Figure 7. Output for comparison tests generated by SPSS.

would be .05 in each of these four comparisons. However, our probability of making *at least one* Type I error across the set of comparisons would be greater than .05. In general, the more comparisons conducted, the greater is the likelihood of making at least one Type I error. Thus, if we wish to make sure that our likelihood of making a Type I error across our entire set of comparisons is no greater than .05, we must use a smaller criteria for determining significance for each comparison. How much smaller is determined, in part, by the total number of comparisons being conducted. In general, the more comparisons that are tested, the smaller the criteria must be.

Two other factors that affect the likelihood of making a Type I error are whether the comparisons are pairwise or complex and whether they were planned in advance of examining the experimental results (i.e., *planned* comparisons) or decided upon after seeing the sample means (i.e., *post hoc* comparisons). Without going into detail, you should realize that the number of potential pairwise comparisons that can be conducted for a given number of groups is limited. In our four group case, there are only six such comparisons possible. However, the potential number of complex comparisons is unlimited since any linear combination of group means is possible. Therefore, there are potentially more opportunities to make a Type I error in complex than pairwise comparisons and, so, more stringent criteria are required for testing complex than pairwise comparisons. Also, post hoc comparisons are at greater risk of leading to Type I errors than planned comparisons. Because post hoc comparisons usually focus on group sample means that appear to have a chance of being statistically significant, they can be affected by chance variation. Planned comparisons, however, are decided upon before looking at the sample means and, so, are not affected. Thus, post hoc comparisons require more stringent criteria than planned comparisons.

Many procedures are available for making the necessary adjustments to maintain the overall Type I error rate at .05 across a set of comparisons. Some of the most popular procedures are the Bonferroni, Tukey HSD, Scheffe, and Dunnett procedures, all of which are available in SPSS. The decision about which procedure to use depends on the three factors mentioned above. Although a thorough discussion of these procedures is beyond the scope of this chapter, we provide some brief suggestions for choosing amongst the possibilities in Table 7 below. For a more in depth discussion of the need for, and use of, these procedures, see Maxwell and Delaney (2004).

Table 7. Recommended procedures for conducting multiple comparisons.

Procedure	**When to use**
Bonferroni	Conducting planned comparisons (pairwise and/or complex), and using it provides more power than would using alternative acceptable procedures
Tukey HSD	Conducting all possible pairwise comparisons if planned, or conducting any number of pairwise comparisons if post hoc
Dunnett	Conducting planned pairwise comparisons in which one control group is compared to every other group
Scheffe	Conducting at least one post hoc complex comparison, or conducting several planned complex comparisons and using it provides more power than would using alternative acceptable procedures

6. Final Considerations

6.1 One-way, Between-subjects ANOVA versus Other Types of ANOVA

As we've seen, ANOVA is used to compare group means on a quantitative dependent variable. When the groups differ along just one factor or independent variable (e.g., by the type of bat used to hit the balls), we say that we are conducting a *one-way* ANOVA because the groups are distinguished from one another in just one way. However, groups could be distinguished along more than one factor. For example, groups might differ with respect to both type of bat and length of bat. This, then, would be referred to as a *two-way* ANOVA since the groups differ in two ways. In general, designs with two or more independent variables are referred to as *factorial* ANOVAs.

Also, each sample unit in an experiment may be observed in only one of the groups involved, or they may be observed in each of the groups. For instance, in our example each baseball was hit by only one of the four types of bat. ANOVA applied to such designs is called *between-subjects* because the comparisons are between different groups of subjects (or balls in this case). However, we could have designed our experiment differently such that each baseball was hit by all four types of bat in random order. ANOVA applied to this second type of design is called *within-subjects* because now comparisons are made within each subject. Within-subjects ANOVAs are also referred to as *repeated-measures* ANOVA because sample units are measured repeatedly, under each group treatment. Although this chapter has focused on one-way, between-subjects ANOVA, you should realize that these other varieties of ANOVA exist, and that the logic on which they are based is similar to the logic that we've been discussing.

6.2 Assumptions of One-way, Between-subjects ANOVA

In order for the probability values associated with the F-ratio of one-way, between-subjects ANOVA to be valid, certain assumptions must be met. The first assumption is that the populations of dependent variable scores from which the sample data were obtained must be normally distributed. In our example, this means that the distances travelled by all balls (not just the sample of balls included in our study, but *all* balls) hit by normal wood, genetically engineered wood, aluminum, and graphite bats must be normally distributed. Unfortunately, we can never be certain if this assumption is met, since we never observe entire populations. However, we can safely make this assumption if one of two situations occurs—either the sample distributions are roughly normally distributed and/or our sample sizes are greater than or equal to 30. When distributions are not normal, ANOVA may still be used although it may not be as powerful as alternative non-parametric tests. The second assumption is that the population variances of the dependent variable scores within each group must be the same. Back to our example, this means that random factors must have the same size effect on balls hit with normal wood, genetically engineered wood, aluminum, and graphite. Again, we cannot know this for sure, but we can estimate based on the sample variances. Unequal group variances are most problematic when the group sample sizes are also unequal. Maxwell and Delaney (2004) suggest consideration of the groups with the largest and smallest sample variances. As long as the sample variance divided by the sample size of the group with the largest variance is no more than four times larger than that of the group with the smallest variance, then even if population variances are not equal, it will not seriously affect the reliability of ANOVA. If they do differ by more than four then there are correction procedures that can be used with ANOVA (e.g., the Greenhouse-Geisser approach). The third assumption is that observations are independent. This means that observations (i.e., balls being hit in our example) are unrelated to one another, both within and between groups. If observations between groups are related, as would occur if the same ball was hit with each of the four types of bat then a within-subjects ANOVA would be more appropriate. Dependence of observations within groups, however, is a much more serious experimental design problem that cannot be handled by ANOVA.

Although we've mentioned these assumptions last in this chapter, in practice one should always perform a check on them first before proceeding with an ANOVA. For a more thorough discussion of the assumptions of ANOVA, multiple comparison procedures, and other varieties of ANOVA, see Maxwell and Delaney (2004), and for a good illustration of how SPSS can be used to conduct ANOVAs, see Field (2009).

Self-Check #4

1. A new engine is being developed for a particular model of car. To test its energy efficiency, 5 cars with the new engine and 5 cars with the old engine are each driven for one week. At the end of the week, the fuel mileage (measured in miles per gallon) of each car is determined, and is shown below.

	New	Old
	33	26
	30	29
	22	34
	38	18
	27	23
Mean:	30	26
Standard Deviation:	6.04	6.04

a) Approximately how much does the gas mileage of cars with the *same* type of engine differ?
b) What are some potential reasons why cars with the same type of engine would not have the same gas mileage?
c) On an average, how much does the gas mileage of cars with different types of engines differ?
d) Is it possible that the difference noted in part c is due to the factors identified in part b (rather than due to type of engine)?
e) If in reality the new engine is no more fuel efficient than the old engine, how likely would it be to conduct a test like this and find a difference between the two groups of cars that is as large as, or larger than, observed here? Explain how you know.
f) Given your answer to part e, what can you conclude about the new engine?

2. Suppose you are hired to test the battery life of three different smartphones. To do so, you loop the same video on several of each of the three types of smartphone and record the elapsed time (in minutes) before each phone's battery is completely drained. An analysis in SPSS generates the results shown below.

Oneway

Descriptives

battery life

	N	Mean	Std. Deviation	Std. Error	95% Confidence Interval for Mean		Minimum	Maximum
					Lower Bound	Upper Bound		
Papr	7	668.4286	41.97959	15.86679	629.6039	707.2532	590.00	720.00
Yuphone	8	420.2500	36.54254	12.91974	389.6997	450.8003	367.00	491.00
Swerve	6	623.6667	46.37097	18.93087	575.0033	672.3300	555.00	690.00
Total	21	561.0952	121.14161	26.43527	505.9522	616.2382	367.00	720.00

ANOVA

battery life

	Sum of Squares	df	Mean Square	F	Sig.
Between Groups	262833.262	2	131416.631	77.121	.000
Within Groups	30672.548	18	1704.030		
Total	293505.810	20			

Figure 8. One-way, between-subjects ANOVA comparing mean battery life of three types of smartphone.

a) Which value provides an estimate of the effect of uncontrolled random factors such as manufacturing errors, environmental conditions (e.g., temperature), errors in measurement of elapsed time, etc.?
b) The between-groups mean square, MS_B, estimates the effect of what factor or factors?
c) Describe what the F-statistic indicates.
d) Explain the meaning of the value labelled "Sig.".
e) What can you conclude based on the value described in part d?
f) What additional comparisons do you think would be of interest to your employer? Describe them, both in words and by writing the coefficients that would allow you to test them.

3. Scientists interested in the health risks associated with diet measured the cholesterol (in mmol/L) in four groups of individuals: Vegetarians, Vegans, Meat-eaters, Fish-eaters. Data, as well as some histograms and a graph from SPSS, are provided below.

Diet			
Vegan	**Vegetarian**	**Fish-eater**	**Meat-eater**
4	4	4	6
3	7	4	5
6	5	5	5
1	4	4	5
6	3	6	7
2	7	4	5
6	5	5	6
4	6	5	6
5	4	6	4
5	5	7	7
4	5	6	5
4	5	5	6
5	5	6	5
4	5	5	6
4	5	6	4
4	4	5	6
6	4	6	5
4	5	6	5
4	4	6	6
3	6	5	6
4	5	4	5
4	5	4	5
4	4	6	5
4	4	4	5
5	4	4	7
6	5	5	4
3	6	4	5
5	5	6	4
4	5	6	4

7	6	5	7
6	3	6	5
4	6	5	8
5	4	5	5
3	6	4	5
1	7	5	5
4	6	5	6
7	4	8	4
5	5	3	4
4	3	3	6
3	4	4	3

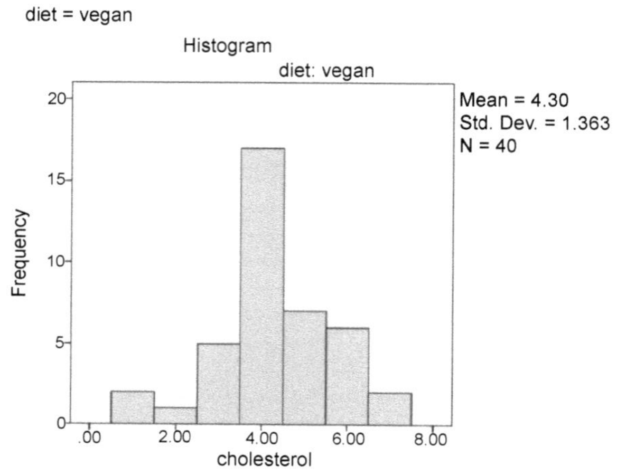

Figure 9. Distribution of Vegan cholesterol.

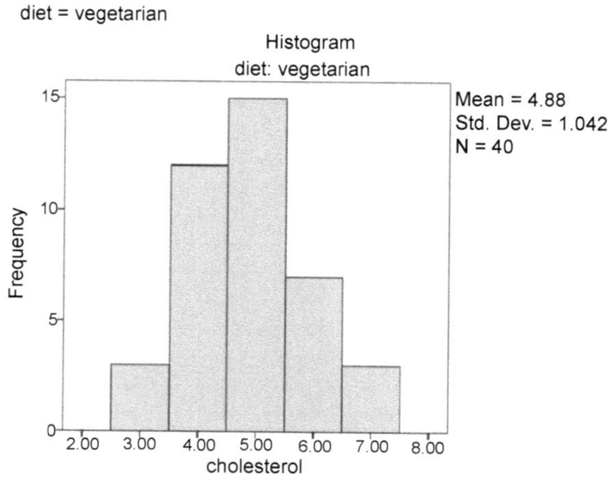

Figure 10. Distribution of Vegetarian cholesterol.

diet = fish eater

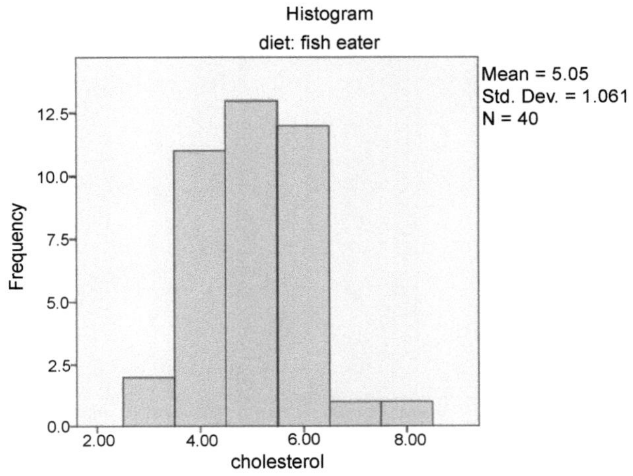

Figure 11. Distribution of Fish eater cholesterol.

diet = meat eater

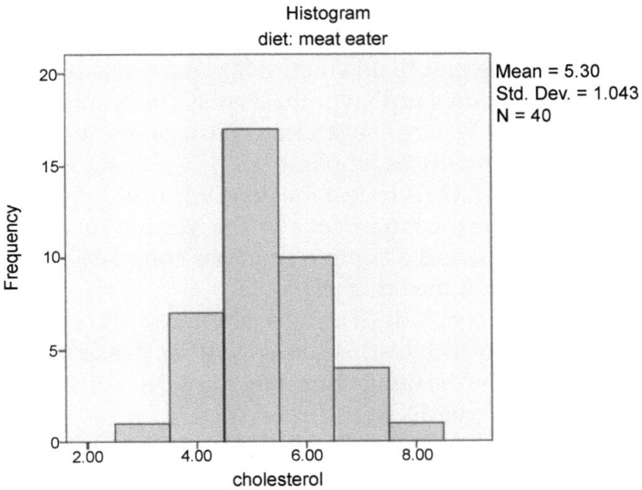

Figure 12. Distribution of Meat eater cholesterol.

Graph

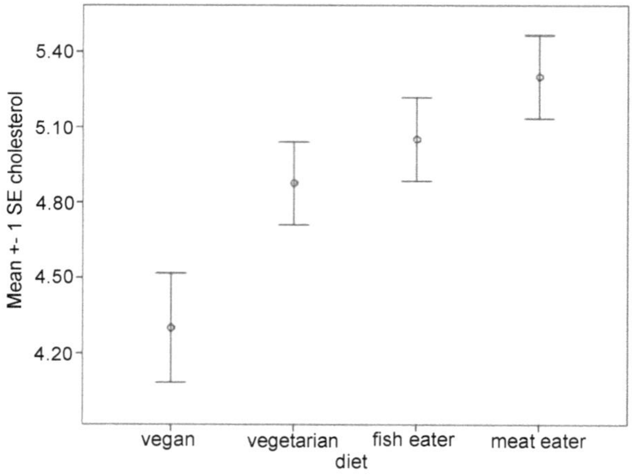

Figure 13. Mean cholesterol by Type of diet.

a) Looking at scores and/or the histograms, does it appear that the normality and equality of variances assumptions required for ANOVA have been met? How do you know?
b) Looking at the graph, do you think that there is sufficient evidence to reject the omnibus null hypothesis that the average cholesterol in the population of Vegans, Vegetarians, Fish eaters, and Meat eaters is the same? Explain why or why not.
c) Conduct an ANOVA to test the omnibus null hypothesis in part b, as well as pairwise comparisons of the Vegetarian diet versus each of the other diets, and a complex comparison of Meat-eaters versus the average of all of the other diets.
d) What can you conclude (if using a criterion of .05 or less for determining significance in each test)? Be as specific as possible.
e) What would be the danger in using a criterion of .05 in each of the four comparisons tested?
f) If, instead, you had used a criterion of $.05/4 = .0125$ or less for the comparisons, how would the conclusions have been different?

7. References

Field, A. 2009. Discovering Statistics Using SPSS (3rd ed.), CA, USA: Sage, Thousand Oaks.

Maxwell, S.E. and H.D. Delaney. 2004. Designing Experiments and Analyzing Data: A Model Comparison Perspective (2nd ed.), New York, USA: Taylor & Francis.

Obrecht, N.A., G.B. Chapman and R. Gelman. 2007. Intuitive *t* tests: Lay use of statistical information. Psychonomic Bulletin & Review 14: 1147–1152.

Trumpower, D.L. 2013. Formative use of intuitive analysis of variance. Mathematical Thinking and Learning, 15: 291–313.

Trumpower, D.L. and O. Fellus. 2008. Naïve statistics: Intuitive analysis of variance. *In:* V. Sloutsky, B. Love and K. McRae [eds.]. Proceedings of the 30th Annual Conference of the Cognitive Science Society. Washington, DC, USA: Cognitive Science Society, pp. 499–503.

CHAPTER 8

Factor Analysis

*Marta B. Quaglino** and *José A. Pagura*

1. Introduction

Multivariate data analysis techniques denominated factor techniques are designed to create representations from simultaneous information on a set of variables in a condensed and synthesized way, optimizing certain aspects in such representations. The objectives that can be covered with these techniques are varied, including identifying similarities and differences among units of analysis, understanding the differences among the represented groups of subjects, analyzing the causes that produce associations among the responses or measurements, studying the structure of the relationship among groups of variables or measurements, identifying subgroups of units or variables due to its similarities, deriving objective rules of classification that would allow us to assign subjects to classes or populations known in advance, etc.

Factor methods always begin with the availability of a wide range of measures registered from a vast set of subjects (countries, regions, cities, companies, products, neighbourhoods, families, people, etc.). These units can represent a sample or the total group or population of interest. The variables can be qualitative characteristics, presence-absence variables, quantitative or ordinals.

The choice of the factor technique depends on the aim of the study and the kind of variables we are analyzing. In any case, the main result will be a set of coordinates that will allow us to visualize the units of analysis and/or the variables in some special charts. These coordinates may also be used to classify the subjects under study (which may result in a possible indicator) or to group them in clusters with similar characteristics. When

Facultad de Ciencias Económicas y Estadística, Universidad Nacional de Rosario (Argentina).
* Corresponding author

applying multivariate methods, it is frequent that the practical use of these techniques be sequential, i.e., the output of one of them is the input for another method.

Particularly, Factor Analysis allows us to analyze the interrelationship system among quantitative variables, so as to discover the presence of underlying complex concepts, non-measurable with a simple indicator, which have produced the associations observed among them (Spearman 1904, Thurstone 1947). Those complex concepts or non-measurable variables are designated as *factors.*

However, it is important to find out if the multivariate techniques are actually useful before using them. In previous chapters, we have learned various statistical methods that let us describe and prove certain hypothesis, involving one or two variables. Now, if we consider that multiple variables are measured simultaneously in each unit, why not to apply those methods repeatedly over each variable? Why is it necessary to use multivariate methods?

The following example is a real case of Factor Analysis application —Advertisement Testing. A leader company in Market Research and Advertising Analysis has information about commercial advertisements. This company has been established in Latin America for more than ten years and has carried out advertising analysis in several countries of the region (Argentina, Brazil, Chile, Colombia, Mexico, Venezuela) and for a wide range of trademarks, gaining a vast experience in the evaluation of commercials. This information comes from surveys of 2381 studies that were carried out between 2000 and 2010. The data collection process is done accurately and systematically, preserving the same criteria to ensure consistency and to allow the historical comparison. The questionnaire contains "filter variables", socio-demographic and attitudinal variables to know the respondents' profile, as well as variables to evaluate the commercials. In the latter case there are specific questions used to measure the effectiveness of the advertisement (Impact variables), plus additional diagnostic questions (Attribute variables), which help to understand the causes for the outcomes of the commercial impact.

Data is aggregated, i.e., each row contains the information of a complete study, summarizing the behaviour of all the people surveyed. The columns of the data matrix contain the variables of the survey and the rows contain each of the tested commercials. The list of variables and labels to be used in this section are:

Attributes variables (as percentages in the top of the scale): *Likeability* (like), *New Brand* (nbrand), *Believable* (beli), *Other Way* (other), *Strong Feeling* (feel), *Entertaining* (enter), *Uniqueness* (uniq), *Irritating* (irri), *Too see again* (again), *Informative* (Info), *For you* (you), *Newness* (new), *Relevant* (rele), *Different* (diff), *Fits with the brand* (fit), *Important* (impo), *Emotive* (emo), *Confusing* (confu), *Aggressive* (agre), *Silly* (silly), *Humour* (humo), *Interesting* (inte), *Name is clear* (clear), *Memorability* (memo), *To win awards* (win), *Be Mentioned* (ment).

Impact variables (continuous variables designed by the company): *Persuasion* (per_I), *Recall* (rec_I), *Recognition* (cog_I), *Brand Association* (bra_I), *Effectiveness* (eff_I).

Some initial univariate or bivariate analysis could involve, for example, the construction of histograms, scatter-plots, t-tests or ANOVA. Figure 1 shows the distribution of the outcomes for one of the variables that evaluate the impact the advertisement produces on people exposed to it: *Persuasion*. The histogram shows that the obtained punctuations are slightly skewed to the right, with a high concentration on the value one hundred, which indicates that the majority of the commercials have an intermediate persuasion level and a small percentage reach a differentiated high impact (Fig. 1).

If the *Persuasion* index distribution is shown by commercial category it is possible to observe that, although the histograms have similar shapes, the highest rates correspond to commercials of the category House (Fig. 2). Do these observed differences from the histograms represent statistically significant differences? An ANOVA test reveals that the means of *Persuasion* are not the same for all the categories (p=0.0001), though not all of them have differences. The category that is different from the remaining ones is House, however the magnitude of the difference among the averages is practically negligible according to the measuring scale (Table 1). The amount of commercials in this study (n = 2381) finds small differences as statistically significant.

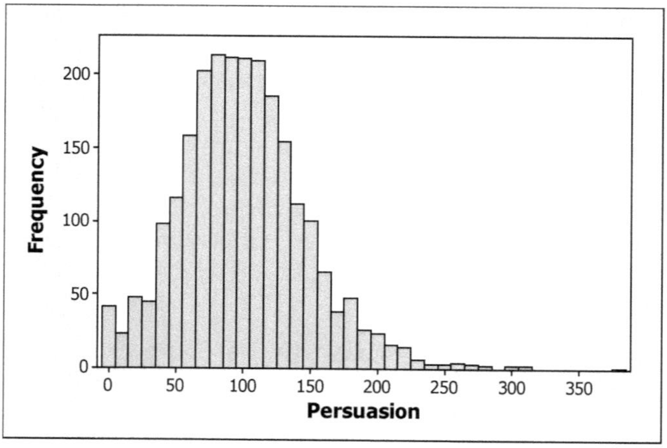

Figure 1. Histogram of *Persuasion*.

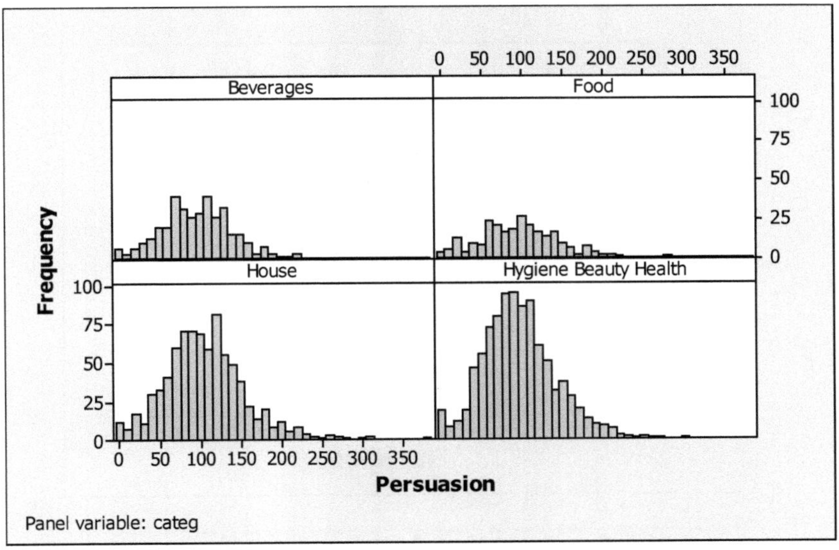

Figure 2. Histogram of *Persuasion* by category of commercials.

Table 1. Mean, St Deviation and individual IC95% of *Persuasion* by Category.

Category	N	Mean	StDev	Lower Bound	Upper Bound
Beverages	363	96.48	41.74	92.17	100.80
Food	249	96.11	47.43	92.19	102.03
House	812	106.25	50.30	102.79	109.72
Hygiene, Beauty, Health	957	96.44	45.08	93.58	99.30

Different peculiarities may be observed if other characteristics are analyzed. As an example, some results are shown for the *Recognition* Index, *Entertaining* and *To See Again* (Figs. 3, 4, 5, 6).

Questions for Self-Evaluation

1) Could you identify some differences in respect of the analysis of *Persuasion*?

If we are interested in bivariate correlations, they could be analyzed from the correlation matrix instead of scatterplots, since there are many variables. Main variables of interest are arranged in two groups corresponding to "impact measures" and "commercial characteristics". Hence it could be of interest to analyze the associations "within" and "between" them, leading to a great number of associations to analyze, presented in Tables 2, 3 and 4.

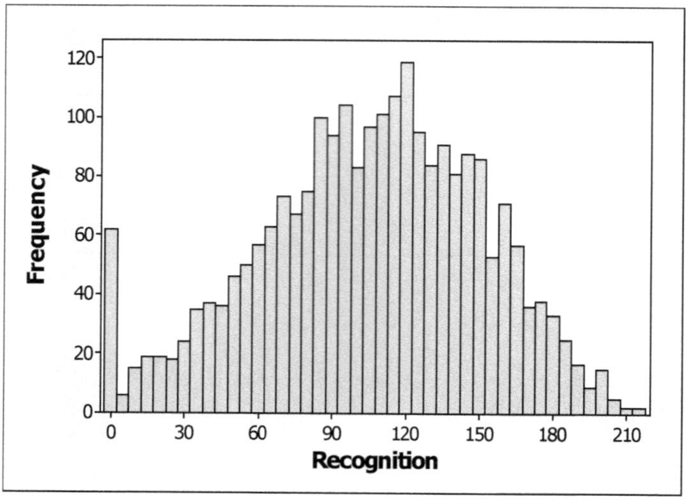

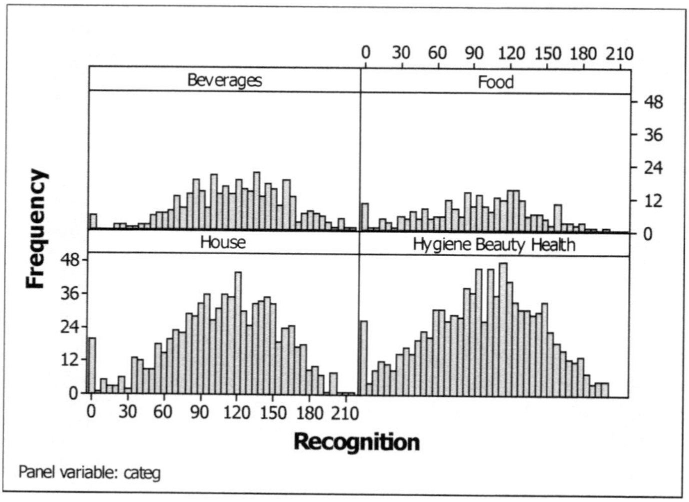

Figures 3, 4. Histogram of *Recognition* for overall commercials and by categories.

Table 2. Pearson correlations within Impact Indexes Group (p-value).

Index	Persuasion	Recall	Recognition	Brand Association
Recall	0.088 (0.000)			
Recognition	0.066 (0.001)	0.727 (0.000)		
Brand Association	0.076 (0.001)	0.865 (0.000)	0.302 (0.000)	
Effectiveness	0.738 (0.000)	0.668 (0.000)	0.490 (0.000)	0.574 (0.000)

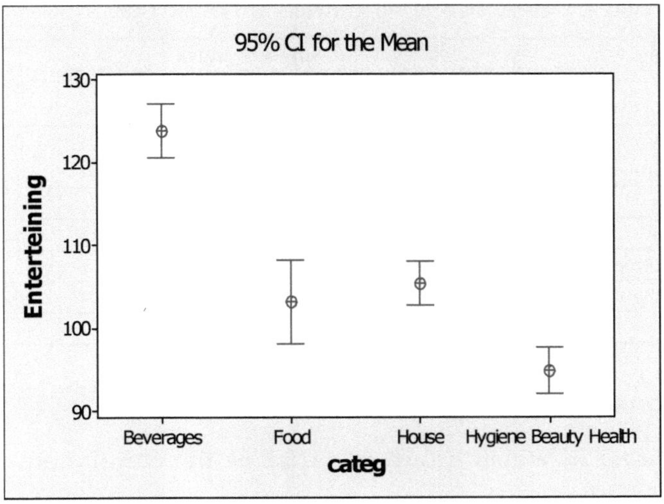

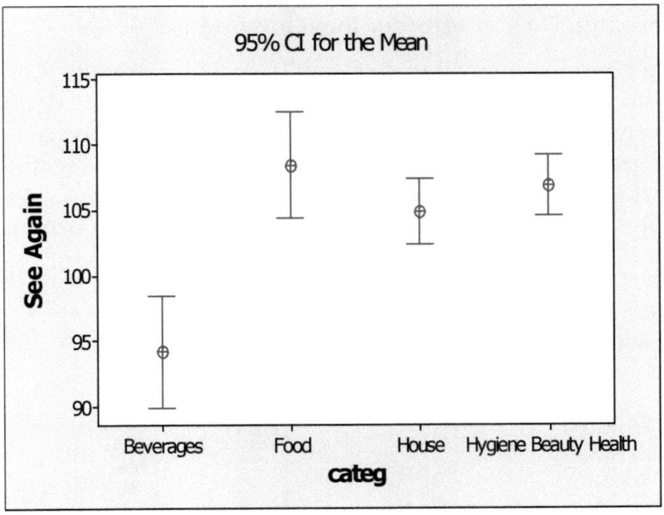

Figures 5, 6. Interval Plots of *Entertaining* and *To see Again* by category of commercials.

Table 3. Pearson correlations within some Commercial's Attributes Group (p-value).

Attributes	Likeability	New Brand	Believable	Other way	Strong Feeling
New Brand	0.568 (0.00)				
Believable	0.638 (0.00)	0.683 (0.00)			
Other way	0.503 (0.00)	0.562 (0.00)	0,554 (0.00)		
Strong Feeling	0.668 (0.00)	0.733 (0.00)	0.712 (0.00)	0.639 (0.00)	
Entertaining	0.368 (0.00)	0.156 (0.00)	0.245 (0.00)	0.193 (0.00)	0.261(0.00)
....	....	....	....	....	....

Table 4. Correlations between Attributes and Impact Groups (p-value).

Attributes	Impact's Index				
	Persuasion	**Recall**	**Recognition**	**Brand Association**	**Effectiveness**
Likeability	0.120 (0.00)	0.135 (0.00)	0.138 (0.00)	0.095 (0.00)	0.160(0.00)
New Brand	0.130 (0.00)	0.074 (0.00)	0.065(0.00)	0.063 (0.00)	0.133(0.00)
Believable	0.135 (0.00)	0.058 (0.00)	0.052 (0.01)	0.048 (0.02)	0.130(0.00)
Other way	0.115 (0.00)	0.068 (0.00)	0.042 (0.04)	0.070 (0.01)	0.122(0.00)
Strong Feeling	0.153 (0.00)	0.061 (0.00)	0.053 (0.01)	0.050 (0.00)	0.141(0.00)
Entertaining	0.034 (0.10)	0.184 (0.00)	0.214 (0.00)	0.103 (0.00)	0.125(0.00)
.....	....	....	....	....	....

Questions for Self-Evaluation

2) Indicate in which group of variables the correlations are most important.
3) Discuss the p-value. There are correlations of around 0.10 that are significant. Do you consider them important?

If we continue analyzing different aspects separately, more features may be discovered. However this strategy is an arduous task and it could lead to erroneous conclusions. Unlike these partial approaches, Factor Analysis performs a comprehensive study. One of the results that we would get is the graphical representation of Fig. 7, showing the total set of variables on a plane. The axes represent the most important factors or

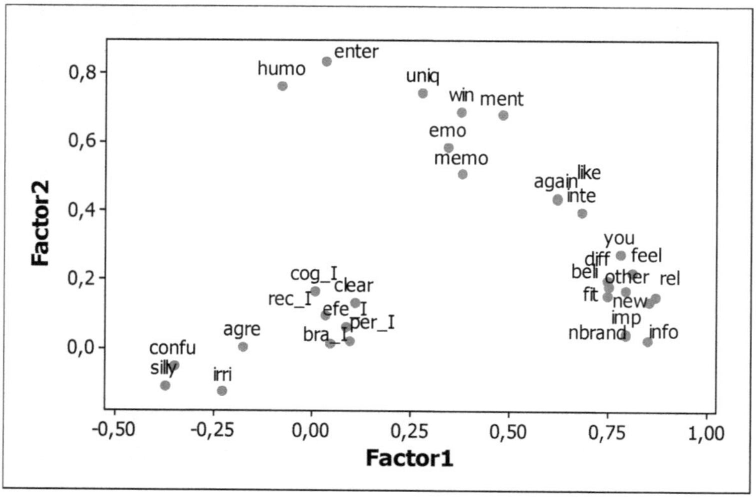

Figure 7. Correlations of variables with Factor 1 and Factor 2.

complex concepts that influence the variables and cause their associations. The coordinates of each variable are the variable-factor correlations. Thus, a simple visual inspection allows us to identify the variables that have stronger associations with each factor by observing the distance to the origin. Further, variables with stronger direct correlations are located nearby, while those with stronger inverse correlations are in opposite directions and the non-associated variables appear on the direction of vectors that form right angles.

Location of the variables in Fig. 7, shows that the most important factor -F1- which causes the associations among variables is a factor that identifies the informative quality of the commercial along with some positive emotional attributes. The furthest variables from the origin in the x-axis direction are: *Relevant* (rele), *Newness* (new), *Informative* (inf), *Strong Feeling* (feel), *Important* (impo), *Different* (diff), *New Brand* (nbrand), *Other Way* (other), *Believable* (beli), *For you* (you), *Fits With the Brand* (fit), *Interesting* (inte), *Likeability* (like) and *To See Again* (again). The negative emotional impact that the commercials cause is scarcely and inversely related to the positive emotional qualities an informative quality. They appear on the other extreme of the *x*-axis and are smaller in absolute value they are near of the origin.

The second factor -F2- is related to entertainment and humour, since it is positively associated with *Humour* (humo), *Entertaining* (enter), *Uniqueness* (uniq), *To win awards* (win), *Be Mentioned* (ment), *Emotive* (emo) and *Memorability* (memo). These variables are the furthest ones from the origin in the *y*-axis direction. These two sets of variables that identify the factors F1 and F2, are hardly associated between them and neither are they with the impact of the commercial indexes, which are grouped around the origin of coordinates. These results will be completed later, after some important concepts are presented, interpreting more factors and getting an ordering of the commercials according to these complex concepts.

Questions for Self-Evaluation

4) Interpret the third and fourth factors using the simple rules we have defined, looking at Fig. 8. Remember that all correlations around the origin are negligible.

How are these results derived? Why does the graphic in the plane show association among variables so simply? How were the coordinates of the variables in the plane generated? That is the key for Factor Analysis. Each axis in the plane represents a new variable (called factor) that is obtained from the originally defined and measured variables. The main aim of the Factor Analysis is to "discover which factors better explain the association among variables". It is supposed that more than one factor could be necessary. How many factors could be defined?

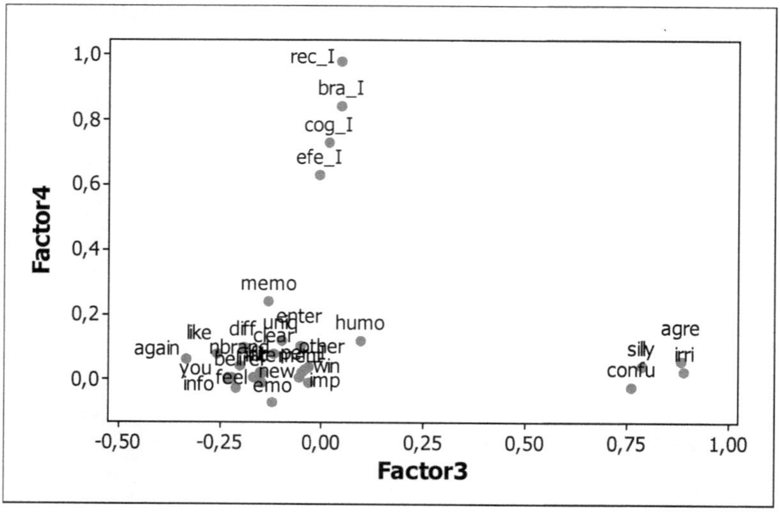

Figure 8. Correlations of variables with Factor 3 and Factor 4.

All these questions could be answered by specifying a statistical model, which begins from a theoretical conception. The model is expressed in a set of equations where parameters or unknown constants, observed variables and new variables or factors are involved. The mathematical model is also accompanied by certain assumptions, which will help to interpret the model.

2. Starting Theoretical Concept

The observed measurements are considered to be the result of certain concepts, variables or complex factors that influence those measurements, probably jointly or by groups. Figure 9 shows this concept assuming that there are six variables and two latent or non-observable factors. As a consequence of the simultaneous influence of the factors over the variables, the associations observed in the correlation matrix are produced. It also be assumed that there exist specific factors which influence only one variable.

Model that represents the theoretical concept

We shall designate X_i, $i = 1,2,\ldots, p$ the observed variables and F_j, $j = 1,\ldots, m$ the not observed latent factors, which are common to all variables. Let e_i, $i = 1,\ldots, p$ specific factors for each variable and a_{ij} certain constants or

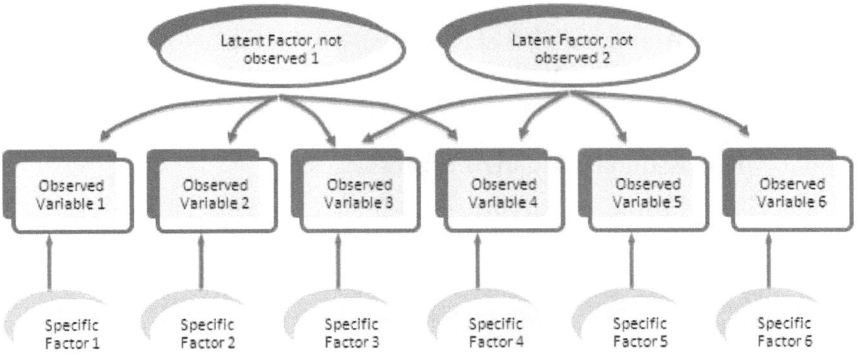

Figure 9. Theoretical Concept of the influence of latent factors on observed variables.

parameters that identify the degree of importance in which each common factor F_j influence on the variable X_i. We set as the factor model the following set of linear equations:

$$X_1 = a_{11}F_1 + a_{12}F_2 + a_{13}F_3 + \cdots + a_{1m}F_4 + e_1$$

$$X_2 = a_{21}F_1 + a_{22}F_2 + a_{23}F_3 + \cdots + a_{2m}F_4 + e_2$$

$$\vdots$$

$$X_p = a_{p1}F_1 + a_{p2}F_2 + a_{p3}F_3 + \cdots + a_{pm}F_4 + e_p$$

These equations could be summarized into the matrix form: $\boldsymbol{X} = \boldsymbol{AF} + \boldsymbol{E}$, where $\boldsymbol{X}$ and $\boldsymbol{E}$ are p-dimensional column vectors, $\boldsymbol{A}$ is a $p \times m$ matrix and $\boldsymbol{F}$ is a column vector of m components.

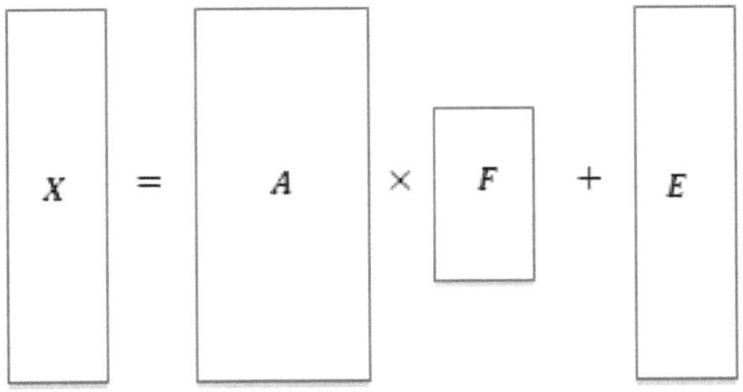

Figure 10. Matrix form of the Factor Model.

For ease of interpretation, we make some assumptions without loss of generality. The variables that represents the latent factors are supposed to be standardized and with null covariance among them[1] and with the specific factors. The specific factors are supposed to be centred and with null covariance among them, though their variances are not necessarily equal. These assumptions can be expressed as:

$$E(F_j) = 0 \quad j = 1, \dots, m$$

$$Cov(F_j, F_k) = \begin{cases} 0 & si\, j \neq k \quad j, k = 1, \dots, m \\ 1 & si\, j = k \quad j, k = 1, \dots, m \end{cases}$$

$$E(e_i) = 0 \quad i = 1, \dots, p$$

$$Cov\,(e_i, F_k) = 0 \qquad j = 1, \dots, p \;\; y\; k = 1, \dots, m$$

In addition, it is possible to work with the original standardized variables, since the interest of the analysis is focused on the interpretation of their association. Remember that the correlation is not affected by linear transformations of variables.

Considering the model with the assumptions, it is possible to deduce the meaning of the parameters a_{ij} and to find a relationship among them and the variances and covariances of the observed variables. To derive this relationship is a simple exercise of application of Expectation, Variances and Covariance moment's properties (proposed for the lector). Results show that:

$$V(X_i) = \sum_{k=1}^{m} a_{ik}^2 + \mathrm{V(e_i)} = \mathrm{h_i^2} + \Psi_i$$

$$Cov(X_i, X_j) = \sum_{k=1}^{m} a_{ik} a_{jk}$$

$$Cov(X_i, F_j) = a_{ij}$$

Therefore, a_{ij} not only indicates the weight or importance of each non-observable factor in the representation of each variable, but also represents exactly the value of the correlation between them (assuming standardized variables). These correlations or "loadings" reproduce the covariances among the variables, but they are not sufficient to explain the variability.

[1]The model can also be stated with correlated factors.

According to the factor model, the variance of each variable is partitioned into two parts, one due to common factors (h_i^2) and another due to the specific factor (Ψ_i). These components are called "communality" and "uniqueness" or "unique variance".

The relationship among the variance-covariance parameters according to the model can be expressed in matrix form as $\boldsymbol{R} = \boldsymbol{AA}' + \Psi$, where Ψ is a diagonal matrix with the uniqueness in the diagonal elements (Fig.11).

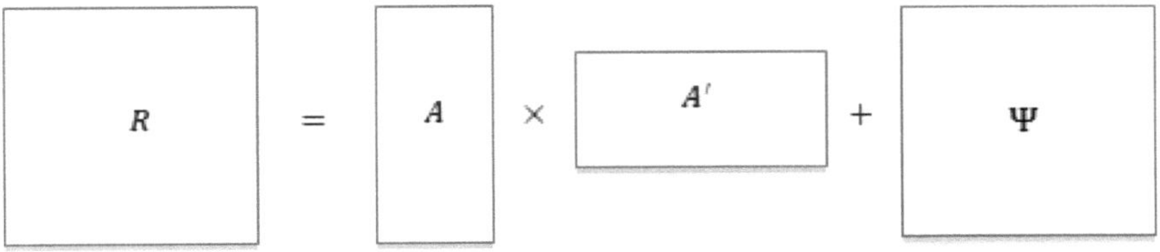

Figure 11. Fundamental Equation of the Factor Analysis.

The meaning of each of the axes in Fig. 7, was interpreted taking into account the model (Fig. 10) and the relationships between the parameters and the correlations (Fig. 11). In applications such as this one, where there are a lot of variables, it is simpler to analyze the numerical values of the coefficients a_{ij} of the model, presenting them in tabular form. Table 5 shows the loadings for the six most important factors, together with the communalities and uniqueness. These last two columns explicit how much of the variability of each variable can be explained by common factors and how much by the specific factor. It should be remembered that these features do not affect the quality of the factor model.

Table 5. Loadings for variables in Advertisement Testing example.

Variable (Label)	Factor						Communality	Uniqueness
	1	2	3	4	5	6		
rel	0,869	0,154	–0,169	0,009	0,058	–0.084	0,8122	0,1878
new	0,854	0,140	–0,147	–0,005	0,009	0.015	0,7719	0,2281
info	0,851	0,027	–0,211	–0,020	0,018	–0.001		0,2280
feel	0,809	0,223	–0,221	0,009	0,055	0.153	0,7569	0,2431
imp	0,795	0,049	–0,056	0,010	0,064	–0.249	0,6427	0,3573
nbrand	0,792	0,041	–0,201	0,047	0,025	0.191	0,6739	
other	0,792	0,172	–0,032	0,043	0,036	–0.154	0,6619	0,3381
you	0,780	0,280	–0,225	0,006	0,013	0.158	0,7385	0,2615
beli	0,750	0,183	–0,232	0,011	0,036	0.195	0,6519	0,3481
diff	0,748	0,200	–0,192	0,102	0,041	–0.093	0,6499	0,3501
fit	0,746	0,156	–0,153	0,026	0,060	0.270	0,6096	0,3904

Table 5. contd....

Table 5. contd.

Variable (Label)	Factor						Communality	Uniqueness
	1	2	3	4	5	6		
inte	0,681	0,402	–0,149	0,021	–0,034	0.055	0,6497	0,3503
like	0,620	0,438	–0,258	0,081	0,015	0.193	0,6502	0,3498
again	0,618	0,440	–0,334	0,066	–0,003	0.086	0,6926	0,3074
enter	0,031	0,840	–0,054	0,108	–0,026	0.159	0,7226	0,2774
humo	–0,080	0,765	0,094	0,121	–0,024	0.208	0,6163	0,3837
uniq	0,276	0,748	–0,100	0,123	0,024	–0.061	0,6630	0,3370
win	0,376	0,691	–0,033	–0,005	0,082	–0.149	0,6271	0,3729
ment	0,479	0,683	–0,051	0,024	0,071	–0.163	0,7053	0,2947
emo	0,343	0,587	–0,120	–0,067	0,068	0.043	0,4871	0,5129
memo	0,376	0,510	–0,132	0,242	–0,058	0.210	0,4821	0,5179
irri	–0,226	–0,123	0,889	0,031	–0,014	–0.038	0,8597	0,1403
agre	–0,176	0,008	0,882	0,061	–0,027	–0.003	0,8155	0,1845
silly	–0,374	–0,106	0,785	0,045	–0,016	–0.014	0,7715	0,2285
confu	–0,351	–0,048	0,761	–0,016	–0,008	–0.145	0,7061	0,2939
rec_I	0,032	0,097	0,046	0,982	0,060	0.022	0,9822	0,0178
bra_I	0,046	0,018	0,044	0,844	0,077	–0.083	0,7230	0,2770
cog_I	0,005	0,167	0,015	0,734	0,003	0.159	0,5677	0,4323
per_I	0,095	0,024	–0,043	0,032	0,974	0.073	0,9628	0,0372
eff_I	0,085	0,067	–0,008	0,633	0,740	0.045	0,9618	0,0382
clear	0,107	0,135	–0,117	0,083	0,090	0.802	0,0588	0,9412

Considering that the coefficients (loadings) are correlations, those with higher absolute values can be selected to identify the ones that correspond to the most influent variables for that factor (for example, $a_{ij} > 0{,}5$). On the graphic, the selection was made considering the furthest variables from the origin, in the direction of each axis or factor. Greatest absolute values of each factor are underlined, and variables are sorted in such a way that those with highest loadings on each factor are grouped. This ordering strategy makes interpretation easier. For the first latent factor, Table 5 shows that the most important correlations are those corresponding to the variables that represents the information that the advertisement brings, while for the second factor the most important ones are those that represents the entertainment and humour attributes. It can be inferred from these results, which are the two most influential factors on the associations between the answers of the respondents: Information and Entertainment. It is also evidenced that these concepts are not associated with the impact on the buying intention.

However, Table 5 contains information about more than two factors, which were not analyzed on Fig. 7. Factor F3 represents the negative emotions (*Irritating* (irri), *Confusing* (confu), *Aggressive* (agre) and *Silly* (silly)) and factor F4 the commercial impact (*Recall* (rec_I), *Recognition* (cog_I), *Brand Association* (bran_I) and *Effectiveness* (eff_I)). The two remaining factors are associated with a small number of variables. *Persuasion* (per_I) and *Effectiveness* (eff_I) are associated with a factor F5, identifying a particular aspect of the commercial's impact. *Name is Clear* (clear) is associated with the factor F6, identifying only one attribute. This is an usual characteristic that appears on applications and it is one of the reason that conduced to not considering those factors in the model.

Questions for Self-Evaluation

5) Complete the gaps in rows 3 and 6 of Table 6. Consider the definition of communality and uniqueness.
6) Which are the best explained variables by common factors? Why?
7) Write two of the estimated equations for the factor model, using the six common factors. How many equations are in the model?
8) Can you identify the value of the common factor F1 for a particular commercial, using algebraic operations with the model equations?

Once the nature of these complex latent factors is interpreted, it would be interesting to identify which are the commercials that show higher rating over each factor. Which have been assessed as the most informative and with positive emotional qualities? Which have been valued as the most entertaining? Which have produced a greater impact on purchase intention? Which were the worst evaluated in these concepts? The answer to these questions requires a measurement of the latent factor on each advertisement. It is not a direct measurement, but it could be obtained using the estimated parameters of the factor model (loadings). Some of these results are presented in Table 6, identifying the country in which the test was done and the category of the advertised product.[2] The factors are new continuous variables that represent complex aspects.

This ordering of the units by the complex concepts discovered through factor analysis could be very meaningful in research.

[2]Name and brand of Commercial are omitted for confidentiality.

Table 6. Highest values of the commercials' latent factors.

F1: Informative - positive feelings			F2: Entertaining - Humor			F3: Negative feelings		
Country	Category	Score	Country	Category	Score	Country	Category	Score
Mexico	Food	2.70	Argentina	Beverages	3.10	Mexico	Food	8.59
Mexico	Food	2.27	Argentina	Food	2.83	Mexico	House	7.54
Argentina	House	2.20	Argentina	Hygiene*	2.46	Colombia	House	7.52
Argentina	Hygiene*	2.12	Argentina	Hygiene*	2.45	Guatemala	House	7.08
Argentina	Hygiene*	2.11	Argentina	Hygiene*	2.40	Mexico	House	6.34
Mexico	Food	1.96	Argentina	Hygiene*	2.21	Brazil	House	6.01

*Hygiene, Beauty, Health

Questions for Self-Evaluation

9) State two or three analysis objectives that could be answered with the results of Table 6.
10) How would you identify the commercials with greater "impact" on the study period?

We have been moving forward in the interpretation of estimated factor models, but without seeing formal procedures to estimate them. Additionally, how many factors should be obtained and interpreted? In the next section we will analyse some estimation methods for a_{ij}.

3. Methods of Identification of Loadings from Data (Estimation)

The parameters a_{ij} of the model, named *loadings*, can be estimated through different methods, some of them statistical and other algebraic. Regardless of the method applied, there is some uncertainty that needs to be solved before using it. If an algebraic method is applied, certain information about the contribution of the factors to the variance of each variable should be given. When applying a statistical method for estimation, in addition, it is necessary to define the number of common factors that will be considered on the model previously. Solutions require some knowledge of linear algebra. For its application to real problems, the use of a computer will be essential given the magnitude of calculations that are needed. The results of this chapter have been obtained using Minitab 16 or SAS 9.3.

Algebraic method for model parameters estimation: Principal Factor Method

When the model was defined, we found relationships among the elements of the correlations matrix of the observed variables (known from data) and the loadings and uniqueness (unknown parameters to estimate). The matrix form of these relationships was written as: $\boldsymbol{R} = \boldsymbol{A}\mathbf{A}' + \Psi$. If the diagonal matrix Ψ were known, the matrix $\boldsymbol{A}$ may be obtained by placing in its columns the eigenvectors corresponding to the eigenvalues of matrix $R - \Psi$ ($R-\Psi$ is always symmetrical but it could be positive semi-definite). The alternatives suggested as approximations for the communalities (diagonal elements of the matrix $\boldsymbol{R} - \Psi$) are:

- the squared multiple correlation coefficients of regression models that consider each X_i as dependent variable and the remaining $p - 1$ variables as explanatories.
- the maximum correlation of each X_i variable with the remaining ones, which is the greatest value of the ith row or column of $\boldsymbol{R}$.
- the average of the correlations among each X_i with the remaining.
- the value one, assuming that only the common factors are the ones that contribute to the variance of the variables (i.e., there are no specific factors in the model).

The solution that is reached when using the valued one in the main diagonal is known as Principal Components and is referred to later in Section 6. Whatever the case is, the suggested solutions are approximations because the actual communalities ($h_i^2 = \sum_{k=1}^{m} \alpha_{ik}^2$) depends on the sought solutions. Thus, a solution through an iterative procedure could be necessary, consist of the following steps:

1) Select an approximate solution for the communalities
2) Diagonalize the matrix $\boldsymbol{R} - \Psi$ (Calculate $\boldsymbol{A} \mid \boldsymbol{A}\boldsymbol{A}' = \boldsymbol{R} - \Psi$)
3) Re-calculate the communalities with the values of $\boldsymbol{A}$
4) Go back to step 2) until a convergence criterion is met.

In general, a small number of iterations are needed to reach convergence and, in practice, the same solution is reached from different initial solutions.

Although this algebraic procedure would lead us to obtain the estimated loadings, there are some matters to be defined. Should all the no null eigenvalues be used? Is it convenient to select them according to any special ordering? Which one of the infinite solutions of the eigenvectors should be used?

The Principal Factor method formalizes these concepts. The loadings for each common factor are obtained maximizing its contribution to the total variability. The factor model establishes that the variance of each variable is produced due to the contribution of the common factors and a specific factor. Thus, the total variability can be expressed as:

$$\sum_{i=1}^{p} Var(X_i) = \sum_{i=1}^{p} (h_i^2 + \Psi_i) = \sum_{i=1}^{p} \left(\sum_{k=1}^{m} a_{ik}^2 + \Psi_i\right) = \sum_{k=1}^{m} \left(\sum_{i=1}^{p} a_{ik}^2 + \Psi_i\right)$$

Let V1 the portion of the variability that is explained by F_1, that is $V_1 = \sum_{i=1}^{p} a_{i1}^2$. The vector $A'_1 = (a_{11}, a_{21}, \ldots, a_{p1})$ is sought so that V_1 is a maximum that satisfies the fundamental identity $\boldsymbol{R} - \Psi = \boldsymbol{AA'}$, which implies $p(p + 1)/2$ restrictions, p for the diagonal elements and $(p^2 - p)/2$ for the non-diagonal elements. As we have already pointed out, the diagonal elements of $\boldsymbol{R} - \Psi$ need to be known to find the solution through the search of conditional extremes, but they depend on the solution. Thus it would be necessary to apply an iterative procedure.

The solutions of the equations obtained by equating to zero the p derivatives of function to be maximized with respect to the p unknown values in A_1 show that the coefficients of F1 for each variable ($a_{i1} = Corr\ (F1, Xi)$, $i = 1 \ldots p$) are the components of the eigenvector associated with the greatest eigenvalue of $\boldsymbol{R} - \Psi$, say λ_1, normalized to the norm λ_1.

Why do we take the greatest eigenvalue? The answer is that the greatest eigenvalue is selected since the eigenvalue is exactly the contribution of the factor to the total variability, so it coincides with V_1, the function to be maximized.

Following the same procedure to estimate the loadings for the factor F2, the restriction should be established that it must not be correlated with F1 and must have the maximum contribution to the total variability, after the contribution of F1. The coefficients of F2 for each variable are then found and they are the components of the eigenvector associated with the second greatest eigenvalue of $\boldsymbol{R} - \Psi$, λ_2-normalized and so on until the required "*m*" factors are obtained. The procedure can be done using all the positive eigenvalues of $\boldsymbol{R} - \Psi$, but considering that the objective of Factor Analysis is to find a simplified explanation of the causes that produce the associations among variables, we should retain as few factors as possible.

This quantity could be identified with an additional calculus. For example, we can fix a small m, form the $\boldsymbol{A}$ matrix with $A_1, A_2, \ldots, A_m$ as successive columns, make the product between $\boldsymbol{A}$ and its transposed ($\boldsymbol{AA'}$) and compare the non-diagonal elements (estimations of the correlations through the model) with the homologous elements of the **R** or $\boldsymbol{R} - \Psi$ matrix (actual associations of the variables calculated from data). If those elements are similar, no other factor is added. This concept will be seen on Section 4: Criteria to define how many factors should be retained in the analysis.

We shall now see this capacity of reproduction of the original correlations in the Advertisement Testing example. If we adopt a model with six factors, the residual matrix $R - AA'$ is shown in Table 7 (a part of the table is shown).

Table 7. Residual correlations in Advertisement Testing example.

Variables (Labels)	like	nbrand	beli	other	feel	enter	uniq	irri	Again
like	0.312	–0.036	–0.007	–0.048	–0.020	–0.073	–0.039	0.025	0.055
nbrand	–0.036	0.289	–0.005	–0.054	0.006	0.050	0.002	0.002	–0.004
beli	–0.007	–0.005	0.310	–0.051	–0.020	0.021	0.007	0.003	–0.020
other	–0.048	–0.054	–0.051	0.314	–0.027	0.043	–0.002	–0.040	–0.055
feel	–0.020	0.006	–0.020	–0.027	0.219	0.011	0.019	–0.002	–0.012
enter	–0.073	0.050	0.021	0.043	0.011	0.252	–0.015	–0.006	–0.037
uniq	–0.039	0.002	0.007	–0.002	0.019	–0.015	0.333	0.035	–0.034
irri	0.025	0.002	0.003	–0.040	–0.002	–0.006	0.035	0.139	0.023
again	0.055	–0.004	–0.020	–0.055	–0.012	–0.037	–0.034	0.023	0.300
info	–0.047	–0.001	–0.007	–0.055	–0.037	0.042	0.028	0.012	–0.029
....	....	....	....	....	....	....	....	....	....

Questions for Self-Evaluation

11) What is your opinion about the magnitude of the residual correlations in Table 7? Are all the associations explained equally by the model?
12) If some diagonal elements (residual variance) are not small, does it cause concern about the goodness of the model? Why?
13) If four factors, rather than six, are included in the model, how would the residuals be affected? Could they remain unchanged compared with the six factors model?
14) Do the estimators obtained by the Principal Factor method require any distributional assumption for the vector of variables?
15) Optional: Establish the function to be maximized according to the Principal Factor method.

Maximum likelihood method

The maximum likelihood method is one of the most known and applied methods for parameter estimation in statistical inference, provided there is a probability model for the distribution of sample observations. For Factor Analysis, it is necessary to make the assumption that the multivariate observations come from a multivariate normal distribution where the variance-covariance matrix Σ is full rank and the number m of factors is pre-set. The model establishes that $\Sigma = AA' + \Psi$. In this case, the information contained in the sample covariance matrix S is sufficient for factor parameters estimation. Expressing the likelihood function of S in terms of its density function (Wishart distribution), the function to maximize with respect to the $p(m + 1)$ parameters of A and Ψ is:

$$ln\,f(\boldsymbol{S}) = L(\mathbf{A}, \boldsymbol{\Psi}) = \text{cte} + \frac{1}{2(n-p-1)} ln|\boldsymbol{S}| - \frac{1}{2n} ln|\boldsymbol{\Psi} + \boldsymbol{AA'}| - \frac{1}{2n} tr(\boldsymbol{\Psi} + \boldsymbol{AA'})^{-1}\mathbf{S}$$

The system of maximum likelihood equations is obtained from the set of derivatives of L and their equalization to zero. Since the system has no explicit solution (Morrison 2004, Johnson and Wichern 2007), it is necessary to use numerical methods. Maximum likelihood solution satisfies the property of invariance, whereby solutions for R are obtained from the transformation of the solutions for Σ.

The ML estimation method has some advantages. A goodness of fit hypothesis test of the factor model can be used to find the appropriate number of factors in the model, confidence intervals for the estimated loadings can be calculated and any additional assumptions on the nullity of some parameters a_{ij} can be added under researchers' assumptions, i.e., set on a prior structure of the associations between latent factors and variables to test its adequacy (confirmatory Factor Analysis). We will return to these issues later on.

The loading estimations for the Advertisement Testing example using ML method with VARIMAX rotation (obtained using SAS 9.3) are presented in Table 8. The interpretation of the factors is simplified by the rotation. The Standard Errors (SE) are added to the table to allow us to evaluate the estimation error. A solution with three factors is chosen due to a numerical convergence problem that appears when selecting more than three factors, producing a particular case, estimated communalities greater than or equal to one (known as Heywood or Ultra-Heywood case). The ML method is very susceptible to this particularity. Sometimes wrong specifications about prior communalities or number of common factors (too many or very few) can be the reason for this situation.

The SAS sentences to obtain the results exposed in Table 8 are:

```
proc factor data=advertisement
n=3 method=ML Heywood se reorder rotate=varimax
run;
```

The option "Heywood" maintains the loading estimations bounded (not greater that one) and the option "se" indicates the standard error estimation. The "method=ML" option produces ML estimation, "n=3" specifies the number of common factors, "reorder" indicates that the loadings are shown grouped by magnitude in the output and "rotate=varimax" is added to simplify the interpretation.

The standard errors of Table 8 can be used to calculate a confidence interval for the loadings. For example, $a_{ij} \pm 2\ SE$ will give an approximate range of 95% of confidence confirming whether the loadings are important or not. According to these criteria $a_{15,2}$, $a_{16,2}$ y $a_{18,3}$ may be accepted null because the 95% approximated confidence interval covers the zero.

Table 8. ML Loadings (SE) for variables in Advertisement Testing example.

Rotated Factor Pattern Estimate/StdErr			
	Factor1	*Factor2*	*Factor3*
RELEVANT	0.86718 0.00560	0.16217 0.01333	0.04807 0.01468
INFORMATIV	0.86262 0.00578	0.06009 0.01367	0.02013 0.01507
IMPORTANT	0.84624 0.00629	0.16511 0.01369	0.03975 0.01493
STROG_FEEL	0.82645 0.00694	0.26800 0.01365	0.05212 0.01470
FOR YOU	0.80047 0.00777	0.31528 0.01385	0.04511 0.01477
NEW BRAND	0.79183 0.00806	0.12165 0.01466	0.08193 0.01564
BELIEVABLE	0.76767 0.00874	0.23403 0.01480	0.05159 0.01555
NEWNESS	0.73848 0.00978	0.05763 0.01552	0.03614 0.01635
DIFFERENT	0.73747 0.00962	0.21764 0.01527	0.10529 0.01586
FIT_WITH BRAND_	0.73355 0.00975	0.22933 0.01534	0.06848 0.01591
OTHER WAY	0.72307 0.01008	0.18804 0.01558	0.06555 0.01617
SEE_AGAIN	0.67442 0.01140	0.44175 0.01465	0.07638 0.01530
INTERESTIN	0.67249 0.01150	0.40670 0.01513	0.04932 0.01565
LIKEABILIT	0.65573 0.01186	0.43962 0.01488	0.10240 0.01544
PERSUAS_I	0.15025 0.02076	0.02559 0.02041	0.09121 0.02108
AGRESSIVE	–0.42633 0.01717	–0.02798 0.01853	0.08433 0.01885

Table 8. contd....

Table 8. contd.

Rotated Factor Pattern Estimate/StdErr			
	Factor1	*Factor2*	*Factor3*
IRRITATING	–0.47339 0.01621	–0.14895 0.01869	0.05625 0.01840
CONFUSING	–0.54603 0.01475	–0.10480 0.01766	0.00901 0.01782
SILLY	–0.58493 0.01384	–0.11484 0.01728	0.04807 0.01758
ENTERTEINI	0.02464 0.01730	0.87773 0.00704	0.08182 0.01408
HUMOUR	–0.12124 0.01751	0.79848 0.00917	0.09320 0.01498
UNIQUENESS	0.28658 0.01684	0.68848 0.01162	0.11416 0.01530
BE MENTION	0.47312 0.01573	0.57188 0.01411	0.06586 0.01587
TO WIN	0.37820 0.01722	0.55751 0.01467	0.04721 0.01648
EMOTIVE	0.37349 0.01767	0.50967 0.01576	–0.00245 0.01700
MEMORABI	0.38060 0.01706	0.50112 0.01572	0.19867 0.01678
NAME CLEAR	0.15257 0.02045	0.20518 0.02106	0.11074 0.02026
RECALL_I	–0.02629 0.01381	0.11450 0.01192	0.99308 0.00410
BRAN_A_I	–0.00877 0.01506	0.02255 0.01387	0.86817 0.00608
RECOGN_I	–0.02755 0.01598	0.19596 0.01559	0.70796 0.01020
EFFECTIV_I	0.09139 0.01630	0.07810 0.01576	0.66636 0.01154

Questions for self-Evaluation

16) Identify the most important loadings and interpret the ML estimated first factor (Table 8).
17) Identify at least three loadings that, according to their SE value, could be considered important (greater than 0.4).

There are other estimation methods that will not be seen in this chapter, as the Harris' Noniterative Canonical Factor Analysis, Alpha Factor Analysis, Unweighted least squares, etc. These different methods will not necessarily bring the same results. However, if the sample size and the number of variables are large, the solutions will be similar (Browne 1968, Linn 1968).

Rotated solutions

A particularity of the solutions for the loadings in the factor model is that they are not unique. Once a particular solution is found, any orthogonal rotation is still a solution. This particularity of the Factor Analysis allow as to find a representation of the solution that is easy to interpret in terms of the factors. These rotations are obtained by introducing additional assumptions, which also makes it possible to achieve a unique solution. The criteria to rotate the solution are based on principles of "simple solution structure", which means that the small loadings are enlarged and each factor is associated with less variables. It is done by differentiating the loadings so that some of them are large and other small (Thurstone 1947). Based on the criterion that is used to perform the rotation, it has a different name. There are orthogonal and oblique rotations. Among the orthogonal ones, VARIMAX attains the simplification of the factors when there are too many variables, because it requires the variability among the factors' loading to be maximum. Another orthogonal rotation criterion used is QUARTIMAX, which is preferable if there are numerous factors, because it maximizes the variability of the loadings for each variable. In all cases, the optimality criterion is translated into a mathematical function to be satisfied and the solutions are expressed in terms of the rotation angles.

Oblique rotations are less attractive because the resulting factors are correlated, they can however be chosen when an interesting interpretation is not achieved. The interpretation of oblique rotations must consider the matrix of associations among factors.

Among other methods available are HK rotation, PROMAX, PROCUSTES; etc.

The loadings, in the factors found as solution for the analysis of the commercials—Advertisement testing, were estimated by the Principal Factor and ML methods and then rotated by the VARIMAX criterion (Tables 5 and 8). This rotation allowed us to identify the variables with highest correlations easily with each factor. Solutions without rotating (Table 9) do not show the associations as clearly, even if we only interpret loadings greater than 0.50. In general, the solutions when applying algebraic methods are characterized by showing the first factor with high loadings in almost all of the variables.

Table 9. Loadings for variables in Advertisement Testig example.

Variables (Labels)	Method of Principal Factor pre-rotation					
	Factor 1	**Factor 2**	**Factor 3**	**Factor 4**	**Factor 5**	**Factor 6**
Rele	0.86052	–0.09896	–0.05272	0.10603	0.01475	0.12655
New	0.85858	–0.07359	0.00681	0.07499	–0.01290	0.12417
Info	0.84543	–0.15333	–0.09217	0.25996	–0.00752	–0.07106
Feel	0.82350	–0.16070	–0.06940	0.24638	–0.03698	0.03617
Impo	0.82334	0.04005	0.08836	–0.10637	–0.03761	–0.01181
Nbrand	0.80960	0.09021	0.08472	–0.07537	–0.01396	0.10443
Other	0.80079	–0.11071	–0.06788	0.06642	–0.00252	0.16595
You	0.79426	–0.25409	–0.15315	0.22353	–0.04192	0.02141
Beli	0.78961	0.03201	0.13284	0.09175	–0.04318	0.02138
Diff	0.77675	–0.05039	–0.09403	0.17185	–0.05184	–0.10790
Fit	0.76981	–0.16288	–0.17961	0.14171	–0.04440	0.19298
Inte	0.76898	–0.07354	–0.08886	0.11120	0.02053	0.25630
Like	0.73763	–0.06573	–0.04658	0.34826	–0.02943	–0.11404
Again	0.69230	0.26539	0.33787	0.07676	0.08751	–0.23407
Enter	0.68890	–0.16763	–0.12436	0.38733	–0.01315	–0.19125
Humo	0.60284	0.32456	0.13454	–0.11731	–0.11370	0.11288
Uniq	0.60265	0.28129	0.37494	0.03802	0.11981	–0.22524
Win	0.57971	0.39537	0.35990	–0.10763	0.02871	–0.18026
Ment	0.57195	0.17360	0.32660	–0.07689	0.13068	–0.04575
Emo	–0.58483	0.29558	0.20648	0.50073	0.06270	0.02285
Memo	–0.62072	0.33182	0.14123	0.48103	0.03442	0.15438
Irri	0.13662	0.75828	–0.51432	–0.03032	–0.34579	–0.06394
Agre	0.14120	0.62026	–0.31025	–0.10954	–0.27427	0.07110
Silly	0.09790	0.60877	–0.49471	0.02969	–0.28975	–0.14202
Confu	0.24072	0.56545	0.48158	–0.19943	0.04327	0.09174
rec_I	0.41009	0.51432	0.49181	–0.26869	0.03429	0.01011
bra_I	0.20606	0.60860	–0.61877	–0.03494	0.39860	–0.08983
cog_I	–0.54229	0.31685	0.13836	0.64509	0.03734	0.17355
pers_I	–0.43957	0.39667	0.20440	0.62189	0.02714	0.18905
effe_I	0.17750	0.20512	–0.40417	–0.02519	0.85378	–0.04160
Clear	0.26466	0.16069	–0.05648	–0.31156	0.10226	0.70393

Questions for Self-Evaluation

18) Prove that if A is a solution, the product $A\Gamma$ is also a solution of the system, being Γ an ortogonal matrix of order p.
19) In Table 9, the communalities and uniqueness have not been added. Why? Could they be different from those shown in table 6?
20) What interpretation would you give to the Factor F2 if you decide to analyze the solution without rotating?
21) And to Factor F3?
22) What concerns does this suggest when analyzing data with this particularity of the Factor Analysis?

In the following example of educational research—Grammatical Construction, different estimation criteria and rotation methods are compared. In this case, the results are obtained with MINITAB 16 (Stat>Multivariate>Factor Analysis).

In grammar, we refer to "relative sentences or clauses" to expressions of the type "I greeted the man who I was presented to", "I met a man who wasn't there", "the child who was given the prize was thrilled". In the process of learning the language, these relative sentences constitute an area of syntactic knowledge that is first acquired orally and then developed further in writing, since some types of these clauses occur only in written production. The management of this aspect of the syntax is linked to the overall development of the language. Language teachers from a school in the city of Rosario, Argentina, studied the process of construction of relative clauses in child grammar, among children of different ages (from 8 to 12 years old). They designed a set of tests to capture every child's ability to construct relative sentences. Two of them consisted of requesting a spontaneous writing, narration and description, and analyzing the number of words used in each one of them (*Narration* and *Description*) and the total number of related sentences used in both (*Free Relatives*). The other two tests consisted of exercises to complete sentences: In one of them, three incomplete relative sentences with increasing order of difficulty were presented, measuring the number of completed correct structures (*Test 1*). Another test presents ten paragraphs with two incomplete structures each and they register the percentage of completed correct structures (*Test 2*).

Information about age (*Age*) and the highest level of education achieved by one of their parents (*Father's education*), encoded in categories that range from 1 to 6, where higher value represents higher education level reached, was added to the data set of the 41 children.

The MINITAB window with the selected options (Principal Components) is (Fig. 12):

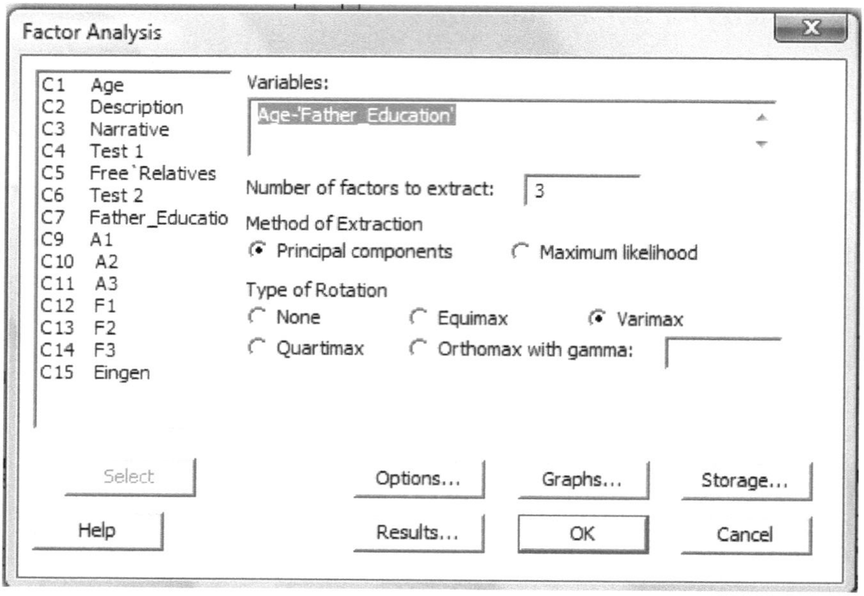

Figure 12. The MINITAB window with the selected options.

The correlation matrix denotes that the variables are associated and a Factor Analysis is appropriate to analyze data in an integrated approach. In principle, three factors are assumed to be the causation of the correlations, so the model is estimated by using $m = 3$. The estimated loading using Principal Factor method and VARIMAX criterion are shown in Table 10. The ML estimations after a QUARTIMAX rotation are shown in Table 11. The loadings $a_{ij} > 0.40$ are underlined.

Table 10. Principal Component Factor Analysis. Grammatical Structures.

Factor Loadings and Communalities							
	Unrotated				Varimax Rotated		
Variable	Factor1	Factor2	Factor3	Communality	Factor1	Factor2	Factor3
Age	0,865	-0,136	-0,046	0,770	0,854	-0,112	0,166
Description	-0,069	0,956	-0,020	0,920	-0,100	0,954	0,023
Narrative	-0,004	0,954	-0,060	0,914	-0,028	0,955	0,002
Test 1	0,863	0,002	-0,193	0,783	0,883	0,036	0,033
Free`Relatives	0,483	0,436	-0,063	0,427	0,465	0,450	0,089
Test 2	0,958	-0,068	-0,140	0,942	0,964	-0,035	0,104
Father_Education	0,408	0,098	0,905	0,995	0,164	0,043	0,983
Variance	2,817	2,048	0,885	5,750			
% Var	0,402	0,292	0,126	0,821			

Tabla 11. Maximum Likelihood Factor Analysis. Grammatical Structures.

	Factor Loadings and Communalities *						
	Unrotated				Quartimax Rotation		
Variable	Factor1	Factor2	Factor3	Communality	Factor1	Factor2	Factor3
Age	1,000	-0,000	0,000	1,000	0,921	0,157	0,355
Description	-0,125	-0,989	0,007	0,994	0,001	-0,993	0,086
Narrative	-0,127	-0,896	-0,156	0,843	0,048	-0,914	-0,076
Test 1	0,580	-0,009	-0,751	0,901	0,813	0,027	-0,489
Free`Relatives	0,292	-0,303	-0,166	0,205	0,367	-0,265	-0,012
Test 2	0,847	-0,006	-0,489	0,956	0,962	0,091	-0,151
Father_Education	0,270	-0,079	-0,071	0,084	0,284	-0,040	0,040
Variance	2,243	1,879	0,860	4,983			
% Var	0,320	0,268	0,123	0,712			

* NOTE * Heywood case

Both estimation methods lead to consistent interpretations. We only notice a difference in loadings of F1 and F2 on the variable Free's Relatives. This variable is key in this study, so it is preferred to use the solution of Table 10.

The loading plot according to the Principal Factor method on the first two factors (Fig. 13) provides an important visual aid to identify the interpretation of factors.

Teachers may deduce from these results that the ability of pupils to build relative sentences is strongly and directly related to the age and barely related to the educational level of parents (F1). There is also another

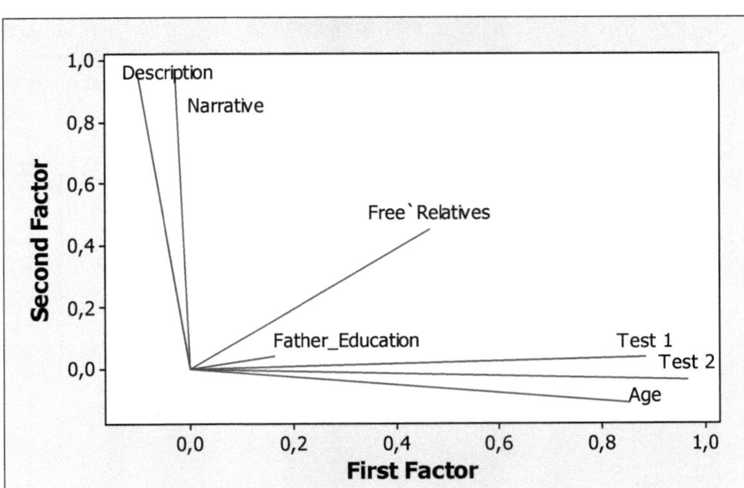

Figure 13. Loading Plot for the Grammatical Construction example.

latent factor (F2) that represents the ease to write essays and that could be identified as the capacity for using the written language. This capacity is linked to the use of related sentences in the spontaneous writing. Figure 14 shows the comparative magnitudes of the eigenvalues of $R-\Psi$ (Scree Plot).

The scatter plot of the two first factors is shown in Fig. 15. The variables (loadings) and pupils (scores) are represented simultaneously. Although we have not yet formalized how to obtain the scores of the latent variables for each pupil, we mentioned that it is possible.

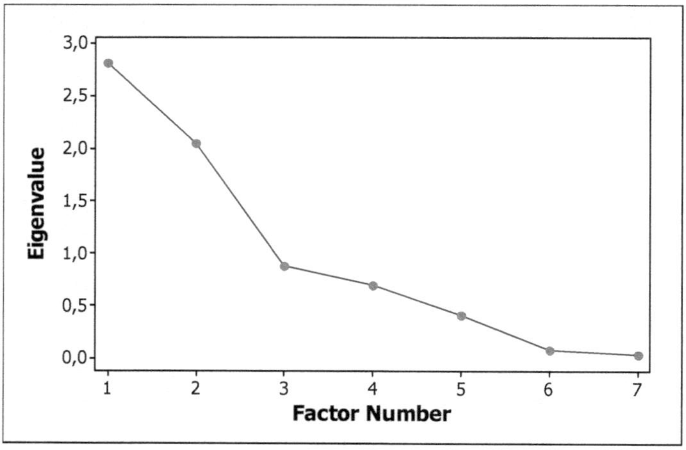

Figure 14. Scree Plot for the Grammatical Construction example.

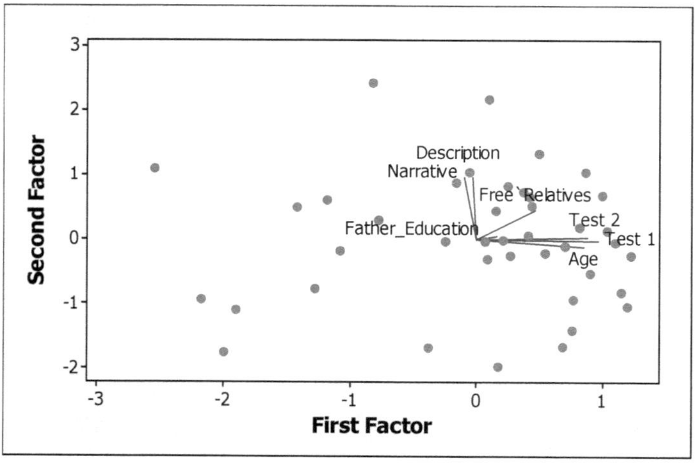

Figure 15. Biplot of variables and subjects for the Grammatical Construction example.

Questions of Self-Evaluation

23) Analyze whether the rotation facilitates the interpretation or not, in both methods applied.
24) If the interpretation of the rotated loadings in Table 9 were not satisfactory, is it possible to apply the same rotation criterion to obtain a new easily and more interpretable expression?
25) The penultimate row of Tables 10 and 11 contains values that correspond to variances. Which variables do such variances correspond to?
26) The last row of Table 9 contains percentages of variances. On what total are the percentages calculated?
27) Repeat the theoretical outline of the factor model (Fig. 9) in this application, identifying by name the complex factors, variables measured and the associations among them according to estimates in Table 10.
28) How would you incorporate factor 3 in a chart similar to Fig. 13?
29) What concept represents latent factor 3? Would you retain it in the analysis? Why?
30) What significance could the Scree Plot have in the application in Fig. 14?
31) Discuss the utility teachers could give to the representation of Fig. 15.

4 Criteria to Decide the Number of Common Factors to Retain in the Analysis

In previous sections, we have seen how to obtain different solutions and interpret the factors, although we have not yet formalized the criteria used to select the number of factors to retain in the analysis. It is necessary to keep in mind that the objective of the multivariate analysis is to obtain a summary of the complex information, and because of that we should try to retain a small number of factors, but still sufficient to explain the structure of associations among the variables in an adequate way. Among the procedures that will guide the decision of how many factors should be explained, there are heuristic methods as well as statistical methods.

Reproduction of observed correlations

The proposed model of Factor Analysis establishes a relationship between latent common factors and observed variables, which is translated into a fundamental relationship between observed correlations and correlations estimated by the model. That relationship is $\boldsymbol{R} = \boldsymbol{AA}' + \Psi$. The information in the first member of this equation comes from observation and that in the second member comes from the abstraction of the model. If the model is well built, choosing an appropriate number of factors, both matrices should have very similar non-diagonal elements (correlations). The difference $R - AA'$ is therefore calculated and the matrix is verified by checking that it is approximately diagonal.

Calculation of an adequacy measure based on residual correlations (RMSE)

If there are many variables, it is often confusing to deduce by observing if the residual correlations are sufficiently small. The overall Root Mean Square Error (RMSE) is the mean of the off-diagonals elements of the residual matrix and allows for making the decision based on a single number. It is defined as:

$$RMSE = \sqrt{\sum_{i=1}^{p} \sum_{\substack{j=1 \\ j \neq i}}^{p} \frac{res_{ij}^2}{p(p-1)}}$$

and it is suggested to use 0.05 as a breakpoint.

Table 12 shows RMSE values for models with different number of common factors in Advertisement Testing and Grammatical Structures examples. It was used for the Principal Factor method.

According to these results four or five common factors would be necessary in the Advertisement Testing example. For the Grammatical Structure example the number of factors suggested by RSME is four, although only two of them are retained according to the Scree Plot and the interpretation of the factors. The criteria for making a decision about the number of factors must be considered simultaneously.

Table 12. Root Mean Square Off-Diagonal Residuals Overall.

Examples	Factors in the model (m)				
	1	2	3	4	5
Advertisement Testing	0.13071	0.09663	0.07449	0.05072	0.04228
Grammatical Structure	0.23102	0.08540	0.07356	0.046072	0.011244

Mean square residuals by variables are used as indicators, so as to notice which are the associations that are best explained by the model. Table 13 shows those values for the models with 1, 2, 3, 4 and 5 factors. Due to the great number of variables involved in the Advertisement Testing example, we only show the results for the Grammatical Structure example.

Table 13. Root Mean Square Off-Diagonal Residuals by variable. Grammatical Structures.

m	Variables						
	Age	**Descrip.**	**Narrative**	**Test 1**	**Free**	**Test 2**	**Education**
1	0.10846	0.38929	0.38751	0.10108	0.1857	0.08317	0.09348
2	0.08322	0.05829	0.07047	0.10107	0.10939	0.07396	0.09005
3	0.08231	0.05785	0.06739	0.08745	0.11020	0.05507	0.01981
4	0.08098	0.03303	0.01925	0.07983	0.00785	0.01035	0.01722
5	0.00771	0.01763	0.01759	0.01025	0.00119	0.00991	0.00014

Questions for Self-Evaluation

32) Analyze the information in table 12. Which example shows the major improvements by increasing the number of factors?
33) Analyse what additional information the RSME provide you by variable in the Grammatical Structure example.
34) Some statistical software adds a measure similar to RSME, based on correlations among variables to fixed values of the factors (i.e. partial correlations corresponding to a conditional distribution). Which values of these averages would indicate a good model? Small or large ones? Why?

Relative magnitudes of the eigenvalues of R−Ψ

According to the Principal Factor method, the solutions for each factor can be obtained from the eigenvectors associated to each eigenvalue λ_i of $\boldsymbol{R} - \boldsymbol{\Psi}$. Their contribution to the reconstruction of the variables' total variance is decreasing according to the magnitude of λ_i. So, it is preferable to retain the factors that contribute the most to the total variance (though initially the goal is not to reproduce the variability with few factors, but to reproduce correlations with few factors). It is recommended then to retain the factors that make a differentiated contribution. A graphical inspection can be made (Scree test), plotting the eigenvalues in descending order of magnitude and selecting those with an important value, in comparison with the others. Those over an approximately horizontal line are discarded.

The Scree Plot for the Advertisement Testing example is shown in Fig. 16. This plot is part of an output of SAS 9.3 and it is accompanied by a cumulative graph that allows for the easy identification of the proportion of total variability explained by the sequence of factors.

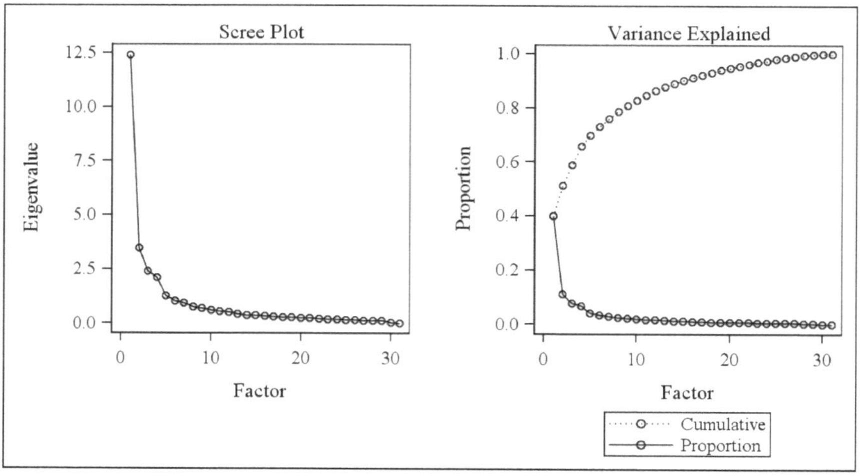

Figure 16. Scree Test for Advertisement Testing example.

Hypothesis test on the number of common factors

If the loadings are estimated by the ML method, it is possible to test the hypothesis that a certain number m of factors are acceptable (Barttlet's Chi Square). Testing the hypothesis does require the assumption of multivariate normality. Using the likelihood ratio test we can test the null hypothesis:

$$H_0\colon \boldsymbol{\Sigma} = \boldsymbol{A}\,\boldsymbol{A}' + \boldsymbol{\Psi}\,, \qquad rank(\boldsymbol{A}) = m\ (known).$$

Under H_0 and for large n the likelihood ratio statistic

$-2\log(\lambda) = (n-1)\ tr(\boldsymbol{AA}' + \boldsymbol{\Psi})^{-1}\boldsymbol{S} - ln|(\boldsymbol{AA}' + \boldsymbol{\Psi})^{-1}\boldsymbol{S}| - p$, has an approximate chi-square distribution with ½($(p-m)^2 - p - m$) degrees of freedom. The value of this statistic is often used as an indicator of the model goodness. If two models are acceptable ($p > \alpha$, the null hypothesis not rejected), the model with the lowest value of the Chi Square statistic is preferred.

Other criteria for model comparison

The Akaike's information criterion (AIC) is a useful indicator to compare models that can be used with ML estimations. It was presented by Akaike in 1973 as a way of comparing models as regards the adequate number of parameters to be included in the model. It was adapted for use in Factor Analysis by the same author in 1987. The logarithm of the likelihood function is used in its calculation. The model that leads to the lowest value of the indicator is chosen as the most adequate.

Other indicators defined with the same purpose are Schwarz's Bayesian Criterion (SBC, Schwarz 1978) and Tucker and Lewis's Reliability Coefficient (Tucker and Lewis 1973).

Table 14 shows the results of Barttlet's test and of these four last indicators, for the model with ML estimations of three factors (showed in Table 9).

Table 14. Goodness of fit indicators for the ML estimated model in Advertisement Testing example with 3 factors.

Significance Tests Based on 2381 Observations			
Test	**DF**	**Chi-Square**	**Pr > ChiSq**
H0: No common factors	465	70255.1705	<.0001
HA: At least one common factor			
H0: 3 Factors are sufficient	375	22955.8228	<.0001
HA: More factors are needed			

Chi-Square without Bartlett's Correction	23083.526
Akaike's Information Criterion	22333.526
Schwarz's Bayesian Criterion	20167.797
Tucker and Lewis's Reliability Coefficient	0.599

The results of the tests in Table 14 show that the hypothesis "No common factors" is rejected, indicating that it is proper to estimate a factor model, and the hypothesis "3 common factors are sufficient" is also rejected. Thus, more than three factors should be used to explain the associations.

Questions for Self-Evaluation

35) Can you use the values of indicators in Table 14 (Chi-Square, Akaike, Schwarz or Tucker and Lewis) to evaluate the model? How can you do it? Why?

We shall see now another example -Quality Honey- and consider the results together to understand how the Factor Analysis discovers interesting matters. Argentina is one of the main honey exporter countries. It has achieved a high level of competitiveness in the apiculture field, regarding performance per beehive and honey production technologies. Nevertheless, it is still recognized in the international market as a provider of honey without differentiation, in times when consumers appreciate classified products. Honey is a natural product that offers several attributes that can be appreciated by the senses. The wide range of flowers that bees are attracted to causes a multiplicity of flavors and aromas which are specific of each source. In studies on the quality of honey in the pampean phytogeographic region in Argentina, eighty one samples of honey from beehive were collected, taking into account 15929 IRAM Quality Standards. Beekeepers take the samples of honey as samples of the global average harvest for the year. On each sample the following continuous variables, among others, are measured:

Physicochemical characteristics: *Color*, expressed in mm Pfund; *Humidity*, expressed as a percentage of the weight of honey; *Ash*, measured in μS/cm and pH of *Acidity*.

Sensorial characteristics: *Sweet*, *Persistence*, *Granularity* and *Intensity*, they all are measured in a quantitative scale and evaluated by a sensorial panel composed by seven members, who were trained according to IRAM 20005 Quality Standards, applying the quantitative descriptive sensory analysis technique.

Pollinic characteristics: *Eucalyptus*, *Trefoil*, *Trifolium*, *Lotus*, *Melilotus* y *Medicago sativa*, all of them measured as the percentage of pollen grains in a sample of honey analyzed under a microscope.

Honey samples are classified into different categories according to pollinic variables. This analysis considers the following classes: *"Eucalyptus Honey"* if its composition contains at least 70% of eucalyptus pollen grains and *"Trefoil Honey"* if in its composition there are Trifolium, Lotus, Melilotus and Medicago sativa pollen, which together reach a minimum value of 45%.

The aim of this analysis is to search for latent factors that explain the stronger associations between physicochemical and sensorial characteristics, then associating them with the categories of honey.

Data was processed with SAS 9.3. The sentences used and the main results obtained are shown below.

```
ods graphics on;
proc factor data=honey
corr MSA
n=3 method=prinit priors=SMC rotate=varimax reorder
residual
plot=(scree preloadings loadings) plots=loadings(vector);
run;
ods graphics off;
```

Sentences in "proc factor" establish that the use of the iterative principal factor method for estimation (method=prinit) include three common factors in the solution (n=3), use multiple correlation coefficients as initial communalities (priors=SMC), show the partial correlation matrix (corr) and a Factor Analysis adequacy measure (MSA), rotate solutions by Varimax method (rotate=varimax), present the rotated solutions grouped in classes according to its magnitude (reorder) and show the residual correlation matrix of the fitted model (residual). Output will contain the scree plot and loadings plot for both rotated and non-rotated loadings (plot=(scree preloadings loadings)). Commands "ods graphics on" and "ods graphics off" are used to improve the graphics quality. Convergence was achieved after four iterations.

Output SAS. Honey Quality example

Correlations								
	humid	*Color*	*ash*	*acidity*	*sweet*	*persis*	*granu*	*intens*
Humid	1.00000	0.43061	−0.04836	0.57628	0.22710	0.29101	0.12986	0.25238
Color	0.43061	1.00000	0.41922	0.56599	0.29868	0.31904	0.08195	0.80011
Ash	−0.04836	0.41922	1.00000	0.11669	0.24753	0.03624	0.13109	0.45002
Acidity	0.57628	0.56599	0.11669	1.00000	0.17765	0.10914	0.19565	0.54286
Sweet	0.22710	0.29868	0.24753	0.17765	1.00000	0.54055	0.13731	0.06089
Persis	0.29101	0.31904	0.03624	0.10914	0.54055	1.00000	0.22891	0.11271
Granu	0.12986	0.08195	0.13109	0.19565	0.13731	0.22891	1.00000	0.20006
Intens	0.25238	0.80011	0.45002	0.54286	0.06089	0.11271	0.20006	1.00000

Partial Correlations Controlling all other Variables								
	humid	*color*	*ash*	*acidity*	*sweet*	*persis*	*Granu*	*Intens*
Humid	1.00000	0.25643	−0.16004	0.46177	−0.01155	0.14516	0.05645	−0.20425
Color	0.25643	1.00000	0.16804	0.12387	0.18984	0.24550	−0.29583	0.71170
Ash	−0.16004	0.16804	1.00000	−0.15947	0.30147	−0.20662	0.11577	0.22137
Acidity	0.46177	0.12387	−0.15947	1.00000	0.15315	−0.22498	0.13554	0.26024
Sweet	−0.01155	0.18984	0.30147	0.15315	1.00000	0.46144	0.03405	−0.30324
Persis	0.14516	0.24550	−0.20662	−0.22498	0.46144	1.00000	0.24537	−0.05465
Granu	0.05645	−0.29583	0.11577	0.13554	0.03405	0.24537	1.00000	0.24486
Intens	−0.20425	0.71170	0.22137	0.26024	−0.30324	−0.05465	0.24486	1.00000

Kaiser's Measure of Sampling Adequacy: Overall MSA = 0.62282735							
humid	*color*	*ash*	*acidity*	*sweet*	*persis*	*granu*	*intens*
0.665242	0.655177	0.632612	0.715675	0.545685	0.552703	0.438404	0.605342

Prior Communality Estimates: ASMC							
humid	*color*	*ash*	*acidity*	*sweet*	*persis*	*granu*	*intens*
0.516333	0.903213	0.408206	0.629816	0.486330	0.491141	0.205816	0.871951

Preliminary Eigenvalues: Total = 3.85723104 Average = 0.48215388				
	Eigenvalue	*Difference*	*Proportion*	*Cumulative*
1	2.66742565	1.77675123	0.6915	0.6915
2	0.89067441	0.26338715	0.2309	0.9224
3	0.62728726	0.43494918	0.1626	1.0851
4	0.19233808	0.13505469	0.0499	1.1349
5	0.05728338	0.19010870	0.0149	1.1498
6	-.13282532	0.04142099	–0.0344	1.1154
7	-.17424630	0.09645982	–0.0452	1.0702
8	-.27070612		–0.0702	1.0000

3 factors will be retained by the NFACTOR criterion
convergence criterion satisfied

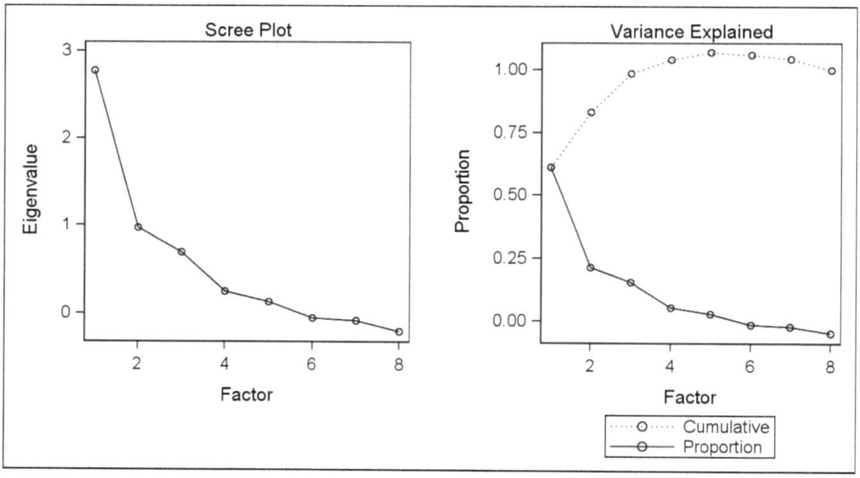

Rotated Factor Pattern			
	Factor1	Factor2	Factor3
Intens	0.82663	0.41874	–0.02181
Color	0.69455	0.49806	0.25970
Ash	0.62256	–0.10440	0.15583
Humid	–0.03519	0.74849	0.25566
Acidity	0.28492	0.73759	0.07221
Sweet	0.12542	0.05566	0.73821
Persis	0.02827	0.15191	0.72299
Granu	0.12080	0.12603	0.19687

Variance Explained by Each Factor		
Factor1	Factor2	Factor3
1.6668460	1.5806411	1.2692189

Root Mean Square Off-Diagonal Residuals: Overall = 0.04216345							
humid	*color*	*ash*	*acidity*	*sweet*	*persis*	*granu*	*intens*
0.014597	0.047642	0.040853	0.037366	0.036623	0.049691	0.060487	0.034707

The off-diagonal elements of the correlation matrix that shown in the SAS Output are medium. The highest are around 0.60. Some of the correlations are low. These values decrease for most of the variables when partial correlations are calculated, which points out that the Factor Analysis is adequate. Kaiser's Measure Sampling adequacy measure (MSA) is 0.6228, greater than 0.50, which confirms what we have observed in the correlations. The MSA per variables show that some of the variables will be better explained than others (MSA for *Acidity*=0.72 and MSA for *Granularity*=0.44).

The Scree Plot in the SAS Output reveals that three important factors accumulate a high percentage of the total data variability. A sharp bend occurs at the fourth eigenvalue, thus indicating that each of the remaining factors does not have important contributions. The observation of residual correlations confirms this finding. Although this matrix of partial correlations is not shown in the output, the RSME summarizes its information. The RSME value is small (0.0422), thus concluding that the three factor model has been sufficient for explaining all the variables. The smallest mean residual correlation by variable was 0.0145 for *Humidity* and the highest one was 0.06 for *Granularity*.

The VARIMAX-rotated loadings show that F1 is highly correlated to two physicochemical characteristics, *Color* and *Ash*, and to a sensorial one, *Intensity*. The honey samples with higher F1 values will present intense flavor and color, and high values of *Ash* (F1 is a factor of "intensity").

F2 shows the highest associations with *Humidity* and *Acidity* (it can be identified as "acidity-humidity") and F3 is highly associated with *Sweet* and *Persistence* (identified as "sweet-persistence").

Questions for Self-Evaluation

36) Look through the SAS program and identify the graphs that were omitted in the output which would have been useful for interpreting the factors.
37) Calculate communalities and specificities for each variable according to the three latent factors model.
38) Write one of the estimated model equations (for example, for variable *Sweet*).

39) Does the output give you any information about the associations between latent complex factors and *"Eucalyptus"* and *"Trefoil"* honey type? What procedure would you suggest to answer the question about honey type characterization?

5. Evaluation of Factors on Individuals

An important matter for complementing the factor analysis, as it was shown in the practical applications, is to measure latent variables over each unit of analysis. This would be a trivial task if the loading matrix $\boldsymbol{A}$ in the factor model would be invertible. Nevertheless, the factor model establishes a relationship between the observed measurements (X) and the latent or non-observed ones (F) that can be used to deduce the values of F over each individual, once the matrix $\boldsymbol{A}$ is estimated.

Least squares method

Consider an estimated factor model with m factors. We will call F_i the vector of latent variables (scores) for the individual i. According to the factor model, the variables X_j with $j = 1, \ldots, p$ can be represented through a linear combination of the F_k factors with $k = 1, \ldots, m$: $\widehat{X_{ij}} = \sum_{k=1}^{m} a_{jk} F_{ik}$ $i = 1, \ldots, n$. The objective is to find F for the individual i so that:

$$\min_F \sum_{j=1}^{p} (X_i - \hat{X}_i)^2 = \min_F \sum_{j=1}^{p} \left(X_i - \sum_{k=1}^{m} a_{jk} F_{ik} \right)^2$$

The values of factors that satisfy this condition are called "least squares estimates" and they will be studied in the Multiple Regression chapter. The system of partial derivatives that must be set to zero in order to satisfy the necessary conditions to find extremes, are expressed in a matrix form. The solutions for the m common factors for the individual i are expressed as:

$$\widehat{\boldsymbol{F}}_i = (\boldsymbol{A}'\boldsymbol{A})^{-1}\boldsymbol{A}'\boldsymbol{X}_i$$

Thus, for each individual $i = 1, \ldots, n$, the m rows of the product matrix $(\boldsymbol{A}'\boldsymbol{A})^{-1}\boldsymbol{A}'$ contain the coefficients of linear combinations of the known X_{ij} variables, which estimate the m latent factors.

Regression method

This method proposes a multiple regression model in which factors are explained by the observed variables, that is: $F_k = \sum_{j=1}^{p} b_{kj} X_j$, $k = 1, \ldots, m$. If coefficients b_{kj} were known for each individual $i = 1, ..., n$, the factor F_{kj} could be estimated from the known X_{ji}. Coefficients b_{kj} are deduced so that:

$$\min_b \sum_{i=1}^{n} (F_{ki} - \hat{F}_{ki})^2 = \min_b \sum_{i=1}^{n} \left(F_{ki} - \sum_{j=1}^{p} b_{kj} X_{ji} \right)^2$$

The solution of the normal equations leads to a least squares estimator of the vector of coefficients $\boldsymbol{\beta}_k$: $\hat{\boldsymbol{\beta}}_k = (\boldsymbol{X}'\boldsymbol{X})^{-1} \boldsymbol{X}' \boldsymbol{F}_k$, where $\boldsymbol{X}$ is the nxp matrix of observed values. These estimators are expressed as functions of the non-observable factors, however, although the vector $\boldsymbol{F}_k$ is unknown for every individual, the product $\boldsymbol{X}'\boldsymbol{F}_k$ is actually known from the model.

If we are working with standardized values of $\boldsymbol{X}$ and $\boldsymbol{F}$ (indeed this is the condition we imposed to specify the model) then $(\boldsymbol{X}'\boldsymbol{F}_k)/\boldsymbol{n} = \boldsymbol{A}$ since every component of the vector is:

$$\frac{1}{n} \sum_{i=1}^{n} X_{ji}\ F_{ki} = \widehat{Corr}\left(X_j\ F_k\right) = a_{jk}, \qquad j = 1, .., p$$

Then,

$$\hat{\boldsymbol{\beta}}_k = (\boldsymbol{X}'\boldsymbol{X})^{-1} \boldsymbol{X}'\ \boldsymbol{F}_k = \left(\frac{\boldsymbol{X}'\boldsymbol{X}}{n}\right)^{-1} \frac{\boldsymbol{X}'\ \boldsymbol{F}_k}{n} = \boldsymbol{R}^{-1}\ \boldsymbol{A}_k$$

where $\boldsymbol{A}_k$ is the $\boldsymbol{k}$th column of $\boldsymbol{A}$.

Thus, the m columns of the product $\boldsymbol{R}^{-1}\ \boldsymbol{A}$ contain the weights or coefficients for the linear combinations of the variables, which give the scores of each factor F_k, $k = 1, ..., m$.

The statistical software allows us to obtain and store the coefficients of the linear combinations as well as the values of the scores over each individual. In Minitab this can be done with: Stat>Multivariate>Factor Analysis>Storage. In SAS, the following options must be written in the proc factor: OUT= "sas data set" for storage the scores and OUTSTAT="sas data set" for storage the coefficients.

Questions for self-evaluation

40) Answer again question 39, from the point of view of section 8.5.
41) Look through the figures in this chapter and identifiy those in which was necessary to estimate the factor scores on the individuals to make that graphical representation.

42) Which values would you use to obtain the valorization of a factor over an individual? Would you use the standardized or non-standardized values?
43) Look at the Fig. 17, which shows the eighty one samples of honey from the Quality Honey example in the scatter plot with factor 1 and factor 2, each one with its label indicating if the sample was categorized as *"Eucalyptus"* (label=E) or *"Trefoil"* (label=T) according to the measurements of pollen. State a description for *"Eucalyptus"* honey and *"Trefoil"* honey.
44) Have you included the variables *"Sweet"* and *"Persistence"* in your description of honeys? Why?

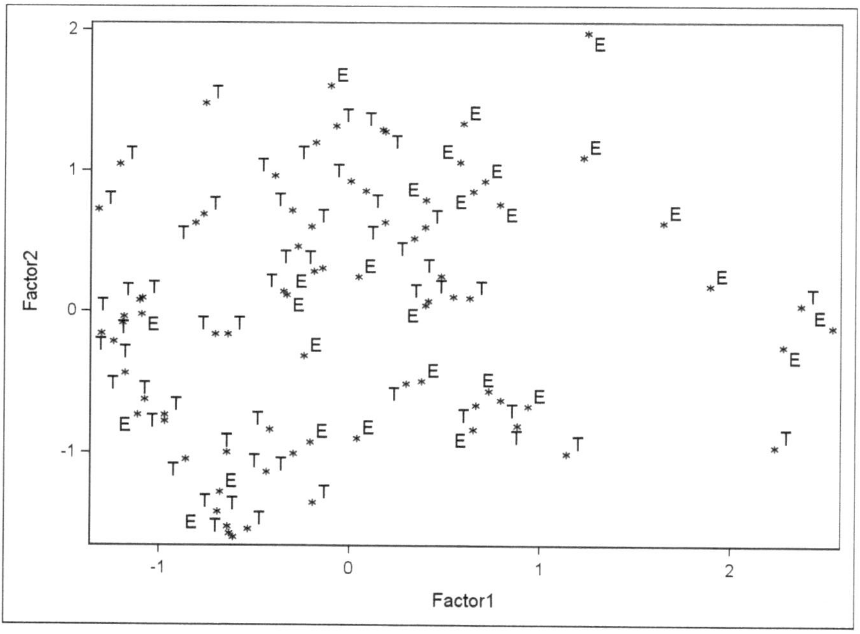

Figure 17. Honey samples on F1 and F2 by pollinic type.

6 Principal Components as a Particular Case of Factor Analysis

A particular case of Principal Factor method occurs when communalities are set to one and only one step of the iterative process is used. This approach leads to the same solutions as another widely used multivariate technique called Principal Components (Pearson 1901, Hotelling 1933). The aim of this technique is to identify the factors that explain the observed differences among individuals on whom a set of quantitative variables have been measured, which is different from the objective of Factor Analysis. This means that the focus of this technique is the study of similarities and differences among rows of the data matrix (individuals) while Factor Analysis focuses its attention on the relationship among columns (variables). Despite this, both techniques have something in common.

The principal components model defines new variables or factors as linear combinations of the original ones so that they explain in the best possible way the total variability of the observed data. Thereby, by retaining a small number of these new variables, much of the variability among individuals would be explained in a condensed form. The model is expressed as:

$Y_k = c_{k1}X_1 + c_{k2}X_2 + \cdots + c_{kp}X_p$, $k = 1,\dots,p$, such that
$V(Y_1) \geq V(Y_2) \geq \cdots \geq V(Y_p)$

Considering the whole group of components, the matrix form of the model can be expressed as $=\boldsymbol{C'X}$, where $\boldsymbol{Y}$ and $\boldsymbol{X}$ are p-dim vectors and $\boldsymbol{C'}$ is a pxp matrix, whose rows contain the coefficients for the linear combination of each new variable Y_k, $k = 1,\dots,p$.

If a small number of components (m) explain a considerable part of the total variability, the original information contained in a p-dimensional space can be explained by a reduced m-dimensional space (with m<p) with little loss of information. This is the reason why Principal Components Analysis is known as a dimension reduction technique.

Schematically:

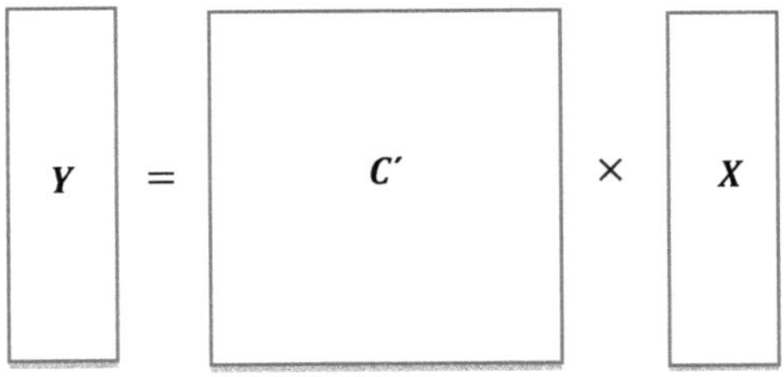

Figure 18. Principal Components model in matrix form

It is worth noting that in Principal Components Analysis the new variables $\boldsymbol{Y}$ are defined by the observed variables $\boldsymbol{X}$, opposite to Factor Analysis in which the latent factors $\boldsymbol{F}$ are discovered by the data or proposed by a theoretical concept and then confirmed by the data.

The coefficients matrix C that satisfies the condition on the variances is an orthogonal matrix that diagonalizes Σ, the variances and covariances matrix of the variables. This means that $\Sigma = C\Lambda C' = C\Lambda^{-\frac{1}{2}}\Lambda^{-\frac{1}{2}}C' = C\Lambda^{-\frac{1}{2}}\left(C\Lambda^{-\frac{1}{2}}\right)'$ where Λ = diag(λ_i) with λi, $i = 1, ..., p$ are the eingenvalues of Σ and C is an orthogonal matrix with the normalized eigenvectors in its columns. In this case Σ is a full-rank matrix and $\lambda i > 0$, $i = 1, ..., p$. The eingevalues are the variance of each new variable Y_i, so, it is necessary to order them so that $\lambda_1 \geq \lambda_2 \geq .... \geq \lambda_p$.

If the original variables are standardized, $\mathbf{\Sigma} = \mathbf{R}$, and $m=p$, we are able to recognize an estimation for the Principal Factor in $C\Lambda^{-\frac{1}{2}}$, according to the factor analysis model: $\mathbf{X} = \mathbf{AF} + \mathbf{E}$. Thus, the interpretation of the latent factors will be similar to that of the principal components. In this case the solution is unique.

Question for self-evaluation

45) Optional: Consider the function to optimize in a Principal Components Analysis. Find the solutions using Lagrange´s multiplier. Note that this solution is unique. Bear in mind that the objective is to obtain new variables as linear combinations of the original $\mathbf{X}$ variables, so as to preserve the differences and similarities among individuals or subjects under analysis.

In the following example—Flexibility, the aim of the analysis is to look for similarities and differences among individuals, rather than identify the factors that cause correlations among variables. In a sports club where swimmers train for competitions, a workout plan was designed to improve the swimmers' flexibility and arthro-muscular elongation. The plan was based on the hypothesis that such physical aptitudes can be improved by doing elongation exercises. A set of measurements were established to evaluate the flexibility of different parts of the body through special appliances. These measurements were registered before and after the training period on twenty athletes. It was also registered the age (in years), sex and whether the swimmer was able to pass a squat test. The most important question to be answered through the analysis of the information was: Did the training plan cause an improvement in the swimmers' flexibility? Principal component analysis may respond to this question and derive other interesting aspects.

The set of variables (labels) used are: *Lumbosacral joint*, with maximum elongation of hamstring muscles (L_S) (+), *Horizontal back* (H_back) (–), *Vertical back* (V_back) (+), *Ventral dorso-lumbar spine* (V_d-l_spine) (–), *Dorsal dorso-lumbar spine* (D_d-l_spine) (+), *Above the shoulder joint* (A_S) (+), *Block ankle joint and foot plantar flexion* (F_plantar) (–), *Block articular ankle and foot in dorsiflexion* (F_dorsif) (+), *Squat test* (Squat), *Age* (age). The signs between brackets indicate whether large magnitudes (+) or small magnitudes (–) of a variable denote greater flexibility. All flexibility measurements are expressed in *cm.*, except the squat test that is a dichotomy measure: passed or failed.

The results obtained with MINITAB 16, are showed below (Stat>Multivariate>Factor Analysis, Method of Extraction: Principal Components, Type of Rotation: None).

Factor Analysis: Flexibility

Principal Component Factor Analysis of the Correlation Matrix

Unrotated Factor Loadings and Communalities

Variable	Factor1	Factor2	Factor3	Factor4	Communality
L_S(+)	0,830	-0,369	0,089	0,008	0,833
H_back(-)	-0,489	-0,657	-0,197	0,077	0,715
V_back(+)	0,407	0,500	-0,405	0,463	0,795
V_dl_spine(-)	-0,729	0,144	-0,352	0,192	0,713
D_dl_spine(+)	0,760	-0,184	-0,278	0,211	0,733
A_S(+)	-0,038	0,131	-0,697	-0,677	0,963
F_plantar(-)	-0,758	0,166	-0,127	0,336	0,731
F_dorsif(+)	0,409	0,767	-0,157	0,012	0,781
Age	0,224	-0,700	-0,485	0,204	0,817
Variance	2,9965	1,9960	1,1740	0,9148	7,0814
% Var	0,333	0,222	0,130	0,102	0,787

Factor Score Coefficients

Variable	Factor1	Factor2	Factor3	Factor4
L_S(+)	0,277	-0,185	0,076	0,008
H_back(-)	-0,163	-0,329	-0,168	0,084
V_back(+)	0,136	0,251	-0,345	0,506
V_dl_spine(-)	-0,243	0,072	-0,300	0,210
D_dl_spine(+)	0,254	-0,092	-0,237	0,231
A_S(+)	-0,013	0,066	-0,594	-0,740
F_plantar(-)	-0,253	0,083	-0,108	0,367
F_dorsif(+)	0,137	0,384	-0,134	0,013
Age	0,075	-0,351	-0,413	0,223

The results shown through the tables and graphics in the MINITAB output are highly informative about the athlete's different characteristics. Two or three factors explain about 70% of the variability among athletes. The first two factors are much more important regarding differentiation (Fig. 19).

The loading plot (Fig. 20) shows that F1 is a differentiation factor identified with "general flexibility". F1 shows high correlations with almost every variable, except *Age* and *Above the shoulder joint* (A_S). The associations are positive with flexibility indicators (+) and negative with flexibility indicators (–). This indicates that high positive values of F1 show a more flexible swimmer in general. High negative values show a less flexible swimmer. This indicator may serve to rank the swimmers according to this concept.

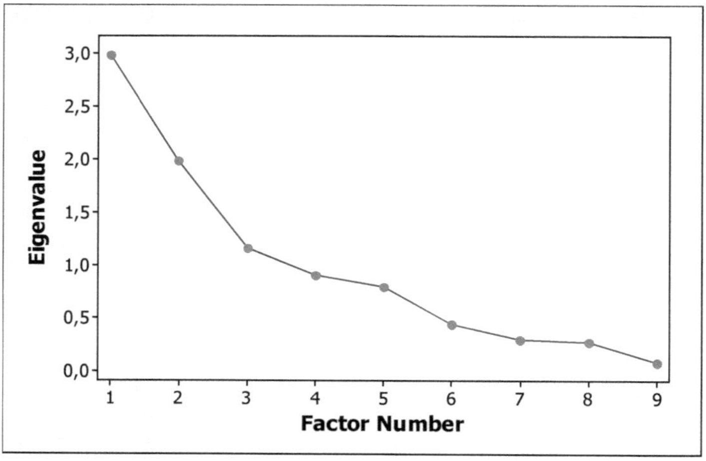

Figure 19. Scree Plot in Flexibility example.

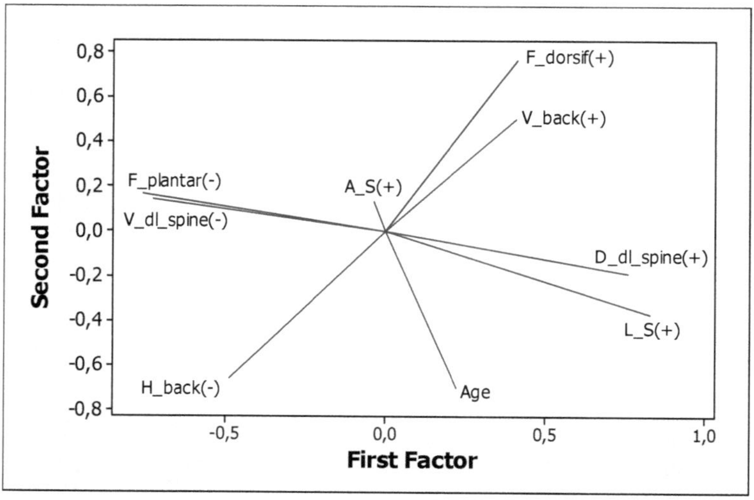

Figure 20. Loading Plot in Flexibility example.

Factor 2 is negatively associated with *Age* and *Horizontal back* (H_back) (–) and positively associated with *Vertical back* (V_back) (+) and *Block articular ankle and foot in dorsiflexion* (F_dorsif) (+). This factor is an indicator of good flexibility in some specifics aspects: *Block articular ankle and foot in dorsiflexion* (F_dorsif) (+), *Horizontal back* (H_back) (–) and *Vertical back* (V_back) (+) and these aptitudes are better in younger swimmers (ranging from 12 to 16). These characteristics may make them more efficient for a specific swimming style.

The loading plot simplifies this interpretation, by observing the location of the variables in the different quadrants.

As additional analysis, Fig. 21 shows the scores on F1 and F2 with a label indicating the sex (F: female and M: male). According to the meaning of F1, the more flexible swimmers are located to the right, and it can be noticed that there are more points labeled "F" on that side of the plane. Moreover, more "M" points are found above the x-axis of the scatterplot, which indicates that men are more flexible in *Horizontal back, Vertical back* and *Block articular ankle and foot in dorsiflexion*. If we repeat this plot (figure is not shown) with the results of the squat test, passed or failed, we may note that the points with label "passed" are concentrated on the right.

Finally, we could focus our interpretation on the objective of the analysis, labeling the points according to whether the observation was before or after the workout program for each swimmer. On Fig. 23 the points for some swimmers were connected to see the pre-post performance, allowing the detection of a pattern of movement going from left to right. This indicates

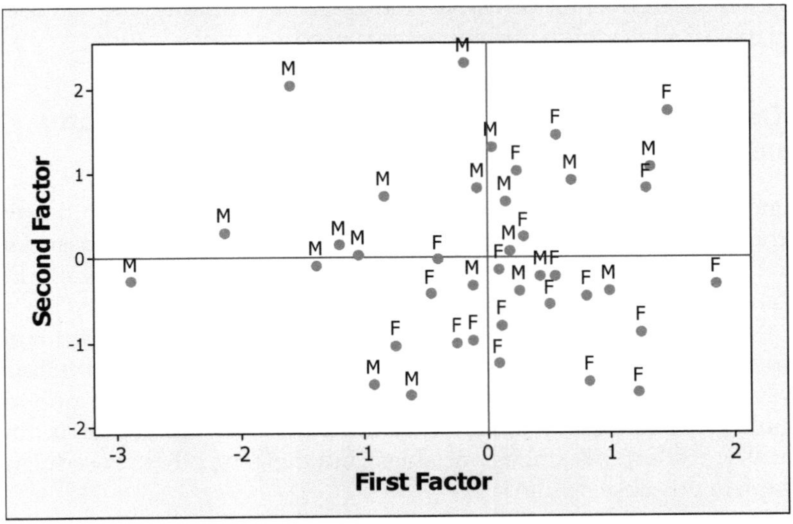

Figure 21. Score Plot for individuals on the plane of Factor 1 and Factor 2.

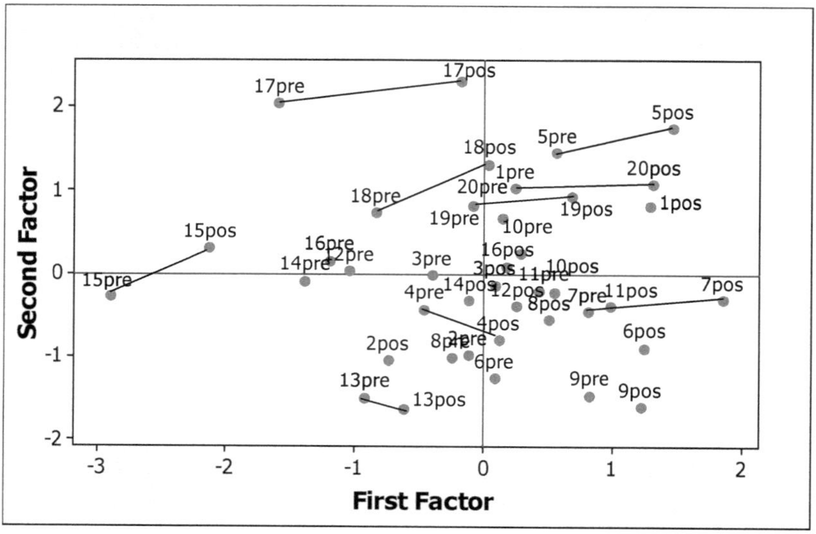

Figure 22. Score Plot of Flexibility by swimmers' pre-post performance.

a flexibility improvement especially in *Horizontal back* (H_back), *Vertical back* (V_back) and *Block articular ankle and foot in dorsiflexion* (F_dorsif) when the trend is to move upwards. These findings are descriptive and were confirmed through multivariate hypothesis tests in the original study.

Interestingly, the first factor (or principal component in this case) can be used as an overall univariate indicator of flexibility, i.e., a summary variable of all flexibility measurements used.

7. Data Evaluation Prior to the Implementation of Factor Analysis

A question we have not done yet, though it corresponds to the initial phases of the analysis, is: When we intend to analyze the structure of associations among a more or less comprehensive set of variables, does the Factor Analysis always provide us with the solution?

Naturally, the answer will be yes provided that such a structure of associations is important. Thus, it is possible to make some preliminary analysis, which will allow us to make this decision. In the previously exposed applications we have examined the correlation matrix (or the pairwise scatterplots among variables), but there are other procedures that complete this observation.

If variables explain each other, the partial correlations, i.e., correlations among variables to fixed values of the other ones, should be small. Thus, the off-diagonal elements of the partial correlation matrix also contains information about whether the application of Factor Analysis is adequate. These correlations is also referred to as negative anti-image correlations.

In either of these cases, if the number of variables is important, it would be difficult to observe all quantities out of the main diagonal simultaneously. Kaiser 1970, Cerny and Kaiser 1977 has defined a global measure called Kaiser's Meyer Sampling Adequacy, MSA. This statistic is a summary of how small the partial correlations are, relative to the original (zero-order) correlations. It is desirable for the MSA to be high. Kaiser, H. and Rice, J. 1974 have established ranges for the MSA values, suggesting that the "mediocre" correspond to those values between 0.60–0.70, "middling" between 0.70–0.80 and "meritorious" between 0.80–0.90. The MSA measure should be equal to 0.5 when the correlation matrix equals the partial correlation matrix.

Another possible evaluation is a statistical test on whether the correlations are null. Bartlett's test on sphericity is applied to test that the correlation matrix is an identity matrix, under the assumption of joint normal distribution of the data. The rejection of this hypothesis indicates that it is adequate to use Factor Analysis technique. This hypothesis test is included by default in SAS output when using the ML method.

Table 15 shows a summary of these measures for all the examples treated in this chapter. Despite obtaining low values in two of the cases, performing a multivariate analysis provide interesting interpretations regarding the objectives.

Tabla 15. MSA in examples of this chapter.

Examples	MSA
Advertisement Testing	0.90755539
Grammatical Structure	0.55323648
Quality Honey	0.62282735
Flexibility	0.51807270

8 Final Comments

Finally, we present some comments concerning general aspects that must be considered when applying this technique.

- Factor Analysis may pursue exploratory or confirmatory aims. In the exploratory approach there is no initial hypothesis regarding the variables that may be associated with the latent factors. Data must identify the association structure. Conversely, in the confirmatory approach, there are theoretical concepts that allow us to assume a structure for the associations between factors and variables prior to estimating the loadings from the observed data, but what is unknown is the strength of those associations. The model adequacy is confirmed after the nonzero loads are estimated. In an exploratory approach, any method of estimation can be used. But, in the confirmatory approach only the ML estimation can be used and it will be necessary to check if the joint normal distribution assumption is adequate.

- The calculation for obtaining the loadings estimations, goodness of fit indicators, or scores, is simplified by computer programmed routines available in statistical software. However, the interpretation of the results will require more time and effort.
- Regarding measurement scales, although the variables are expected to be measured in quantitative continuous scales, variables measured in discrete or ordinal scales could be considered, provided they have a broad range of possible values. It should be noted that, as the analysis takes input as the correlation matrix, these correlation parameters should be interpretable. In case of having ordinal scale variables, Spearman's or Kendall's correlation coefficients should be calculated and used as input.
- The method does not distinguish different roles among explained variables or explanatory variables. Although both types of variables can be considered simultaneously, they act on the analysis without distinction. The role may be taken into account when interpreting the results.
- Factor Analysis is very rich in its results and can be applied in a wide range of practical situations. The analysis can be used to study associations in an integrated approach, to identify variables or latent factors that cause associations, identify differentiation factors among individuals, confirm the choice of indicators selected to approximate the measuring of complex concepts, achieve a ranking of individuals or objects from non measurable factors, graphically synthesize the relationship among latent factors and grouping factors, etc.

In any case, Factor Analysis provides an integrated multivariate data analysis from the most diverse areas of interest, covering various objectives and discovering aspects that hardly arise with other approaches.

Answers for Questions of self-evaluation

1. The distribution of *Recognition* is more symmetrical than the one of *Persuasion*, when full data and categories of both are considered. There are no large outliers. There are atypical data values in the scale lowest values. Distributions by category are homogenous. The average values of the variable *Entertaining* are higher for the Beverages category and

lower for Hygiene, Beauty and Health. *To See Again* averages show no significant differences among House, Food and Hygiene, Beauty and Health categories. They are slightly lower for Beverages.

2. The correlations among the variables of attribute are generally higher. The correlations among impact variables are low for *Persuasion* except for *Effectiveness*. The remaining are correlations ranging from moderate to high.
3. The number of commercials under study is very large and correlations of the order of 0.10 have been detected significantly different from zero. A correlation of that magnitude is very low and does not represent an important association.
4. The F3 factor represents "negative emotions" and F4 the "impact of the commercial". This is deduced from observation of the variables grouped away from the origin on the addresses of the x-axis and y-axis.
5. Communality of *Informative*= $0.854^2 + 0.14^2 + (-0.147)^2 + (-0.005)^2 + 0.009^2 + 0.015^2 = 1 - 0.228$
   Unique Variance of *New Brand*= 1-0.6739=0.3261
6. The variables better explained by the model are *Recall* (rec_I), *Persuasion* (per_I) and *Effectiveness* (eff_I). They have the three highest communalities.
7. rel=0.869 F1+0.154 F2- 0.169 F3+0.009 F4+0.058 F5-0.084 F6+e1
   new=0.854 F1+0.140 F2- 0.147 F3- 0.005 F4+0.009 F5 +0.015 F6+e2
   The factor model has 31 equations, one for each observed variable.
8. It is not possible to obtain F1 from the model. The coefficient matrix A is not invertible.
9. Some objectives may be:
   a. Observing commercial categories considered most "entertaining".
   b. Identifying if a particular commercial was among the best or worst rated with respect to the 'information' provided.
   c. Using scores of "impact" in other studies, to associate this concept with other variables.
10. Valuing to F4 in commercials and doing a ranking.
11. Most of the residual correlations (non-diagonal elements) in absolute value are less than 0.05. Some correlations are estimated by excess and others by default. Not all are equally explained, the values are not identical.
12. A large residual variance does not affect the quality of the model. The variances have contribution of the specific and the common factors. Correlations are only affected by common factors.
13. Residuals will be higher if we consider four factors instead of six, although some residual correlations could remain unchanged if it is that two additional factors do not affect these variables.

14. They do not require distributional assumptions.
15. Conditioned extremes should be used. The function to maximize is:

$$\Phi = V1 - \sum_{i=1}^{p} \lambda i((r_{i,i} - \sum_{j=1}^{m} (a_{ij}{}^{2} + \Psi_{i})) - \sum \sum_{i<j=1}^{p} \delta_{ij} \ (r_{ij} - \sum_{k=1}^{m} a_{ik} \, a_{jk} \ ))$$

16. With the rotated ML estimations, the same interpretation is obtained. High and low associations for F1, F2 and F3 are similar to those found by the method of the Principal Factor.
17. The loadings $a_{5,1}$, $a_{6,1}$ and $a_{7,1}$ are about 0.75 and 0.80. The intervals $a_{i,j}$ $\pm$ 2 SE have the lower limit greater than 0.75, so they are important.
18. Given $A/AA' = R - \Psi$ and the orthogonal matrix Γ, so $B = A\Gamma$ is also a solution, since $BB' = A\Gamma(A\Gamma)' = A\Gamma\Gamma'A' = A(\Gamma\Gamma)'A' = AA'$ because Γ is orthogonal.
19. The pre- and post- rotated communalities and uniqueness of solutions are identical.
20. If we consider the loadings without rotating, the factor F2 is high and positively associated with negative attributes (*Irritating* (irri), *Aggressive* (agre), *Silly* (silly), *Confusing* (confu)) and two variables impact (*Recall* (rec_I) and *Recognition* (cog_I)). F2 is a factor of "memory of commercials which have produced negative emotions". A high value on F2 will represent the commercials most remembered by associating them with negative emotions.
21. If we consider the loadings without rotating, the factor F3 is only associated inversely with two variables: a negative attribute (*Aggressive* (agre)) and an impact attribute (*Brand Association* (bra_I)). F3 is a factor of the association of the commercial with its brand by one of the negative attributes (*Aggressive* (agre)). F3 low values represent commercials associated with such features.
22. Pre and post rotation solutions may be different. They are different readings of the same associations. It is convenient to use the rotated solution because it simplifies the structure of associations and is unique. If different alternatives show differences, the one that matches your interpretation with theoretical knowledge of the analyzed problem should be selected.
23. In the Grammatical Constructions example, the rotated solutions estimated through ML and Principal Factor has the same interpretation than unrotated.
24. No. The solutions that satisfy the rotation criteria are unique.
25. They are the variances of the common factors (F1, F2, F3).
26. The variance percentages are calculated with respect to the total variability of the data. With standardized variables, the sum of the variances is equal to the number of variables (p). In this case the total variance is 7.

27. The estimated model can be represented by the next figure:

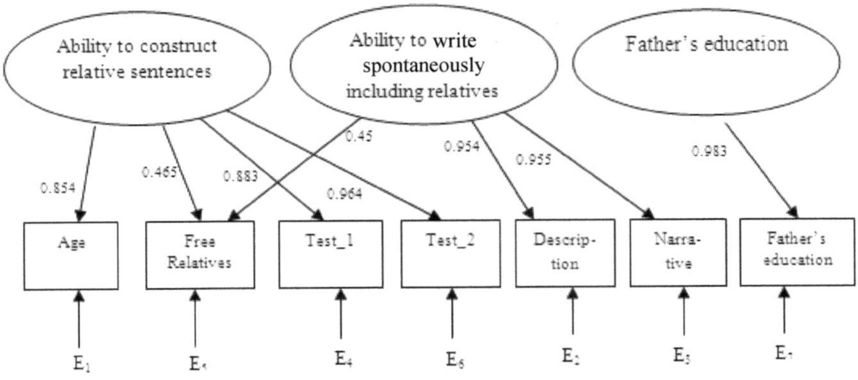

Figure 23

28. A plane with axes F1 and F3 or F1 and F2 could be considered to include a chart with F3. While it could only be considered in a straight line, the representation of the correlations with the variables is not as clear as in a plane.
29. F3 represents only parents education and has a correlation with this variable very close to one (0.983). It is not a complex concept because it repeats the information of the variable education of parents. It is not convenient to keep it in the analysis.
30. Scree Plot is one of the resources to identify how many factors should be retained in the analysis. The decision should be based on several aspects.
31. The figure shows in a summarized way the skills developed so far in the application of written language among the children who participated in the study. This can be communicated to teachers for the design of special activities in order to make progress regarding capabilities in writing for those who have not yet developed themselves.
32. In the Grammatical Constructions example, the decrease of the RMSE is more important only when going from one to two factors.
33. Observing the RMSE by variable in the Grammatical Constructions example, it is possible to identify that the less explained variables by a model with a single factor are *Descriptive* and *Narrative*. Their residuals lower considerably as other factors are added. With two factors, the highest average residuals are about 0.1, for *Test_1* and *Free Relatives*. From the third factor onwards, residuals decrease shortly for all the variables.
34. If averages of correlations to fixed values of the factors (conditional distribution) are calculated, these should be small in order to confirm

the goodness of the model. If the factors are causing the associations, the correlations, once fixed, should disappear or be very small.

35. The values of these indicators may only be used comparatively to choose among alternative models. There are no models for comparison in table 14.
36. Pre and post-rotation loadings plots were ignored. Interpretation was done on the tables with the values of the loadings, obtaining the same information. Loading plot generated by SAS with the option: "plots=loadings (vector)" is shown below (Fig. 24). The circle with radius one allows us to identify the important loadings easily.

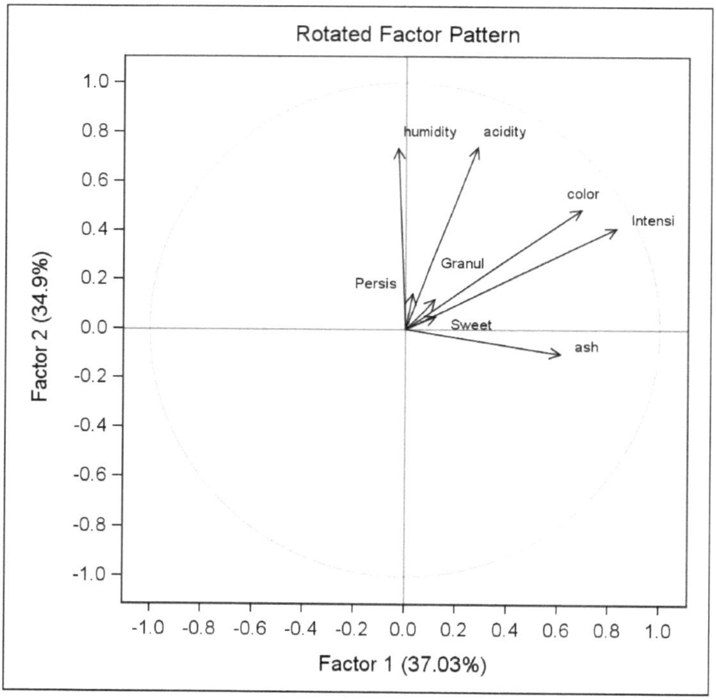

Figure 24. Loading Plot for the Honey Quality Example.

37. Communalities are calculated as the sum of the squared loadings for each variable: *Humidity*=0.627, *Color*=0.798, *Ash*=0.423, *Acidity*=0.63, *Sweet*=0.564, *Persistence*= 0.547, *Granularity*=0.069, *Intensity*=0.859. Total communality is 4.517.

    Uniqueness: *Humidity*=1–0.627=0.373, *Color*=0.202, *Ash*=0.577, *Acidity*=0.37, Sweet= 0.436, *Persistence*=0.453, *Granularity*=0.931, *Intensity*=0.141.
38. *Sweet*=0.12542 F1+ 0.05566 F2- 0.73821 F3 + e1

39. There is no information to associate the factors with the pollen attributes classification. The complex factors of individuals should be evaluated and then descriptive measures considering the defined classes should be calculated. The comparison should demonstrate whether there are differences among the groups or not.
40. Regression or least squares methods explain how to evaluate factors on individuals. Therefore, it is possible to make comparisons between the *Eucalyptus* and *Trefoil* groups, according to the complex concepts that have been found (F1, F2 and F3). It is also possible to generate a graphic for F1 and F2 (or F1 and F3) plotting the samples of honey tagged with the classification (*Eucalyptus, Trefoil*) and the loadings of the model. If classes are grouped in any sector of the plane, the pollen features may be described using the meaning of the factors. If they are mixed and scattered, the concepts that the factors represent are not associated with pollen characteristics.
41. Figs. 15, 17, 21 and 22.
42. The standardized values.
43. The *Eucalyptus* and *Trefoil* samples are grouped in the direction of x-axis (Fig. 17). On the left half plane there are more *Trefoil* samples (T) and on the right half there are more *Eucalyptus* points (E). Therefore *Eucalyptus* honey has a more intense flavor and color and it contains higher levels of minerals (*Ash*). There are no differences between different kinds of honey regarding the concepts represented by F2 (*Granularity* and *Acidity*).
44. The characterization does not include *Persistence* or *Sweet* because these variables are not involved in the construction of the common factors F1 and F2. In order to identify these features the samples should be represented on F3.
45. Step 1: find $\boldsymbol{C_1}'=(c_{11}, c_{12}\ldots c_{1p})$ with module 1 that maximize $\text{Var}(Y_1)=\text{Var}(\boldsymbol{C_1}'\boldsymbol{X})=\boldsymbol{C_1}'\Sigma\boldsymbol{C_1}$. Using Lagrange multipliers: $\Phi=\boldsymbol{C_1}'\Sigma\boldsymbol{C_1}-\lambda(\boldsymbol{C_1}'\boldsymbol{C_1}-1)$. In matrix form, the system of partial derivatives equalized to zero is: $2\Sigma\boldsymbol{C_1}-2\lambda\boldsymbol{C_1}=\boldsymbol{0}$. Then $(\Sigma-\lambda I)\boldsymbol{C_1}=\boldsymbol{0}$ and $\boldsymbol{C_1}'\Sigma\boldsymbol{C_1}=\text{Var}(Y_1)=\lambda$. Therefore $\boldsymbol{C_1}$ is the normalized eigenvector of Σ corresponding to the greatest eigenvalue λ_1.
    Step 2: find $\boldsymbol{C_2}=(c_{21}, c_{22}\ldots c_{2p})$ with module 1 and orthonormal to $\boldsymbol{C_1}$ that maximize $\text{Var}(Y_2)$. Using Lagrange multipliers: $\Phi=\boldsymbol{C_2}'\Sigma\boldsymbol{C_2}-\lambda(\boldsymbol{C_2}'\boldsymbol{C_2}-1)-\upsilon\,\boldsymbol{C_1}'\boldsymbol{C_2}$. In matrix form, the system of partial derivatives equalized to zero is: $2\Sigma\boldsymbol{C_2}-2\lambda\boldsymbol{C_2}-\upsilon\boldsymbol{C_1}=\boldsymbol{0}$. Pre multiplying by $\boldsymbol{C_1}'$ it is deduced that $\upsilon=0$, then $(\Sigma-\lambda I)\boldsymbol{C_2}=\boldsymbol{0}$ and $\boldsymbol{C_2}'\Sigma\boldsymbol{C_2}=\lambda$. Therefore $\boldsymbol{C_2}$ is the normalized eigenvector of Σ corresponding to the second greatest eigenvalue, λ_2. And so on until the p positive eigenvalues of Σ are used.

Proposed Problems

In this section, we present several multivariate problems, stating the main objective that led to the collection of data and synthesizing the obtained information. The files containing the data are on the website. The proposal for the students is to apply Factor Analysis following all the steps presented in the chapter and to prepare a report with the main findings. As a guide for this task, the analysis of problem 1 is also available on the webpage.

1. **Academic Performance.** Data of students in the first three years of high school was collected to make an analysis of academic performance at a school in a city in Argentina. The variables include school year, sex and average marks obtained during the year in each subject, namely: Language and Literature, Geography, History, Biology, Physical Chemistry, Mathematics, Accounting, Accounting Cabinet, Plastic Education, Physical Education, Foreign Language and Theatre. Note that the subjects of the first three years are common regarding the area of knowledge taught, but certainly not with regard to the content.
2. **Tomato fruits phenotypic variability.** Pratta et al. 2011[3] studied the phenotypic variability of the tomato in order to differentiate tomato lines. Seeds of different inbred lines were germinated in seedling trays and the total number of harvested fruits was 6072 (approximately 17 fruits per plant). Fruit traits evaluated by line (average) were: *weight* (W, in g), *diameter* (D, in cm) and *height* (H, in cm) at the moment of harvest in the breaker stage; *shelf life* (SL), in days from harvest at the breaker stage to the beginning of fruit softening; *shape index*, as the H/D ratio; *soluble solids content* (SS, in Brix degrees), as the percentage of glucose and fructose in the homogenized juice from ripe fruits; *pH*, in 100 g of the homogenized juice from ripe fruits diluted in 100 mL of distilled water; *titrable acidity* (TA, in g of citric acid per 100 g of the homogenized juice from ripe fruits); *the reflectance percentage* (L) and the *absorbency index* (a/b ratio, where the absorbency is at 540 nm and b at 675 nm); *firmness* (F): measured with a Shore A (Durofel, Forges-Les-Eaux, France) equipment on two opposite sides of ripe fruits.
3. **Social Vulnerability.** In developing countries, the identification of vulnerability situations and its geographical location is an important analysis for the design of policy actions. In a research on this topic in an Argentine state, census data were considered. The measurements registered on housing clusters correspond to different aspects of vulnerability: activity, education, housing, health and demographic indicators. The list with 36 variables is provided with the file.
4. **Bank Clients.** A financial institution offers a credit card customers' portfolio informing about aspects of profitability, consumption and

[3]Data are used with permission of the authors.

failure to pay on time in order to evaluate the suitability of the purchase of that portfolio. Among all the available information, only the following variables were selected: cardholders' age and sex, number of additional cards, purchase limit, amount to pay in instalments, days late in payment, average account balance for three consecutive months, average consumption, average paid amount, minimum payment and age of the card.

Acknowledgements

The authors wish to thank Nicolás Ballarini, Daniela Dianda, Lucía Hernández and Leandro Kovalevski for helping in the review of the text and the translation into English.

Special thanks to Patricio Giuliani, Celina Beltrán, María Cristina Ciappini, María Susana Vitelleschi, Juliana Merello and Nicolás Renzi for letting us use their data for the examples of this chapter.

We also thank to Six Soft (Minitab Argentina) and SAS Institute Argentina SA for the Cooperation Agreements that keep for years with the School of Statistics at the National University of Rosario and allowed us to make the analysis in this chapter.

References

Akaike, H. 1987. Factor Analysis and AIC. Psychometrika, 52: 317–332.

Browne, M.W. 1968. A Comparison of Factor Analytic Techniques. Psychometrika, 33: 267–334

Cerny, C.A. and Kaiser, H.F. 1977. A study of a measure of sampling adequacy for factor-analytic correlation matrices. Multivariate Behavioral Research, 12(1): 43–47.

Hotelling, H. 1933. Analysis of a Complex of Statistical Variables in Principal Components. Journal of Educational Psycology, 24: 417–441, 498–520

Johnson, R. and Wichern, D. 2007. Applied Multivariate Statistical Analysis. Pearson Prentice Hall. 6th Edition

Kaiser, H.F. 1958. The Varimax Criterion for Analytic Rotation in Factor Analysis. Psycometrika, 23: 187–200.

Kaiser, H.F. 1970. A Second Generation Little Jiffy. Psychometrika, 35: 401–415.

Kaiser, H.F. and Rice, J. 1974. Little Jiffy, Mark IV. Educational and Psychological Measurement. 34: 111–117.

Linn, R.L. 1968. A Monte Carlo Approach to the Number of Factors Problems. Psychometrika, 33: 37–71.

Morrison, D. 2004. Multivariate Statistical Methods. Duxbury Advanced Series. 4th Edition

Pearson, K. (1901). On Lines Planes of Closest Fit to a System of Points in Space. Philosophical Magazine, 2: 557–572.

Pratta, G., Rodriguez, G., Zorzoni, R., Valle, E. and Picardi, L. 2011. Phenotypic and molecular characterization of selected tomato recombinant inbred lines derived from the cross *Solanumly copersicum* × *S. pimpinellifolium*. Journal of Genetics, 90, 2: 229–237

Schwarz, G. 1978. Estimating the Dimension of a Model. Annals of Statistics, 6: 461–464.

Spearman, C. 1904. General Intelligence, Objectively Determined and Measured. American Journal of Psychology, 15: 201–293

Thurstone, L. 1947. Multiple Factor Analysis. Chicago. University of Chicago Press.

Tucker, L. and Lewis, C. 1973. A Reliability Coefficient for Maximum Likelihhod Factor Analysis. Psycometrika, 38: 1–19.

CHAPTER 9

Discriminant Analysis

T. Ramayah,[1,*] *Joshua Ignatius,*[2] *Jasmine Yeap Ai Leen*[3] and *Lo May Chiun*[4]

Introduction

Business management and the social sciences often rely on a set of statistical methods to support the observations made. However, the challenge is in selecting an appropriate statistical technique for analyzing the relationship among variables. The 'correct test' depends on a couple of things, ranging from the number of observations to the data structure (which includes the level of measurement). For instance, a common test for assessing the extent of a group of independent variables influencing another dependent variable is the Multiple Regression Analysis (MRA). The prerequisite condition for the MRA is that the level of measurement of all the variables must be continuous.

However, there is also a multivariate method to address the situation where the dependent variable is categorical in nature. This situation may arise in a number of different areas. A marketing firm may be interested in analyzing key features that influence buyers, while in the healthcare industry, risk classification of patients will allow insurance firms to package their portfolios better. The banking industry may be more concerned with features that determine the credit worthiness of their clients. The logistic regression analysis and the discriminant analysis are appropriate statistical techniques to analyze the above situations. For the purpose of this chapter, we will be focusing on the latter.

[1] School of Management, Universiti Sains Malaysia.
[2] School of Mathematical Sciences, Universiti Sains Malaysia.
[3] School of Management, Universiti Sains Malaysia.
[4] Faculty of Economics and Business, Universiti Malaysia Sarawak.
* Corresponding author

Discriminant analysis is generally used when the researcher wants to classify a set of objects into 2 or more distinct groups. When the focus is only on a two-group classification it is called the two-group Discriminant Analysis (DA) and if the focus extends to more than 2 groups, it is called the Multiple Discriminant Analysis (MDA).

Hair et al. (2010) specifies the purpose of discriminant analysis as follows:

1. To determine statistically whether there are differences in the average score of a variable for 2 groups and in the case of more than 2 groups *a priori* should be set.
2. To determine which of the variables accounts for most of the differences in the average score in the case of the 2 or more groups.
3. To establish the number and composition of dimensions of discrimination between the groups based on a set of independent variables.
4. To establish procedures to classify objects into groups based on their scores on the independent variables.

We will look into the methodology of Discriminant Analysis in the next section.

Method

Discriminant, or Discriminant Function Analysis is a parametric technique, which determines the weights of quantitative variables or predictors that can best discriminate between 2 or more groups of cases; and this must meet the 'better than chance' requirement (Cramer 2003).

In Discriminant Analysis, the variable generated by the composite cannot be used as a predicted score on the dependent variable. It is merely a discriminant function score that is fed into the computation of predicted probability of a case being classified in a category of the dependent variable. This predicted probability is later utilized to generate a predicted category for each case.

The maximum number of discrimination functions considered is the smaller value between the number of predictors, or one less the number of groups. For example, if you have 2 groups then the number of discriminant function is (2-1).

$$Z_{jk} = a + W_1 X_{1k} + W_2 X_{2k} + \ldots + W_n X_{nk}$$

Where:

Z_{jk} = discriminant Zscore of discriminant function j for object k

a = intercept

W_i = discriminant coefficient for the independent variable i

X_{ik} = independent variable i for object k

The number of discriminant equations required for classification will vary, depending on the number of groups that we have. If we have only 2 groups then we will need one equation, whereas if there are more than 2 groups the number of equations needed can be calculated as $(n - 1)$, with n being the number of groups.

Some Issues of Concern

Some of the issues that need to be looked at when running discriminant analysis are discussed below (see Ramayah et al. 2004, 2006, 2010).

1. **Sample size**
   In Discriminant Analysis there are two concerns about sample size which are the overall sample size and the sample size per individual group. The ratio of samples to independent variables should be at least 5:1 which is the lower threshold. The higher threshold is 20:1 which is very difficult to achieve. Also for each individual group the sample size should be at least 20.
2. **Division of sample**
   The preferred means of validating the Discriminant Analysis is to split the sample into 2 sub-samples; one is called the estimation sample and the other one is called the holdout sample. This method is called the split-sample validation or cross-validation (Cohen 1988, Perreault et al. 1979). No definite guidelines have been established for the split. Many researchers split one half for the estimation and the other half for the validation. Some researchers have used the 60:40 or even the 75:25 split depending on the overall sample size.
3. **Assumptions**
   Like many other parametric analysis techniques, Discriminant Analysis has several assumptions that need to be addressed (Leech et al. 2008) as follows:
   i. Relationship between all pairs of predictors must be linear.
   ii. Multivariate normality must exist.
   iii. Population covariance matrices of predictors must be equal across all groups.

   Discriminant analysis is however, fairly robust to these assumptions, although violation of multivariate normality may affect accuracy of estimates of the probability of correct classification (Leech et al. 2008). Remedies may be possible through transformations to reduce disparities among the covariance matrices (Hair et al. 2010).

4. **Predictive accuracy**

   Since the basic idea of the Discriminant Analysis is about classification accuracy, once the analysis is done we need to test whether the model we developed is actually performing better than chance. To do that there are 3 criteria that can be used: the maximum chance criterion, the proportional chance criterion and the Press Q statistics.

   i. Maximum chance criterion

      The probability we get from arbitrarily assigning all the objects to the largest group.

      C_{MAX} = Size of the largest group

   ii. Proportional chance criterion

      $C_{PRO} = p^2 + (1 - p)^2$

      Where:

      p = proportion of individuals in group 1

      1 – p = proportion of individuals in group 2

   iii. Press Q statistic

      This is the statistic that can be calculated using the formula given below.

      Press Q

      $$\text{Press Q} = \frac{[N - (n*k)]^2}{N(k-1)}$$

      Where: $Q \sim \chi^2$ with 1 degree of freedom

      N = Total sample size

      n = number of observations correctly classified

      k = number of groups

5. **Explanatory power**

   Part of any multivariate analysis is to make an assessment of how good a model is. In regression analysis we look at the R^2 to see how good the explanatory power is. Values closer to 1 indicate good explanatory power whereas values closer to 0 indicate weak explanatory power. In Discriminant Analysis we can look at the canonical correlation value which when squared is similar to the R^2.

6. **Important discriminating variables**

   One of the reasons for doing the Discriminant Analysis is to see if the independent variables are useful in classifying the objects into the correct group or not. One of the tests of whether there is a model or not is to look at the value of Wilks' Lambda. If the Wilks' Lambda is not significant **($p > 0.05$)**, we can conclude that the particular variable is not a good discriminator and if it is significant **($p < 0.05$)** then we can conclude that the particular variable is a good discriminator.

Self-Evaluation Exercises

1. Multivariate Statistical Analysis allows us to test the effects of three or more variables concurrently.
   1. True 2. False
2. Discriminant Analysis predicts a categorical dependent variable based on a linear combination of independent variables.
   1. True 2. False
3. Which type of analysis attempts to predict a categorical dependent variable?
   a. Regression Analysis
   b. Cluster Analysis
   c. Factor Analysis
   d. Discriminant Analysis
4. What value will you look at to assess whether a particular variable is significant or not?
   a. F-value
   b. Wilks' Lambda
   c. t-values
   d. R^2
5. What test will you look at to see if all the independent variables are useful in classifying the objects into the correct group or not?
   a. F-test
   b. Wilks' Lambda
   c. t-values
   d. R^2

Answers to Self-Evaluation Exercises

1. True
2. True
3. D
4. A
5. B

A Two-Group Discriminant Example

To illustrate how this analysis can be applied, we consider the case of the decision to adopt an Enterprise Resource Planning (ERP) program among Malaysian companies. The survey consists of 344 firms. From the literature, we have identified 5 predictors of adoption, *Top Management Support* (TOPMGMT), *Size of Company* (SIZE), *Organization Culture* (CULTURE), *Perceived Benefits* from ERP adoption (BENEFITS) and *Resistance to Change* (RESISTANCE). The dependent variable is measured using 2 categories

(1= Non-Adopters, 2 = Adopters). The variables used are also summarized below:

ADOPTION	1 = NO, 2 = YES
TOPMGMT	Top management support
SIZE	Size of company (number of employees)
CULTURE	The culture of the organization
BENEFITS	The expected perceived benefits
RESISTANCE	The resistance to change among the employees

We asked the respondents whether they have adopted ERP to which they responded 1=NO and 2=YES. Hence, the 2-group Discriminant Analysis can be used to see how good is our classification accuracy by using the 5 variables TOPMGMT, SIZE, CULTURE, BENEFITS and RESISTANCE to classify them into the correct categories. The analysis is carried out using SPSS and the steps involved are shown below.

Step 1: To start click **Analyze > Classify > Discriminant** as shown below (Fig. 1).

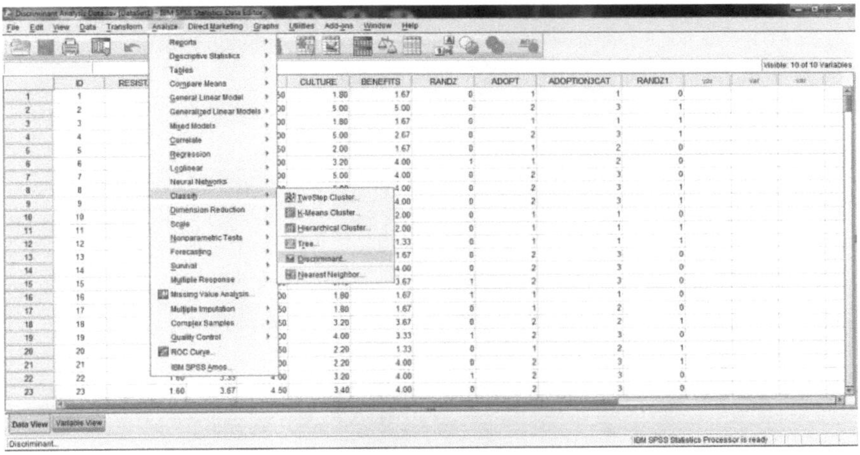

Figure 1. Setting up the analysis in SPSS.

Step 2: Click the dependent variable (ADOPT) into the **Grouping Variable** tab, then click the 5 independent variables into the **Independents** tab. Next, click in RANDZ into the **Selection Variable** tab, and set the value to 0 as shown below (Fig. 2).

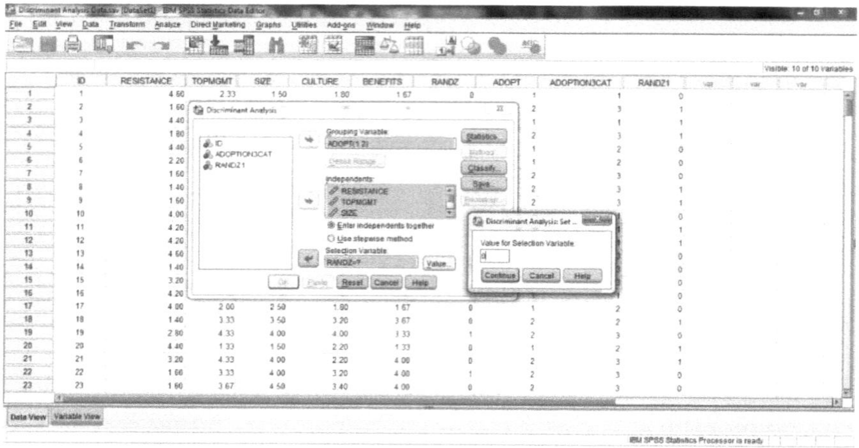

Figure 2. Selecting the analysis and the holdout group.

Step 3: Click the **Statistics** tab, select all the features under **Descriptives** and also **Function Coefficients**.(Fig. 3).

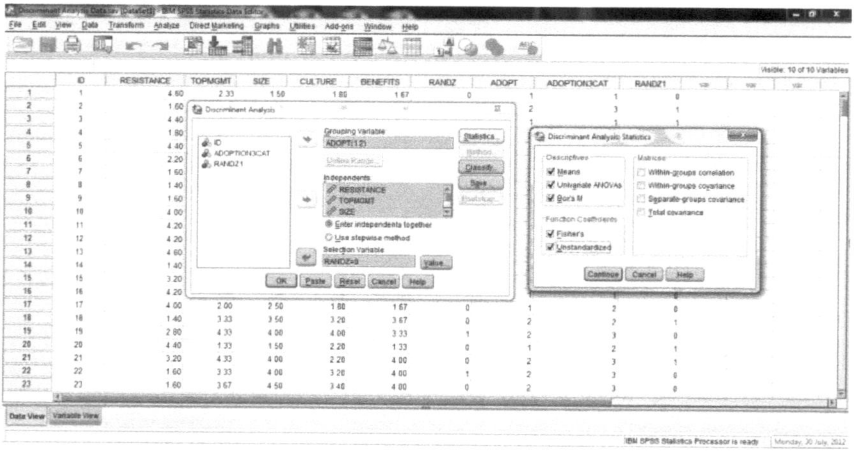

Figure 3. B Selecting additional options.

Step 4: Click the **Classify** tab and under the **Display** column select **Summary table** and also **Leave-one-out classification** (Fig. 4).

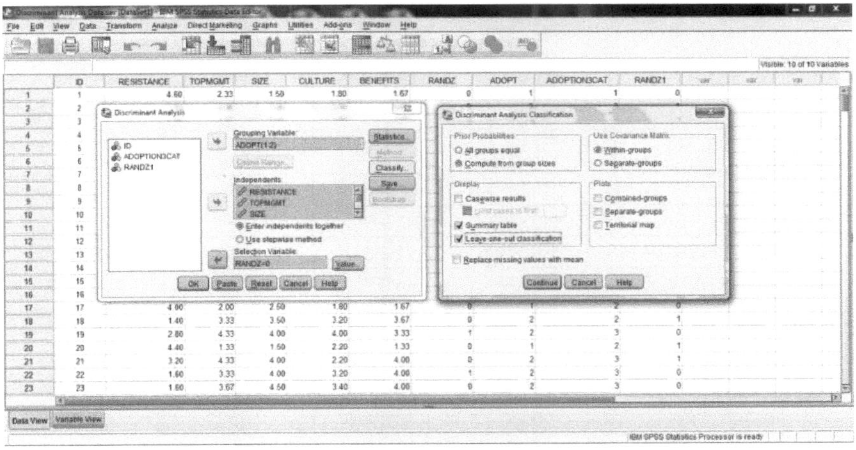

Figure 4. B Selecting Classification options.

Once we have executed this analysis we will get a long list of tables that can be used for interpretation. We will look at the important ones that need to be interpreted in this analysis.

The first table (Table 1) gives us the case processing summary indicating that the total number of cases we have in the data set is 344. The summary also indicates that the sample has been split into the analysis sample (222) and the holdout sample (122). The holdout sample was split according to a 65:35 ratio split. The ratio is not fixed but most researchers suggest that the analysis sample is bigger than the holdout sample. The holdout sample will be used to assess the predictive accuracy of the model developed using the analysis sample.

Table 1. Case Processing Summary

Analysis Case Processing Summary			
Unweighted Cases Valid		N 222	Percent 64.5
Excluded	Missing or out-of-range group codes	0	.0
	At least one missing discriminating variable	0	.0
	Both missing or out-of-range group codes and at least one missing discriminating variable	0	.0
	Unselected	122	35.5
	Total	122	35.5
Total		344	100.0

The next table (Table 2) summarizes the mean and standard deviation for the 2 groups, i.e., those who adopt and those who do not adopt.

Table 2. Descriptive Statistics

Group Statistics					
Adoption Status		Mean	Std. Deviation	Valid N (listwise)	
				Unweighted	Weighted
No	TOPMGMT	2.1036	.56038	119	119.000
	SIZE	217.2269	62.5845	119	119.000
	CULTURE	2.0538	.41309	119	119.000
	BENEFITS	1.8683	.71993	119	119.000
	RESISTANCE	4.0739	.44177	119	119.000
Yes	TOPMGMT	4.3366	.60768	103	103.000
	SIZE	414.077	62.7856	103	103.000
	CULTURE	4.2369	.76786	103	103.000
	BENEFITS	4.0000	.68121	103	103.000
	RESISTANCE	1.7456	.88215	103	103.000
Total	TOPMGMT	3.1396	1.25846	222	222.000
	SIZE	308.5586	116.5829	222	222.000
	CULTURE	3.0667	1.24656	222	222.000
	BENEFITS	2.8574	1.27519	222	222.000
	RESISTANCE	2.9937	1.34822	222	222.000

Table 3 presents the univariate ANOVA test and the test indicates that the 2 groups differ significantly ($p < 0.01$) for all the 5 independent variables. Those who adopt showed higher mean values as compared to those who do not adopt for TOPMGMT, SIZE, CULTURE, BENEFITS whereas for RESISTANCE, the mean value was higher for those who do not adopt as compared to those who adopt (see Table 2 for mean values).

Table 3. Test of Equality of Group Means

Tests of Equality of Group Means					
	Wilks' Lambda	F	df1	df2	Sig.
TOPMGMT	.213	810.518	1	220	.000
SIZE	.288	544.600	1	220	.000
CULTURE	.234	721.144	1	220	.000
BENEFITS	.302	508.727	1	220	.000
RESISTANCE	.255	643.011	1	220	.000

Table 4 reports the test for homogeneity of covariance which is one of the assumptions that needs to be tested for the discriminant analysis. The test indicates that there is a significant difference in the covariance of the 2 groups ($F = 8.228$, $p < 0.01$). As stated by Meyers et al. (2006), it should be noted that Discriminant Analysis is robust to the violation of homogeneity of covariance assumption, provided the data do not contain extreme outliers (z scores $> \pm 4$).

Table 4. Test of Equality of Group Covariance

Test Results		
Box's M		126.526
F	Approx.	8.228
	df1	15
	df2	186359.724
	Sig.	.000

Tests null hypothesis of equal population covariance matrices.

One of the important values that we would like to see is this value called canonical correlation and the Wilks' Lambda (see Table 5 and Table 6). The canonical correlation which measures how strong a relationship is with a value of more than 0.5 indicating strong relationship. Also when we square the canonical correlation value we will get the percentage of variance explained in the dependent variable which in this case is 0.835 (0.914^2), something that indicates that the explanatory power is extremely good.

Table 5. Canonical Correlation

Eigenvalues				
Function	Eigenvalue	% of Variance	Cumulative %	Canonical Correlation
1	5.051[a]	100.0	100.0	.914

a. This is part of the SPSS software output.

The Wilks' Lambda tests the eigenvalue of the discriminant function when there is more than 1 function. In this case since we have only 2 groups we have only 1 function and the eigenvalue is significant ($\chi^2 = 391.561, p< 0.01$).

Table 6. Wilks' Lambda

Wilks' Lambda				
Test of Function(s)	Wilks' Lambda	Chi-square	df	Sig.
1	.165	391.561	5	.000

The next three tables (Table 7, 8 and 9) give us the discriminant function coefficients. The standardized canonical discriminant function coefficients are like the beta weights (standardized) in regression analysis. This will tell us the relative importance of the variable in predicting adoption versus non-adoption of ERP. As can be seen in Table 7 the most important variable in enhancing adoption is top management support followed by the size of the company, which means companies that are bigger in size (number of employees) and companies with a stronger top management support adopt ERP systems. The most important factor for non- adoption is the resistance to change. The companies that did not adopt the ERP system reacted as such due to their strong resistance to change from their legacy system to the ERP system.

Table 7. Standardized Coefficients

Standardized Canonical Discriminant Function Coefficients

	Function 1
TOPMGMT	.432
SIZE	.238
CULTURE	.194
BENEFITS	.187
RESISTANCE	−.239

The next table on structure matrix (Table 8), presents the simple (zero-order) correlation of each variable with the discriminant function.

Table 8. Structure Matrix

Structure Matrix

	Function 1
TOPMGMT	.854
CULTURE	.806
RESISTANCE	−.761
SIZE	.700
BENEFITS	.677

- Pooled within-groups correlations between discriminating variables and standardized canonical discriminant functions
- Variables ordered by absolute size of correlation within function.

The last table (Table 9) is the canonical discriminant function coefficients which display the unstandardized discriminant function coefficients similar to the regression analysis (unstandardized beta). These coefficients are used to develop the discriminant equation as follows:

$$Z_{jk} = -4.195 + 0.742(\text{TOPMGMT}) + 0.004(\text{SIZE}) + 0.321(\text{CULTURE}) + 0.266(\text{BENEFITS}) - 0.350(\text{RESISTANCE})$$

Table 9. Unstandardized Coefficients

Canonical Discriminant Function Coefficients

	Function 1
TOPMGMT	.742
SIZE	.004
CULTURE	.321
BENEFITS	.266
RESISTANCE	–.350
(Constant)	–4.195

Unstandardized coefficients

Functions at group centroids display the mean for the groups which are in standardized form (*z score*) which is based on the weighted linear combination making up the discriminant function. These values will be used to calculate the cutting score which is used to establish a cutting point for the purpose of classification.

Table 10. Group Centroids

Functions at Group Centroids	
Adoption Status	Function 1
No	–2.082
Yes	2.405

Unstandardized canonical discriminant functions evaluated at group means.

The cutting score is needed to establish a cutting point as we can see from Table 10, if the variate calculated is less than –2.082 then a company will be classified as a non adopter whereas if the variate is greater than 2.405 a company will be classified as an adopter. As can be seen clearly, there is a gap between the 2 values. This is where the cutting score is needed to patch the gap. If the two groups are of equal size then the best cutting score will be half way between the two values. If the group sizes are not equal then we will need to use the weighted average to calculate the cutting score.

Since the two group sizes are not equal, we will use the formula shown below in the calculation.

$$Z_{CS} = \frac{N_A Z_B + N_B Z_A}{N_A + N_B}$$

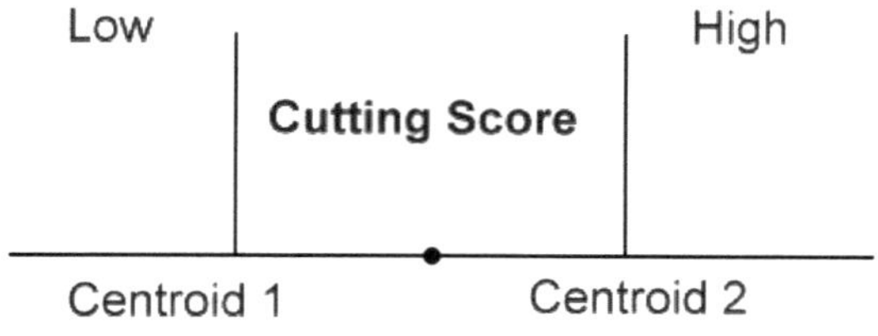

Figure 5. Cutting score

Where:

Z_{CS}	=	optimal cutting score between group A and B
N_A	=	number of observations in group A
N_B	=	number of observations in group B
Z_A	=	centroid for Group A
Z_B	=	centroid for Group B

$$Z_{cs} = \frac{103(-2.082)+119(2.405)}{103+119} = 0.323$$

So the analysis will classify any company with a variate of less than 0.323 as a non adopter of ERP and any company with a variate more than 0.323 will be classified as an adopter.

Example: Let's say you are give the values below for the 5 independent variables from a particular company, we can use them in the calculation of the variate and then, based on the cutting score, decide whether the particular company will be classified as an adopter or a non-adopter.

TOPMGMT	4
SIZE	300
CULTURE	4
BENEFITS	3
RESISTANCE	2

Using the equation:

$$Z_{jk} = -4.195 + 0.742(\text{TOPMGMT}) + 0.004(\text{SIZE}) + 0.321(\text{CULTURE}) + 0.266(\text{BENEFITS}) - 0.350(\text{RESISTANCE})$$

$$Z_{jk} = -4.195 + 0.742(4) + 0.004(300) + 0.321(4) + 0.266(3) - 0.350(2)$$

$Z_{jk} = 1.355$ (> 0.323, therefore it will be classified as **ADOPT**)

Once we have finished the interpretation, we need to interpret the classification accuracy. As shown in Table 11 we have 3 classification hit ratios, the first one is the classification with all the data that we have which is called the original classification. The second one is called the cross-validated classification. In the cross-validated classification the analysis will leave one case out at a time and run the analysis. The third one is the hit ratio when we use the holdout sample.

Table 11. Classification Results

Classification Results[a,b,d]						
	Adoption Status			Predicted Group Membership		Total
				No	Yes	
Cases Selected	Original	Count	No	114	5	119
			Yes	4	99	103
		%	No	95.8	4.2	100.0
			Yes	3.9	96.1	100.0
	Cross-validated[c]	Count	No	114	5	119
			Yes	4	99	103
		%	No	95.8	4.2	100.0
			Yes	3.9	96.1	100.0
Cases Not Selected	Original	Count	No	51	5	56
			Yes	1	65	66
		%	No	91.1	8.9	100.0
			Yes	1.5	98.5	100.0

a. 95.9% of selected original grouped cases are correctly classified.
b. 95.1% of unselected original grouped cases are correctly classified.
c. Cross validation is done only for those cases in the analysis. In cross validation, each case is classified by the functions derived from all cases other than that case.
d. 95.9% of selected cross-validated grouped cases are correctly classified.

Next what we need to do is to assess the hit ratio which is actually a measure of the classification accuracy that may range from 0% of all respondents wrongly classified to 100% of all respondents correctly classified. We also need to make sure that the hit ratio that we get from the analysis is better than chance which indicates that our model works well in predicting the group membership. To assess this, we need to assess the maximum chance criterion, the proportional chance criterion and the third is to assess the significance using a Press Q test. The calculation is given as follows:

1. Maximum chance
   C_{MAX} = Size of the largest group
   Example:
   Estimate Sample 119/222 = 0.536 **(see Table 11)**
   Holdout Sample 66/122 = 0.541

2. Proportional chance

   $C_{PRO} = p^2 + (1-p)^2$

   where:

   p = proportion of individuals in group 1

   1 – p = proportion of individuals in group 2

   Example:

   Estimate Sample $(119/222)^2 + (113/222)^2 = 0.502$

   Holdout Sample $(56/122)^2 + (66/122)^2 = 0.503$

3. Press Q

$$\text{Press Q} = \frac{[N - (n^*k)]^2}{N(k-1)}$$

Where: Q ~χ^2 with 1 degree of freedom

N = Total sample size

n = number of observations correctly classified

k = number of groups

Press Q Calculation for **Estimate Sample (see Table 11)**

$$\text{Press Q} = \frac{[222 - (213^*2)]^2}{222(2-1)} = 187.46$$

Press Q Calculation for **Holdout Sample**

$$\text{Press Q} = \frac{[122 - (116^*2)]^2}{122(2-1)} = 99.18$$

The values calculated are summarized in Tables 12, 13 and 14.

Table 12. Classification Accuracy (Estimate Sample).

Measure	Value	Hit Ratio for Estimate Sample
Maximum Chance Proportional Chance	53.6 50.2	95.9 95.9
Press Q Table Value Press Q Calculated Value	6.635 187.46**	

** p< 0.01

Table 13. Classification Accuracy (Leave One Out).

Measure	Value	Hit Ratio for Leave One Out Sample
Maximum Chance Proportional Chance	53.6 50.2	95.9 95.9
Press Q Table Value Press Q Calculated Value	6.635 187.46**	

** p< 0.01

Table 14. Classification Accuracy (Holdout Sample).

Measure	Value	Hit Ratio for Holdout Sample
Maximum Chance	54.1	95.1
Proportional Chance	50.3	95.1
Press Q Table Value		6.635
Press Q Calculated Value		99.18**

** $p < 0.01$

Based on the calculation we can conclude that the model that we developed can predict the group membership greater than the chance criterion of maximum chance (53.6%) and also proportional chance (50.2%) for the estimate and the leave one out classification (95.9% see Table 11 (d)) whereas for the holdout sample the hit ratio was greater than the chance criterion of maximum chance (54.1%) and also proportional chance (50.3%). The Press Q is also significant at $p < 0.01$, thus we can conclude that the model we developed is good. Hair et al. (2010) have also suggested a more stringent criterion which is 25% higher than the chance criterion which for Proportional chance would have been 1.25*0.503 = 0.629 whereas the proportional chance 1.25*0.541 = 0.68. The hit ratios exceeded even the more stringent criterion established by Hair et al. (2010).

Self-Evaluation Exercises

1. If a company manager would like to differentiate between an employee who will use the new system to be implemented by the company or would not use the new system, which analysis should he choose?

   a. Factor Analysis
   b. Multidimensional scaling
   c. MANOVA
   d. Discriminant Analysis

The next few questions are based on the data of an internet portal service provider who would like to assess the satisfaction of their users with regards to the service they have provided. The variables they have used as predictors are as follows:

CORE	Core services provided
SUPPORTING	Supporting services provided
VALUE	Perceived value of the services provided
SATISFACTION	1= Not Satisfied; 2 = Satisfied

A 2 group Discriminant Analysis was run and based on the output please answer the questions that follow.

2. Given the table below, can you suggest which variables would be able to differentiate a satisfied user from a dissatisfied user based on the descriptives provided?

Group Statistics					
Satisfaction Level		Mean	Std. Deviation	Valid N (listwise)	
				Unweighted	Weighted
Not Satisfied	CORE	4.9413	1.01992	352	352.000
	SUPPORTING	3.5928	.56283	352	352.000
	VALUE	3.1581	.79125	352	352.000
Satisfied	CORE	5.4466	.82363	287	287.000
	SUPPORTING	3.9721	.55540	287	287.000
	VALUE	3.5714	.73289	287	287.000
Total	CORE	5.1682	.96936	639	639.000
	SUPPORTING	3.7632	.59010	639	639.000
	VALUE	3.3438	.79218	639	639.000

3. Given the output below, if you are asked to test whether all the independent variables can help to differentiate the 2 groups of users, how will you decide and justify your decision?

Wilks' Lambda				
Test of Function(s)	Wilks' Lambda	Chi-square	df	Sig.
1	.843	108.570	3	.000

4. Which variables are significant discriminators of the 2 groups (**Satisfied/ Not Satisfied**) based on the output given below.

Tests of Equality of Group Means					
	Wilks' Lambda	F	df1	df2	Sig.
CORE	.933	45.986	1	637	.000
SUPPORTING	.898	72.666	1	637	.000
VALUE	.933	46.070	1	637	.000

5. Using the output given below determine whether the covariance of the 2 populations are equal or significantly different.

Test Results		
Box's M		23.665
F	Approx.	3.928
	df1	6
	df2	4164087.738
	Sig.	.001

Tests null hypothesis of equal population covariance matrices.

6. How much of the variance in the dependent variable can the three independent variables explain?

Eigenvalues				
Function	Eigenvalue	% of Variance	Cumulative %	Canonical Correlation
1	.186[a]	100.0	100.0	.396

a. First 1 canonical discriminant functions were used in the analysis.

Answers to Self-Evaluation Exercises

1. D
2. By looking at the mean values, it can be seen that the mean values are higher for satisfied users for all the three independent variables as compared to dissatisfied users. Thus the 3 independent variables are possible significant differentiators of the level of satisfaction.
3. The Wilks' Lambda value of 0.843 is significant at $p< 0.01$. Thus we can conclude that the 3 independent variables are useful discriminators of the satisfaction level of the internet portal users.
4. All three independent variables are significant predictors of satisfaction as the p-values are all < 0.05.
5. Box's M is significant ($F = 3.928$, $p < 0.01$), thus it can be concluded that the covariance of the 2 populations are not equal.
6. The canonical correlation is 0.396, if you square it. It will give a value of 0.157 which translates to about 15.7% variance explained.

A Three-Group Discriminant Example

We will use the same data set used for the two group Discriminant Analysis but the dependent variable is measured using 3 categories (1= Non-Adopters, 2 = Plan to Adopt and 3 = Adopters). The variables used are also summarized below:

ADOPTION	1 = NO, 2 = PLAN and 3 = YES
TOPMGMT	Top management support
SIZE	Size of company (number of employees)
CULTURE	The culture of the organization
BENEFITS	The expected perceived benefits
RESISTANCE	The resistance to change among the employees

We asked the respondents whether they have adopted ERP to which the answers elicited was 1=NO, 2= PLAN and 3 = YES. So the 3-group Discriminant Analysis can be used to see how good our classification accuracy is by using the 5 variables TOPMGMT, SIZE, CULTURE, BENEFITS and RESISTANCE to classify them into the correct categories. The analysis is carried out using SPSS and the steps involved are similar to the 2 group example. The only changes are in terms of the dependent variable and the variable used to split the group (see below for the change into 3 types of adopter categories under **Grouping Variable,** Fig. 6).

Table 15 gives us the case processing summary indicating that the total number of cases we have in the data set is 344. The summary also indicates that the sample has been split into the analysis sample (213) and the holdout sample (131). The holdout sample was split according to a 60:40 ratio split. The ratio is not fixed but most researchers suggest the analysis sample to be bigger than the holdout sample. The holdout sample will be used to assess the predictive accuracy of the model developed using the analysis sample.

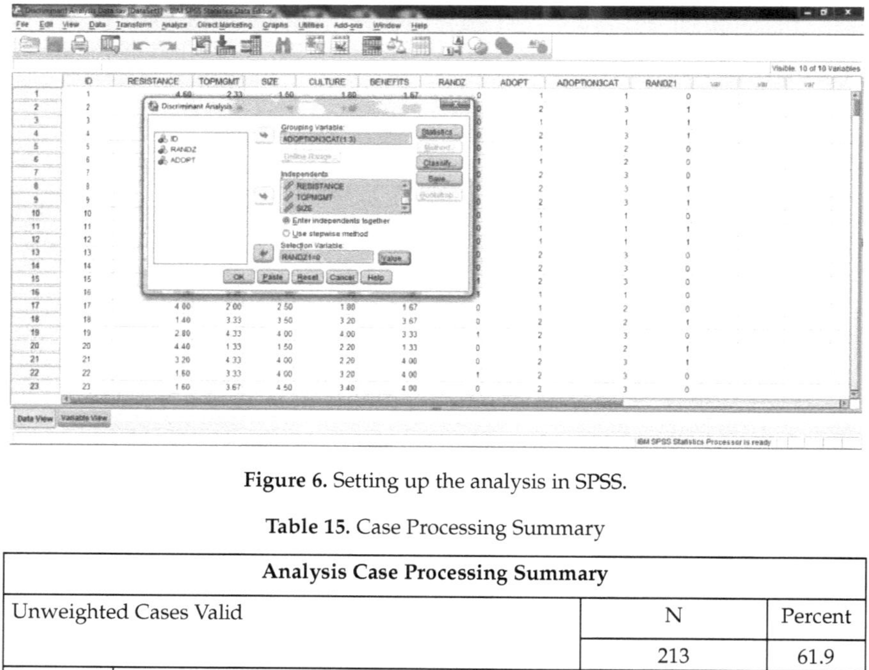

Figure 6. Setting up the analysis in SPSS.

Table 15. Case Processing Summary

Analysis Case Processing Summary			
Unweighted Cases Valid		N	Percent
		213	61.9
Excluded	Missing or out-of-range group codes	0	.0
	At least one missing discriminating variable	0	.0
	Both missing or out-of-range group codes and at least one missing discriminating variable	0	.0
	Unselected	131	38.1
	Total	131	38.1
Total		344	100.0

Table 16 summarizes the mean and standard deviation for the 3 groups, i.e., those who adopt, plan to adopt and do not adopt.

Table 16. Descriptive Statistics

Group Statistics					
Adoption Level		Mean	Std. Deviation	Valid N (listwise)	
				Unweighted	Weighted
No	TOPMGMT	2.1293	.38379	49	49.000
	SIZE	192.857	48.947	49	49.000
	CULTURE	1.9061	.27719	49	49.000
	BENEFITS	1.8435	.63136	49	49.000
	RESISTANCE	4.2041	.15540	49	49.000
Plan	TOPMGMT	2.4229	.76449	67	67.000
	SIZE	256.716	73.286	67	67.000
	CULTURE	2.3463	.73326	67	67.000
	BENEFITS	2.1841	.99372	67	67.000
	RESISTANCE	3.7104	.83214	67	67.000
Adopted	TOPMGMT	4.3368	.62267	97	97.000
	SIZE	415.98	61.891	97	97.000
	CULTURE	4.1588	.76482	97	97.000
	BENEFITS	3.9210	.66801	97	97.000
	RESISTANCE	1.7299	.89118	97	97.000
Total	TOPMGMT	3.2269	1.19886	213	213.000
	SIZE	314.554	114.613	213	213.000
	CULTURE	3.0704	1.21271	213	213.000
	BENEFITS	2.8967	1.22336	213	213.000
	RESISTANCE	2.9221	1.34436	213	213.000

Table 17 presents the univariate ANOVA test and the test indicates that the 3 groups differ significantly ($p< 0.01$) for all the 5 independent variables with those who adopt showing higher mean values as compared to those who do not adopt and those who plan to adopt for TOPMGMT, SIZE, CULTURE, BENEFITS whereas for RESISTANCE, the mean value was higher for those who do not adopt as compared to those who adopt and those who plan to adopt.

Table 18 reports the test for homogeneity of covariance which is one of the assumptions that needs to be tested for the Discriminant Analysis. The test indicates that there is a significant difference in the covariance of the 3 groups ($F = 9.178$, $p< 0.01$). As stated by Meyer et al. (2006), it should be noted that discriminant analysis is robust to the violation of homogeneity of covariance assumption, provided the data do not contain extreme outliers (z scores >± 4).

Table 17. Test of Equality of Group Means.

Tests of Equality of Group Means					
	Wilks' Lambda	F	df1	df2	Sig.
TOPMGMT	.272	281.090	2	210	.000
SIZE	.301	244.271	2	210	.000
CULTURE	.306	238.411	2	210	.000
BENEFITS	.401	157.019	2	210	.000
RESISTANCE	.321	221.797	2	210	.000

Table 18. Test of equality of group covariance.

Test Results		
Box's M		286.020
F	Approx.	9.178
	df1	30
	df2	84498.997
	Sig.	.000

Tests null hypothesis of equal population covariance matrices.

The canonical correlation is presented in Table 19. Since we have 3 groups we will get 2 discriminant functions. The 1st function has an explanatory power of 0.771 (0.878^2) which is high whereas the 2nd function explains an additional 0.055 (0.234^2) explanatory power. The total combined explanatory power is 0.826.

Table 19. Canonical Correlation

Eigenvalues

Function	Eigenvalue	% of Variance	Cumulative %	Canonical Correlation
1	3.361[a]	98.3	98.3	.878
2	.058[a]	1.7	100.0	.234

a. First 2 canonical discriminant functions were used in the analysis.

The Wilks' Lambda tests the eigenvalue of the discriminant function when there is more than 1 function. In this case since we have 3 groups we have 2 functions. Referring to Table 20, the eigenvalues are significant for the 1st function ($\chi^2 = 318.035, p< 0.01$) and also for the 2nd function ($\chi^2 = 11.694, p< 0.05$).

Table 20. Wilks' Lambda

Wilks' Lambda				
Test of Function(s)	Wilks' Lambda	Chi-square	df	Sig.
1 through 2	.217	318.035	10	.000
2	.945	11.694	4	.020

The next three tables (Tables 21, 22 and 23) give us the discriminant function coefficients. The standardized canonical discriminant function coefficients are like the beta weights (standardized) in regression analysis. This will tell us the relative importance of the variable in predicting adoption versus non-adoption of ERP. When there are 2 or more functions, a composite or summary measure is useful in describing the contributions of a variable across all significant functions (Hair et al. 2010). Thus, the potency index which is the relative measure among all variables is indicative of each variable's discriminating power (Perreault et al. 1979).

Table 21. Standardized Coefficients

Standardized Canonical Discriminant Function Coefficients

	Function	
	1	2
TOPMGMT	.475	.975
SIZE	.373	−1.110
CULTURE	.139	.049
BENEFITS	−.074	.199
RESISTANCE	−.256	.152

The next table on structure matrix (Table 22) presents the simple (zero-order) correlation of each variable with the discriminant function.

Table 22. Structure Matrix.

Structure Matrix

	Function	
	1	2
TOPMGMT	.891*	.411
SIZE	.829*	−.509
CULTURE	.822*	.035
RESISTANCE	−.793*	−.012
BENEFITS	.667*	.171

- standardized canonical discriminant functions
- Variables ordered by absolute size of correlation within function.

* Largest absolute correlation between each variable and any discriminant function

Table 23 presents the canonical discriminant function coefficients. It displays the unstandardized discriminant function coefficients similar to the regression analysis (unstandardized beta). These coefficients are used to develop the discriminant equation and are as follows:

Function1: $Z_{jk} =$ –3.676 + 0.756(TOPMGMT) + 0.006(SIZE) + 0.206(CULTURE) –0.096(BENEFITS) – 0.335(RESISTANCE)

Function 2: $Z_{jk} =$ –1.023 + 1.551(TOPMGMT) -0.018(SIZE) + 0.073(CULTURE) + 0.256(BENEFITS) + 0.199(RESISTANCE)

Table 23. Unstandardized Coefficients.

Canonical Discriminant Function Coefficients

	Function	
	1	2
TOPMGMT	.756	1.551
SIZE	.006	–.018
CULTURE	.206	.073
BENEFITS	–.096	.256
RESISTANCE	–.335	.199
(Constant)	–3.676	–1.023

Unstandardized coefficients

As there are **2 functions,** we need to calculate a **Potency Index** to see which variable has the most impact. This composite measure is useful only in depicting the relative position (such as the rank order) of each variable (Hair et al. 2010). The Potency Index is calculated by a 2 step process as follows (Hair et al. 2010):

Step 1: Calculate potency value for each variable

a. First calculate the squared discriminant loadings.
b. Then weigh them by their relative contribution of the discriminant function

We will first calculate the relative eigenvalue measure using the formula below:

$$\text{Relative eigenvalue of discriminant function } j = \frac{\text{Eigenvalue of discriminant function } j}{\text{Sum of eigenvalue across all significant functions}}$$

Next the potency value of each variable is calculated as follows:

$$\text{Potency value of variable } i \text{ on function } j = (\text{Discriminant loading}_{ij})^2 \times \text{Relative eigenvalue of function } j$$

Step 2: Calculate a composite potency index across all significant functions.
The composite potency index is calculated as follows:

Composite potency of variable *i*	=	Sum of potency of variable *i* across all significant discriminant functions

Table 24 shows the calculated **Potency Index values.**

Functions at group centroids display the mean for the groups which are in standardized form (*z score*) which is based on the weighted linear combination making up the discriminant function. These values will be used to calculate the cutting score which is used to establish a cutting point for the purpose of classification. Since we have 3 groups we will get 2 functions (see Table 25).

Once we have finished the interpretation, we need to interpret the classification accuracy. As shown in the following table (Table 26) we have 3 classification hit ratios, the first one is the classification with all the data that we have which is called the original classification and the second one is called the cross-validated classification. In the cross-validated classification the analysis will leave one case out at a time and run the analysis. The third one is the hit ratio in which we use the holdout sample.

Next, we need to assess the hit ratio. The hit ratio is actually a measure of the classification accuracy which may range from 0% of all respondents wrongly classified to 100% of all respondents correctly classified. We also need to make sure that the hit ratio that we get from the analysis is better than chance which would indicate that our model works well in predicting the group membership. To assess this we need to assess the maximum chance criterion, the proportional chance criterion and the third is to assess the significance using a *Press Q* test. The calculation is given as follows:

1. Maximum chance

   C_{MAX} = Size of the largest group

   Example:

   Estimate Sample 97/213 = 0.455 **(see Table 26)**

   Holdout Sample 62/131 = 0.473

2. Proportional chance

   $C_{PRO} = (p_1)^2 + (p_2)^2 + (p_3)^2$

   where: p_1 = proportion of individuals in group 1

   p_2 = proportion of individuals in group 2

   p_3= proportion of individuals in group 3

   Example:

   Estimate Sample $(49/213)^2 + (67/213)^2 + (97/213)^2 = 0.359$

   Holdout Sample $(49/131)^2 + (67/131)^2 + (97/131)^2 = 0.371$

Table 24. Calculation of the Potency Index

	Discriminant Function 1				*Discriminant Function 2*				
Independent Variables	**Loadings**	**Squared Loadings**	**Relative Eigenvalue**	**Potency Value**	**Loadings**	**Squared Loadings**	**Relative Eigenvalue**	**Potency Value**	**Potency Index**
TOPMGMT	0.891	0.794	0.983	0.780	0.411	0.169	0.017	0.003	0.783**(1)**
SIZE	0.829	0.687	0.983	0.676	-0.509	0.259	0.017	0.004	0.680**(2)**
CULTURE	0.822	0.676	0.983	0.664	0.035	0.001	0.017	0.000	0.664**(3)**
BENEFITS	0.667	0.445	0.983	0.437	0.171	0.029	0.017	0.000	0.438**(5)**
RESISTANCE	-0.793	0.629	0.983	0.618	-0.012	0.000	0.017	0.000	0.618**(4)**

Note 1: The Relative Eigenvalues are calculated as follows (see Table 19), Eigenvalue Function 1 = 3.361, Function 2 = 0.058, Total Eigenvalue = (3.361 + 0.058 = 3.419), Relative Eigenvalue Function 1 = (3.361/3.419) = 0.983 and Relative Eigenvalue for Function 2 = (0.058/3.419) = 0.169.

Note 2: The loadings are taken from the structure matrix (see Table 22).

Note 3: The numbers in brackets indicate the ranking of important variables.

Table 25. Group Centroids

Functions at Group Centroids

Adoption Level	Function	
	1	2
No	-2.117	.337
Plan	-1.294	-.309
Adopted	1.964	.043

Unstandardized canonical discriminant functions evaluated at group means

Table 26. Classification Results

Classification Results[a,b,d]

Adoption Level				Predicted Group Membership			Total
				No	Plan	Adopted	
Cases Selected	Original	Count	No	33	16	0	49
			Plan	14	39	14	67
			Adopted	0	4	93	97
		%	No	67.3	32.7	.0	100.0
			Plan	20.9	58.2	20.9	100.0
			Adopted	.0	4.1	95.9	100.0
	Cross-validated[c]	Count	No	33	16	0	49
			Plan	18	35	14	67
			Adopted	0	4	93	97
		%	No	67.3	32.7	.0	100.0
			Plan	26.9	52.2	20.9	100.0
			Adopted	.0	4.1	95.9	100.0
Cases Not Selected	Original	Count	No	12	14	0	26
			Plan	11	26	6	43
			Adopted	0	1	61	62
		%	No	46.2	53.8	.0	100.0
			Plan	25.6	60.5	14.0	100.0
			Adopted	.0	1.6	98.4	100.0

a. 77.5% of selected original grouped cases correctly classified.
b. 75.6% of unselected original grouped cases correctly classified.
c. Cross validation is done only for those cases in the analysis. In cross validation, each case is classified by the functions derived from all cases other than that case.
d. 75.6% of selected cross-validated grouped cases correctly classified.

3. Press Q

$$\text{Press Q} = \frac{[N - (n*k)]^2}{N(k-1)}$$

Where: $Q \sim \chi^2$ with 1 degree of freedom
N = Total sample size
n = number of observations correctly classified
k = number of groups

Press Q Calculation for **Estimate Sample (see Table 27):**

$$\text{Press Q} = \frac{[213 - (165*3)]^2}{213(3-1)} = 186.676$$

Press Q Calculation for **Holdout Sample:**

$$\text{Press Q} = \frac{[131 - (99*3)]^2}{131(3-1)} = 105.176$$

The values calculated are summarized in Tables 27, 28 and 29.

Table 27. Classification Accuracy (Estimate Sample)

Measure	Value	Hit Ratio for Estimate Sample
Maximum Chance Proportional Chance	45.5 35.9	77.5 77.5
Press Q Table Value Press Q Calculated Value	6.635 186.676**	

** $p < 0.01$

Table 28. Classification Accuracy (Leave One Out)

Measure	Value	Hit Ratio for Leave one out Sample
Maximum Chance Proportional Chance	45.5 35.9	75.6 75.6
Press Q Table Value Press Q Calculated Value	6.635 171.127**	

** $p < 0.01$

Table 29. Classification Accuracy (Holdout Sample)

Measure	Value	Hit Ratio for Holdout Sample
Maximum Chance Proportional Chance	47.3 37.1	75.6 75.6
Press Q Table Value Press Q Calculated Value	6.635 105.176**	

** $p < 0.01$

Based on the calculation we can conclude the model that we developed can predict the group membership greater than that of the chance criterion of maximum chance (45.5%) and also proportional chance (35.9%) for the estimate and the leave one out classification whereas for the holdout sample the hit ratio was greater than that of the chance criterion of maximum chance (47.3%) and also proportional chance (37.1%). The Press Q is also significant at $p < 0.01$, thus we can conclude that the model we developed is good. Hair et al. (2010) has also suggested a more stringent criterion which is 25% higher than the chance criterion which for the estimate and leave one out classification would have been 1.25 * 0.455 = 0.569 whereas for the holdout sample 1.25 * 0.473 = 0.591. The hit ratios exceeded even the more stringent criteria established by Hair et al. (2010).

Self-Evaluation Exercises

1. In a 3-group discriminant analysis how many functions are you expected to get?
   _____ functions
2. It is a must to split the analysis sample and holdout sample using a 60:40 ratio.
   a. True b. False

The next few questions are based on the data of an Internet portal service provider who would like to assess the loyalty of their users with regards to the service they have provided. The variables they have used as predictors are as follows:

CORE	Core services provided
SUPPORTING	Supporting services provided
VALUE	Perceived value of the services provided
LOYALTY	1= Not Loyal; 2 = Neither; 3 = Loyal

A 3 group discriminant analysis was run and based on the output. Please answer the questions that follow.

3. From the results given below, are the 2 discriminant functions derived from the analysis significant?

Wilks' Lambda				
Test of Function(s)	Wilks' Lambda	Chi-square	df	Sig.
1 through 2	.890	74.001	6	.000
2	.985	9.335	2	.009

4. How much of variance is explained by each of the functions derived from the analysis?

Eigenvalues				
Function	Eigenvalue	% of Variance	Cumulative %	Canonical Correlation
1	.107[a]	87.9	87.9	.311
2	.015[a]	12.1	100.0	.121

a. First 2 canonical discriminant functions were used in the analysis.

5. Which variables are significant discriminators of the 3 groups (**Not Loyal/Neither/Loyal**) based on the output given below.

Tests of Equality of Group Means					
	Wilks' Lambda	F	df1	df2	Sig.
CORE	.995	1.633	2	636	.196
SUPPORTING	.978	7.239	2	636	.001
VALUE	.918	28.327	2	636	.000

6. Can the model developed by using this analysis predict better than the proportional chance criterion?

Classification Results[a,b,d]								
Loyalty Level				Predicted Group Membership			Total
				Not Loyal	Neither	Loyal	
Cases Selected	Original	Count	Not Loyal	47	105	38	190
			Neither	43	144	59	246
			Loyal	24	74	105	203
		%	Not Loyal	24.7	55.3	20.0	100.0
			Neither	17.5	58.5	24.0	100.0
			Loyal	11.8	36.5	51.7	100.0
	Cross-validated[c]	Count	Not Loyal	37	115	38	190
			Neither	43	144	59	246
			Loyal	24	74	105	203
		%	Not Loyal	19.5	60.5	20.0	100.0
			Neither	17.5	58.5	24.0	100.0
			Loyal	11.8	36.5	51.7	100.0
Cases Not Selected	Original	Count	Not Loyal	17	35	10	62
			Neither	17	36	21	74
			Loyal	4	22	19	45
		%	Not Loyal	27.4	56.5	16.1	100.0
			Neither	23.0	48.6	28.4	100.0
			Loyal	8.9	48.9	42.2	100.0

a. 46.3% of selected original grouped cases correctly classified.
b. 39.8% of unselected original grouped cases correctly classified.
c. Cross validation is done only for those cases in the analysis. In cross validation, each case is classified by the functions derived from all cases other than that case.
d. 44.8% of selected cross-validated grouped cases correctly classified.

Answers for Self-Evaluation Exercises

1. 2 functions
2. B. False
3. Function 1 Chi Square = 74.001 p-value < 0.01
   Function 2 Chi Square = 9.335 p-value < 0.01
   Thus we can conclude that both functions are significant at $p < 0.01$ level.
4. The canonical correlation for function 1 = 0.311, the squared value will give a 0.097 which means 9.7% of the variance in Loyalty can be explained by the first function. The second function has a canonical correlation of 0.121 and the squared value will yield a value of 0.015 which means only 1.5% of the variance in loyalty can be explained by the 2nd discriminant function.
5. Only 2 variables, i.e., Supporting ($F = 7.239$, $p < 0.01$) and Value ($F = 28.327$, $p < 0.01$) are significant discriminators. Core is not a significant predictor of loyalty.
6. The hit ratio for the holdout (unselected) sample is 39.8%, the proportional chance criterion can be calculated as follows:

Not Loyal	=	$(62/181)^2$	=	11.73
Neither	=	$(74/181)^2$	=	16.71
Loyal	=	$(45/181)^2$	=	6.18

   The total of 34.62% is lower than the hit ratio from the analysis indicating that the analysis performs better than the proportional chance criterion.

Conclusion

This chapter has highlighted the use of Discriminant Analysis and how the results can be used in the interpretation of a research project. We have illustrated the basic requirements that need to be met before the analysis can be conducted and once the analysis is done we have also shown how the output can be interpreted. This illustrated guide should be useful for novice researchers as well as seasoned researchers who would like to embark on undertaking this technique of analysis and how best the output from the SPSS analysis can be interpreted as well as presented in standard table forms.

Additional Unsolved Exercises

1. Explain what you understand by the term "standardized canonical discriminant coefficient" and what do we use them for?

2. How do you assess predictive accuracy of the model that you develop in a Discriminant Analysis?
3. What are the assumptions that need to be fulfilled when using the Discriminant Analysis?
4. Discriminant Analysis is essentially a classification technique.
   a. True b. False
5. The M test of the Box is used to test what the assumption is in a Discriminant Analysis.

Case Analysis

A local bank manager would like to develop a model to discriminate their customers who are internet banking users and those who do not use internet banking (IB). The bank has invested a lot of money in developing the internet banking platform and if their clients do not adopt the use of internet banking, the investment will not give them the required returns. The cost to service a customer at the bank physically is increasing and the focus of the bank is to reduce the offline usage and move to an online environment to reduce cost.

The bank manager has identified 4 factors from the literature that would be able to help in discriminating a user form a non-user. The 4 factors are PU (Perceived Usefulness), PEU (Perceived Ease of Use), PBC (Perceived Behavioral Control) and also the RISK (Risk Perception).

Using the data (**Internet Banking.sav**) run a 2 group discriminant analysis using the variables below:

PU	**Perceived Usefulness of Internet banking**
PEU	**Perceived Ease of Use of Internet banking**
PBC	**Perceived Behavioral Control of Internet banking**
RISK	**Perceived Risk of Internet banking**
USER2GRP	1= NON-USER, 2 = USER
RANDZ	This is the selection variable, Set Value = 0

Using the output derived, answer the following questions (justify using the values that you get from the analysis).

1. Can all the four predictors help to discriminate the user category (USER/NON- USER)?
2. How much of the variance in the user category is explained by all the predictor variables?
3. Which variables are significant predictors in discriminating the user category?

4. Which variable is the best discriminator in this case?
5. Calculate the cutting score for the classification of the 2-groups.
6. Can the model predict better than chance? (Need to benchmark)

References

Cramer, D. 2003. Advanced Quantitative Data Analysis. Buckingham: Open University Press.

Cohen, J. 1988. Statistical Power Analysis for the Behavioral Sciences. Lawrence Erlbaum Associates: Hillsdale, NJ.

Hair, J.F., Black, W.C., Babin, B.J. and Anderson, R.E. 2010. Multivariate Data Analysis: A global perspective. Prentice-Hall: Upper Saddle River, New Jersey.

Leech, L.L., Barret, K.C. and Morgan, G.A. 2008. SPSS for Intermediate Statistics: Use and Interpretation. Taylor & Francis Group: USA.

Meyers, L.S., Garnst, G. and Guarino, A.J. 2006. Applied Multivariate Research: Design and Implementation. Sage: Thousand Oaks, CA.

Perreault, W.D., Behrman, D.N. and Armstrong, G.M. 1979. Alternative Approaches for Interpretation of Multiple Discriminant Analysis in Marketing Research. Journal of Business Research, 7: 151–173.

Additional Readings

Ramayah, T., Ahmad, N.H., Halim, H., Mohamed Zainal, S.R. and May-Chiun, L. 2010. Discriminant Analysis: An Illustrated Example. African Journal of Business Management, 4(9): 1654–1667.

Ramayah, T., Jantan, M. and Chandramohan, K. 2004. Retrenchment Strategy in Human Resource Management: The Case of Voluntary Separation Scheme (VSS). Asian Academy of Management Journal, 9(2): 35–62.

Ramayah, T., Md. Taib, F. and Koay, P.L. 2006. Classifying Users and Non-Users of Internet Banking in Northern Malaysia. Journal of Internet Banking and Commerce, August 2006, 11(2), [Online] Available: http://www.arraydev.com/commerce/JIBC/2006-08/Thurasamy.asp.html

CHAPTER 10

Multiple Regression Analysis

María V. López, María C. Fabrizio* and *María C. Plencovich*

1. Introduction

In our professions, we may sometimes have to deal with issues in which there are two or more variables at a stake. For example, in a company that builds denitrification plants for nitrate elimination, we may wonder to what extent the nitrate level of a river may change when the water pH is modified. And if it does, how does it change? And what other variables may influence this change? Can we predict the level of nitrate from a group of variables?

It may be crucial to have an answer to these questions in order to make appropriate decisions about the future of a company, which also involve environmental issues and professional responsibility.

Let us see how these issues affect a company which builds denitrification plants transforming nitrate to non-toxic nitrogen forms at critical points along a river.

Drinking water may contain nitrates in small concentrations, but elevated concentrations of nitrates in surface waters indicate pollution from industrial and/or community sewage. Imagine a situation taking place in Rughipolis, a remote city of a distant country. The Environmental Agency of Rughipolis has decided to hire *Chemaly, Inc.*—a well-known chemical company specialized in designing denitrification plants—to build some plants in several places along the Rughi-Rughi River, which runs through the city. The company wants to design plants of an adequate size to control the level of pollution in the place.

Facultad de Agronomía, Universidad de Buenos Aires (Argentina).
* Corresponding author

John Denitri, the industrial chemical engineer who is responsible for improving processes in the sectors, knows that the key issue for building a denitrification plant is how to estimate the nitrate levels of the river.

Suppose that John wants to predict the nitrate level having only the nitrate values from a sample. What is the best estimation or prediction? Would his prediction be modified knowing other variables, such as pH of water, electrical conductivity, chlorine levels, or oxygen dissolved levels? A precise prediction of the pollution level will allow him to build a plant of adequate size, eliminating health risks for nearby residents and reducing costs.

This chapter seeks to provide readers with the tools needed to solve complex problems with multiple variables, one of which is considered the dependent or response variable, and the others are considered the independent, explanatory, or predictor variables. To do so, multiple regression analysis will be presented as a robust tool to predict, explain and evaluate the nature of relationships among multiple variables. The R software is used in this chapter (R Development Core Team 2008).

2. Prediction with no independent variable

In the Chemaly problem of the denitrification plant the nitrate, NO_3^-, level is the dependent variable, which we call Y. If John wants to predict the NO_3^- level having only a random sample of 30 observations taken from sites along the river (see Appendix A), what will his best prediction be? Is it the mean or the median value? Or is it another measurement from the sample?

The different sampling places will vary in their NO_3^- levels. If all values were constant, he would be absolutely certain that his prediction, the constant, would be perfect. But as all the values are not equal, his prediction will not be perfect. The multiple variables accompanying the process, such as pH, electric conductivity, chlorine concentration, level of oxygen dissolved in water, and many other either unknown or not evaluated variables may cause a joint variation.

However, without having any further information than the nitrate level variable, the best prediction can be obtained from a model that can be written as

$$Y_i = \beta_0 + \varepsilon_i \quad i = 1,2,\dots,n \tag{1}$$

where Y_i is the value for the response variable or dependent variable, nitrate level measured in ppm at the ith observation (with i varying from 1 to 30 in this case), β_0 is the parameter representing the expected value of the response variable and ε_i expressing the difference between the observed value, Y_i and the expected value. ε_i is a random error term with expected value $E(\varepsilon_i) = 0$ and variance $V(\varepsilon_i) = \sigma^2$. It is assumed that these error terms are not correlated, that is $\sigma(\varepsilon_i, \varepsilon_j) = 0$ if $i \neq j$.

According to the model, the response variable Y_i, nitrate level, is the sum of a constant, β_0, and a random error term ε_i. Then, Y_i is a random variable. As you may remember, the mean of the sum is the sum of the means, and that the mean of a constant is the same constant. Then, we may express the expected value of variable Y as:

$$\mu_{Y_i} = E(Y_i) = E(\beta_0 + \varepsilon_i) = E(\beta_0) + E(\varepsilon_i) = \beta_0$$

On the other hand, as the variance of the sum of a constant plus a variable equals the variable variance, the variance of Y is:

$$\sigma^2_{Y_i} = V(Y_i) = V(\beta_0 + \varepsilon_i) = V(\varepsilon_i) = \sigma^2$$

In the case that errors are independent and have a normal distribution, Y_i also has a normal distribution and Y_i are independent observations.

Suppose that the model (1) is valid, how can John find a good estimator for β_0 from a data set containing nitrate levels? He may intuitively choose that value that is closer to the observed values.

Statistics provide a method to answer this question: *the least square method*. This method is so called because of all possible β_0 values, it looks for the value which minimizes the error sum of squares. Thus, John would design all the denitrification plants with the same size, based on this estimated value, assuming that this value is closer to the observed values. In the model (1), he would look for

$$\min_{\beta_0} \sum_i (Y_i - \beta_0)^2$$

Using a derivative operator from Mathematical Analysis, he can infer that the estimator forβ_0, called b_0, which provides that value, is

$$\widehat{\beta}_0 = b_0 = \frac{\sum_{i=1}^{n} Y_i}{n} = \overline{Y}$$

Therefore, he may obtain

$$\widehat{Y}_i = b_0 \quad i = 1,..n$$

where $\widehat{Y}_i$ represents the estimated or fitted value of Yat the i-th observation. As expected, this means that $\widehat{Y}_i = \overline{Y}$. So, in the case, without any explanatory variable the mean is the best predictor of the response variable. John Denitri estimates $\widehat{Y} = 200.29$ ppm as his best prediction for the nitrate level using only one value for estimating the size of all the plants over the river.

How could he determine how precise this estimate is? This precision analysis is based on the difference between the observed value Y_i and the corresponding fitted value $\hat{Y}_i$. This difference is called *residual* and is denoted by e_i:

$$e_i = Y_i - \hat{Y}_i$$

In our model (1), $e_i = Y_i - b_0$.

It can be shown that $\sum_{i=1}^{n} e_i = 0$.

From the residual values, John can estimate how far his prediction is from the observed values and how variable are these differences. He will obviously be much interested in getting small values.

The sum of squares of the deviations between the observed values and the fitted values from the model is a variability measure of the residuals and consequently of the response variable. It is called *error sum of squares* or *residual sum of squares* and is denoted by *SSE*:

$$SSE = \sum_{i=1}^{n} (Y_i - \hat{Y}_i)^2 = \sum_{i=1}^{n} e_i^2$$

This expression will surely remind you of the numerator of the sample variance. If this expression is divided by its degrees of freedom, we get the *error mean square* or *residual mean square*, denoted by *MSE*:

$$MSE = \frac{SSE}{n-1} = \frac{\sum_{i=1}^{n} (Y_i - \hat{Y}_i)^2}{n-1} = \frac{\sum_{i=1}^{n} e_i^2}{n-1}$$

This expression is the estimated variance of the *Y* variable. The degrees of freedom for *SSE* are $n - 1$ since $\hat{Y}_i$ is obtained estimating only one parameter, β_0.

Let's see the data from the observations of the nitrate levels (Table 1). If John builds the plant according to estimated nitrate levels, he will find that in *site 1* the plant will be smaller than needed and that the contamination will not be controlled effectively. However, he will see that in *site 3* the plant will have a larger size than needed and will be more expensive.

Table 1. Nitrate levels. Observed and fitted values, residuals and squared residuals according to model (1).

Observation	Observed nitrate levels (ppm) Y_i	Fitted Nitrate Levels (ppm) $\widehat{Y}_i$	Residual $e_i = Y_i - \widehat{Y}$	Squared residual e_i^2
1	288.6	200.29	88.31	7,798.66
2	213.5	200.29	13.21	174.50
3	128.3	200.29	–71.99	5,182.56
....	...	...	...	...
29	229.5	200.29	29.21	853.22
30	214.8	200.29	14.51	210.54

$$\overline{Y} = 200.29\,ppm \quad \sum_{i=1} e_i = 0 \quad \sum_{i=1}^{30} e_i^2 = 105{,}391.37$$

In this case $MSE = \dfrac{105{,}391.37}{29} = 3{,}634.19$, corresponding to a standard deviation of 60.28 ppm representing 30 percent of the mean value.

If errors follow a normal distribution, John may continue his analysis inferring the β_0 value from data as we may usually estimate the population mean.

Error term $\varepsilon_i = Y_i - E(Y_i)$ should be distinguished from the residual $e_i = Y_i - \widehat{Y}_i$. The former corresponds to a Y_i value deviation to its expected unknown value $E(Y_i)$; therefore, ε_i is also unknown. The latter corresponds to a deviation of an observation Y_i from its fitted value, which can be estimated and therefore known.

Self-evaluation

10.1 Choose the correct answer. *MSE* is the estimator for:

a) Residual sum squares
b) Sample mean
c) Error variance
d) Standard error deviation

10.2 If in any data distribution where at least one observation differs from the rest, the squared differences of the observation values are estimated according to the minimum value instead of the mean, the new sum of squares will be:

a) lower b) higher c) equal d) depends on the sample

10.3 A sample of eight BATTY alkaline batteries was taken and battery run time was recorded in order to predict its life in hours. The following results were recorded: 50, 48, 45, 42, 54, 26, 30, 40. Calculate the sum of squares of deviations with respect to: i) the sample mean, ii) the median, iii) the minimum value. Compare the results obtained.

10.4 Superoxide dismutases (SOD) are antioxidant enzymes that catalyze the reactions to prevent oxidative stress. These enzymes protect cells and organisms. Copper (Cu), iron (Fe), manganese (Mn) and zinc (Zn) oligoelements are part of the SOD protein structure. Deficiency in the mineral consumption or the presence of factors interfering in animal metabolism, especially in grazing cattle, may decrease enzyme activity associated with it, and produce a decrease of weight gain and of fat milk yield and cause other problems. A researcher team was guided by the following hypothesis: "Blood activity of SOD in grazing cattle depends on the concentrations of Cu, Fe and Mn in the feed and may be used as a biological marker of the mineral metabolism". To validate this hypothesis, they selected at random five two-year-old Holstein heifers, grazing mainly on grass, from 15 dairy farms. SOD activity (in Ug^{-1} Hb) was determined from blood samples and each farm average was recorded. Forage samples were also taken from the prairie at the same day. Only data corresponding to Cu (in ppm) are presented in the following table:

Dairy farm	SOD	Cu
1	2,149	53.05
2	350	7.00
3	1,610	49.08
4	1,289	18.94
5	981	14.27
6	2,357	68.30
7	1,467	21.43
8	1,070	10.00
9	2,507	41.60
10	1,681	38.52
11	1,702	43.18
12	1,109	19.25
13	350	7.00
14	1,216	12.93
15	350	7.00

a) What is the best estimate for the Superoxide dismutase (SOD) level in cattle without considering Copper content in grass?
b) What is the estimated variance for SOD levels on dairy farms?

Answers:

10.1 c. $MSE = \hat{\sigma}_{\varepsilon}^2$ because $MSE = \frac{\sum_{i=1}^{n}(Y_i - \hat{Y}_i)^2}{n-1} = \frac{\sum_{i=1}^{n}(Y_i - \overline{Y}_i)^2}{n-1}$

10.2 b. It has been shown that the minimum value of the expression $\sum_{i=1}(Y_i - \beta_0)^2$ is obtained only if $\beta_0 = \overline{Y}$.

10.3 i) $\overline{Y} = 41.875 \Rightarrow \sum_{i=1}^{8}(Y_i - 41.875)^2 = 656.875$

ii) $Y_{.50} = 43.5 \Rightarrow \sum_{i=1}^{8}(Y_i - 43.5)^2 = 678$

iii) $Y_{\min} = 26 \Rightarrow \sum_{i=1}^{8}(Y_i - 26)^2 = 2{,}673$

It is shown that the sum of squares of the deviations reaches its minimum value when deviations are considered with respect to the mean of the sample.

10.4.

a) The best estimate of SOD level is the mean of the entire sample when there is no other information. In this case, it is 1,345.867 Ug^{-1} Hb.
b) Variance estimator is MSE = 471,422.552 $(Ug^{-1}\text{ Hb})^2$.

3. Prediction with one Independent Variable

The Chemaly chemical engineer will surely want to make a more precise prediction about the nitrate level to optimize the design of the denitrification plants. We have seen that the sample mean is the best prediction when there is no further information about other variables that may explain the variability of the response variable that is nitrate levels. However, John knows that the pH of water is an important variable for denitrification since it affects the microflora activity involved in this process. Therefore, John thinks that knowing the pH values at different river points may explain part of the variability of the nitrate levels through a linear relationship between both variables.

The analysis of simple linear regression allows studying the simplest relationship that can be determined between variables. On one hand, we have a response variable Y that in this case is the nitrate level, and only one independent variable, also known as a predictor or explanatory variable, X, that in his case is the pH of water.

In general, to be able to determine the nature of the relationship between variable X and Y (linear, quadratic, or of any other type, and even when there is no relationship), we use a scatter plot. Figure 1 shows the scatter plot corresponding to the pH values and nitrate levels taken at the different sites (see Appendix A for the corresponding values).

John understands that, although the relationship is not perfect, as pH increases, the nitrate levels also increase. There is an upward linear trend in the nitrate level to vary with regard to the pH of water. But this does not indicate a functional relationship since all the points are not aligned. Also, for the same value of the pH of water there are several different observed values of the nitrate levels. Therefore, this may indicate a statistical relationship.

Taking into account these two issues, the tendency of the nitrate levels to vary systematically with pH in a linear way and the nitrate level dispersion for the same pH value, John may represent nitrate levels in the general model the following way:

$$Y_i = \beta_0 + \beta_1 X_i + \varepsilon_i \quad i = 1,2,...n \tag{2}$$

where Y_i represents the value of the response variable or dependent variable, nitrate level measured in ppm at the ith observation (with i varying from 1 to 30 in this case), β_0 and β_1 are the coefficients of the regression function, X_i are the values taken up by the independent variable and ε_i, is the difference between the observed value, Y_i, and the expected value, provided that they follow the same assumptions as in Model (1).

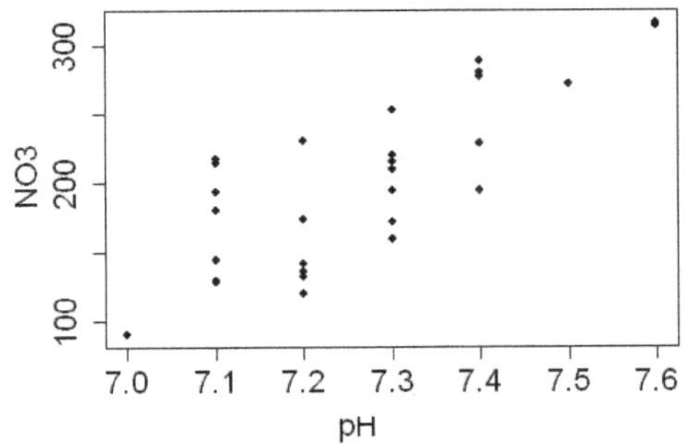

Figure 1. Scatter plot corresponding to the pH values and nitrate levels taken at the different sites plot(nitro$pH,nitro$NO3, pch=20, xlab="pH", ylab="NO3").

Model (2) is called *simple linear regression model*. It is *simple and linear in the parameters* and *linear in the independent variable*. It is simple because it has only one independent variable, and linear in the parameters because in each term there is only one parameter to the first power, and linear in the independent variable because this variable is raised to the first power. So far, suppose that variable X takes fixed values.

The same as in Model (1), Y_i is the sum of two components: a constant $\beta_0 + \beta_1 X_i$, expressing the mathematical function which relates both variables and the random term ε_i, representing the variability in Y not explained by the linear relationship between Y and X. Therefore, Y_i is a random variable.

John will surely be interested in knowing the true nitrate levels on sites with different pH. He will not consider only one constant nitrate value, but will have a different nitrate mean according to the pH value of the site. The expected nitrate level may now be written down according to the pH value as follows:

$$\mu_{Y/x_i} = E(Y_i) = E(\beta_0 + \beta_1 X_i + \varepsilon_i) = \beta_0 + \beta_1 X_i + E(\varepsilon_i) = \beta_0 + \beta_1 X_i$$

According to the relationship $E(Y_i) = \beta_0 + \beta_1 X_i$, parameter β_0 is the expected value of Y_i if $X_i = 0$. We have to be careful with this interpretation because if the model does not include $X_i = 0$, the parameter β_0 will not be meaningful since the relationship between variables may be different in values close to zero. John knows that the pH value is not zero, therefore he is not interested in β_0 interpretation in this case. Parameter β_1 is the slope of the regression function. As a consequence, it indicates the expected change of the nitrate level, Y, associated with a 1 pH unit increase.

As in model (1), the variance of the nitrate levels is:

$$\sigma^2(Y_i) = \sigma^2(\beta_0 + \beta_1 X_i + \varepsilon_i) = \sigma^2(\varepsilon_i) = \sigma^2$$

Note that the variance of Y is the same constant for all levels of X.

Figure 2 shows the regression assuming normality and constant variance for the different levels of the independent variable. We see that for each level of the independent variable, the response variable values are distributed according to a normal probability distribution, whose mean belongs to a regression straight line. The variances of these distributions are the same for all the levels of the variable X, but the expected values differ in their position according to the functional relationship of the variables.

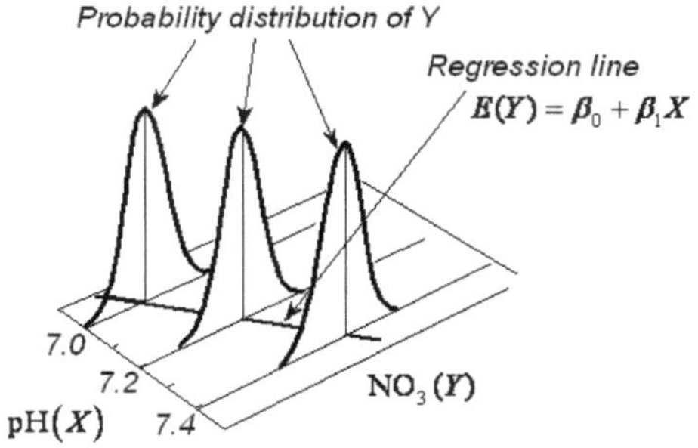

Figure 2. Representation of regression model (2).

3.1 Parameter Estimation of the Regression Function

The Chemaly engineer thinks that Model (2) is correct and wants to use it to predict nitrate levels with more precision than Model (1). He wonders how to get good estimators of the model parameters to get a straight line that fits the data best. Once again the answer is provided by the least square method. As in Model (1), this method determines that estimators of β_0 andβ_1, denoted b_0 and b_1, are the values minimizing the sum of squares of ε_i. Therefore,

$$\min_{\beta_0,\beta_1} \sum_{i=1}^{n} \left[\varepsilon_i^2\right] = \min_{\beta_0,\beta_1} \sum_{i=1}^{n} \left[Y_i - (\beta_0 + \beta_1 X_i)\right]^2$$

To solve this problem, partial derivates are used, obtaining the so called *normal equations system*:

$$\begin{cases} \sum_{i=1}^{n} Y_i = nb_0 + b_1 \sum_{i=1}^{n} X_i \\ \sum_{i=1}^{n} X_i Y_i = b_0 \sum_{i=1}^{n} X_i + b_1 \sum_{i=1}^{n} X_i^2 \end{cases}$$

Solving this system, estimators b_0 and b_1 are obtained.

$$\widehat{\beta}_1 = b_1 = \frac{\sum_{i=1}^{n} (X_i - \overline{X})(Y_i - \overline{Y})}{\sum_{i=1}^{n} (X_i - \overline{X})^2}$$

$$\widehat{\beta}_0 = b_0 = \frac{1}{n}\left[\sum_{i=1}^{n} Y_i - b_1 \sum_{i=1}^{n} X_i\right] = \bar{Y} - b_1 \bar{X}$$

Figure 3 shows part of the results obtained by John Denitri using the R software.

```
Model2 <- lm(NO3~pH, data=nitro)
summary(Model2)
Call:
lm(formula = NO3 ~ pH, data = nitro)
```

Coefficients:

	Estimate	Std. Error	t value	Pr(>|t|)
(Intercept)	–2085.55	334.02	–6.244	9.52e-07 ***
pH	314.56	45.96	6.845	1.95e-07 ***

Signif. codes: 0 '***' 0.001 '**' 0.01 '*' 0.05 '.' 0.1 ' ' 1

Figure 3. R Regression output for Model (2) for nitrate level prediction.

From the computer output, the industrial engineer observes that $b_0 =$ –2,085.55 and $b_1 = 314.56$. The b_0 value does not have a practical use in this particular problem, because there were no water samples with pH equal or close to zero. However, this does not justify cancelling it since it can help prediction. From the b_1 value, John estimates that one additional pH unit in water leads to an increase in the mean nitrate level of approximately 314.6 ppm. Thus, the regression function that estimates is $\widehat{Y}_i = -2,085.55 + 314.56X_i$ $\widehat{Y}_i$ being the estimated mean of the nitrate level for the X_i value of the independent variable pH. Thus, for a pH value $X_i = 7.5$, its corresponding estimated value of nitrate level is $\widehat{Y}_i = -2,085.55 + 314.56(7.5) = 273.7\,ppm$. This value is an estimate of the expected value of the nitrate level for a site having a 7.5 pH and differs from the initial prediction when the pH value was unknown, which was 200.29 ppm. John is careful in predicting only nitrate levels for pH values within the range observed in this study, because at higher pH levels the river denitrification bacteria may be inhibited and the relationship between variables could be different.

Figure 4 shows the scatter plot of the observations, the fitted regression line and the horizontal line representing the mean of the nitrate levels without considering the pH. Note that observations are closer to the fitted regression line than to the horizontal line; hence it represents the best prediction for the nitrate levels at different sites.

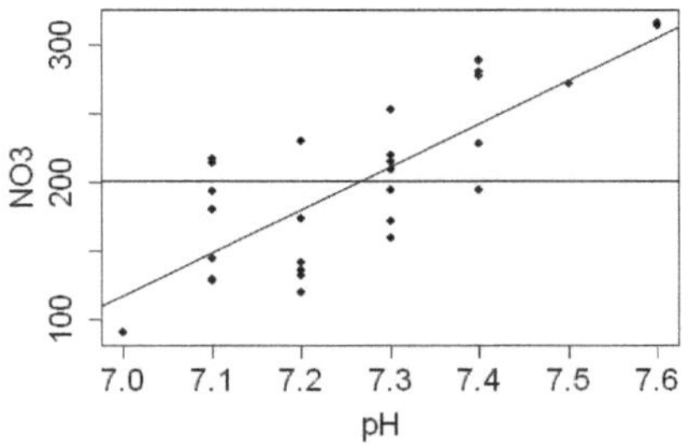

Figure 4. Scatter Plot fitted regression line of pH and nitrate levels and mean of nitrate level without considering pH.

```
plot(nitro$pH,nitro$NO3, pch=20, xlab="pH", ylab="NO3")
abline(Model2)
abline(h=200.29)
```

3.2 Estimation of Error Terms Variance σ^2 and goodness of fit

John now needs a measure to evaluate how far the observed values are from the predicted ones. As he did when he used Model (1), he will calculate residuals:

$$e_i = Y_i - \widehat{Y}_i = Y_i - (b_0 + b_1 X_i)$$

and the *error sum of squares* or *residual sum of squares, SSE:*

$$SSE = \sum_{i=1}^{n} e_i^2 = \sum_{i=1}^{n} (Y - \widehat{Y}_i)^2$$

This *SSE* has $n - 2$ degrees of freedom associated with it, because parameters β_0 and β_1 are estimated for its calculation. Thus, its corresponding mean square, *MSE*, estimator of σ^2, is

$$\widehat{\sigma}^2 = MSE = \frac{SCE}{n-2} = \frac{\sum_{i=1}^{n}(Y_i - \widehat{Y}_i)^2}{n-2} = \frac{\sum_{i=1}^{n}\left[Y_i - (b_0 + b_1 X_i)\right]^2}{n-2}$$

Under Model (2) assumptions, variance σ^2 of error terms ε_i coincides with the variance of the probability distributions of Y_i. A large value of σ^2 would indicate that the observed (x, y)'s are quite spread out about the true regression line, whereas a small value of σ^2 would indicate that observations are very close to the true line. As the true line is unknown, the estimate is based on the extent to which the sample observations deviate from the estimated line. Table 2 shows part of the predictions of nitrate levels and the residuals for each site.

Table 2. Calculations to estimate the variance in the nitrate example using Model (2)

Site	pH levels X_i	Observed nitrate levels (ppm) Y_i	Fitted nitrate levels (ppm) $\hat{Y}_i$	Residual $e_i = Y_i - \hat{Y}$	Squared residual e_i^2
1	7.4	288.6	242.23	46.37	2,150.18
2	7.1	213.5	147.86	65.64	4,308.61
3	7.1	128.3	147.86	–19.56	382.59
.	.	.	.	.	.
.	.	.	.	.	.
29	7.2	229.5	179.32	50.18	2,518.03
30	7.3	214.8	210.78	4.02	16.16

Thus, SSE = 39,423.94 and MSE= 1,408.

John takes a look at his results and smiles. He has improved the estimation of the nitrate level in the water, incorporating the information of the pH values. MSE changed from 3,634.19 to 1,408. We may wonder now, how he can assess the gain in the precision of prediction by the pH incorporation. The total variation of the observations without taking the predictor variable into account is known as the *total sum of squares, SST*. Similarly, the variation in Y when a regression model using the predictor variable X is employed, is measured by SSE. The difference between these sums of squares, known as *regression sum of squares SSR*, is the amount in which the unexplained variation of the nitrate levels decreased, and is a measure of predictive success. Thus, $SSR = SST - SSE$.

However, as the sum of the squares has the original variable units squared, fitness cannot be compared when dependent variables have different units. To get a comparable measure, we may consider that difference as a proportion of the total variability in Y thus getting a measure without units, known as the *coefficient of determination* denoted by R^2:

$$R^2 = \frac{SSR}{SST} = 1 - \frac{SSE}{SST}$$

As $0 \le \frac{SSR}{SST} \le 1$, it follows that $0 \le R^2 \le 1$.

Since $0 \le SSE \le SST$, the extreme values of R^2 are associated to the extreme SSE values in the following way:

1) If $SSE = 0$, then all the residuals are zero; therefore all points are aligned, resulting in perfect fitness. The predictor variable X explains all the variation in Y observations through a linear relationship. Thus, $R^2 = 1$.
2) If $SSE = SST$, the total variability of Y is explained by the error and there is no linear relationship between X and Y; thus X does not explain Y variation through a linear relationship and $R^2 = 0$.

John estimates the *regression sum of squares* from his data,

$SSR = 105{,}391.37 – 39{,}423.94 = 65{,}967.43$ and gets the following R^2:

$$R^2 = \frac{65{,}967.43}{105{,}391.37} = 0.626$$

Note that from this result about 63 percent of the total variability of the nitrate levels is explained by its relationship with the pH according to our model. However, John notices that there is 37 percent of the total variation of nitrate levels that is still unexplained.

We have to be cautious in interpreting a coefficient of determination close to zero. This does not necessarily mean that variables are not related. If R^2 is small, variables may be related through a non-linear model.

Self-evaluation

10.5 Determine if the following statement is true (T) or false (F)
"To get estimators of the regression parameters by least squares method, error distribution should be normal."

10.6 When variation on the response variable not explained by regression is very large, the error sum of squares against the total sum of squares is:

a) Larger b) Equal or smaller c) It is necessary to make calculations.

10.7 Show that $\sum_{i=1}^{n} (\widehat{Y}_i - \overline{Y}) = 0$

10.8 Clover yield responses (kg seed per ha) to phosphate application were studied in an experiment. The data are presented below:

Phosphate (kg per ha)	0	0	0	0	10	10	10	10	20	20
Yield (kg seed per ha)	82	85	80	79	131	125	135	128	154	158
Phosphate (kg per ha)	20	20	30	30	30	30	40	40	40	40
Yield (kg seed per ha)	150	152	148	152	143	153	100	105	92	108

a) Make a scatter plot and then write some comments about it.
b) What percentage of the total variation of seed yield is explained by the linear relationship with the phosphate? Do you think it is adequate?

10.9 Refer to Problem 10.4:

a) Make a scatter plot considering SOD as a response variable and copper soil level as predictor variable. Write some comments about it.
b) Estimate the parameters of the simple linear regression model explaining their meaning in terms of the problem.
c) What SOD level mean will you expect if animals are fed on pastures with 50 ppm copper? And if pastures had 200 ppm copper?
d) Estimate the error mean square and compare it with the one obtained in Problem 10.4.

Answers

10.5 False. The least squares method only uses derivatives without any statistical assumptions.

10.6 b. *SSE* is part of the *SST*, then $SSE \leq SST$.

10.7. $$\sum_{i=1}^{n}(\hat{Y}_i - \overline{Y}) = \sum_{i=1}^{n}\hat{Y}_i - n\overline{Y} = \sum_{i=1}^{n}(b_0 + b_1X_i) - n\overline{Y} = nb_0 + b_1\sum_{i=1}^{n}X_i - n\overline{Y} =$$
$$n(\overline{Y} - b_1\overline{X}) + b_1 n\overline{X} - n\overline{Y} = 0$$

10.8 a) Seed yield is low at low and high phosphate levels, and it is higher for medium phosphate levels. This suggests that the relationship between seed yield and phosphate levels is quadratic.
b) $R^2 = 0.088$ indicating that only 8.8 percent of the total yield variation is explained by its linear relationship with phosphate levels. The R^2 value obtained indicates a poor fit of the data to the linear model. This result is confirmed by the scatter plot and suggests the addition of a quadratic term to the model. Nonetheless, a small R^2 value does not necessarily mean that the variables are not associated with or are weakly related to each other.

10.9 a) As copper level in forage increases, SOD level in blood increases in a linear trend. The variable relationship does not fit perfectly to a function because not all the points are aligned.
b) $b_0 = 506.89$ Ug^{-1} Hb, does not have an interpretation in terms of the problem because the Cu values observed in the sample are not close to zero.
$b_1 = 30.58$ Ug^{-1} Hb per ppm, for each Cu ppm increase in forage, SOD level increases in average 30.58 Ug^{-1} Hb in blood.

c) For 50 Cu ppm: $\hat{Y} = 506.89 + (30.58)(50) = 2,035.89$ Ug^{-1} Hb.
For 200 Cu ppm it is not proper to estimate the SOD value, because that forage values of Cu are far from the Cu value ranges considered. The relationship for values close to 200 ppm may be different.

d) Considering Cu, *MSE* = 113,405.991. Without considering Cu, *MSE* = 471,422.552. The information about the different Cu values in the forage allows reducing the unexplained variation of SOD values in blood.

3.3 From the Sample to the Population

John Denitri has estimated the regression equation from the observations of his sample. However, he is not interested in only the sample but in making inferences for decision making. In studying the relationship among the variable *Y*, nitrate levels and regressor variable *X*, pH, he is mostly interested in checking whether there is a true linear relationship among them. Does *X* provide information to predict *Y*? Do data indicate that *Y* increases linearly as *X* increases in the observation interval? These questions can be answered by a statistical test with the following hypotheses, H_0: $\beta_1 = 0$ o H_a:$\beta_1 \neq 0$.

As mentioned, the total sum of squares can be divided into two sources: one attributed to the regression and the other unexplained by the model corresponding to the random error.

Similarly as you have seen in the previous chapter of ANOVA, we have the equality

$$\sum_{i=1}^{n}(Y_i - \bar{Y})^2 = \sum_{i=1}^{n}(\hat{Y}_i - \bar{Y})^2 + \sum_{i=1}^{n}(Y_i - \hat{Y}_i)^2$$

The term at the left side is the total square sum of squares (*SST*) associated to $n - 1$ degrees of freedom, because in its calculation only one parameter is estimated, the mean of *Y*. The first sum of the second part corresponds to the regression sum of squares (*SSR*), to which a degree of freedom is associated with since there is only one independent variable in the model. The second sum of squares on the right term is the error sum of squares (*SSE*), with $n - 2$ degrees of freedom. By dividing the sum of squares of the second part by its respective degrees of freedom, we obtain the corresponding *MSR*, the *regression mean square*, and the *MSE*, *error mean square* or *residual mean square*:

$$MSR = \frac{SSR}{1} = SSR \qquad \text{and} \qquad MSE = \frac{SSE}{n-2}$$

The selection of the statistical test is based on the expected values of the mean square. Statistical theory provides the following results:

$$E(MSR) = \sigma^2 + \beta_1^2 \sum_{i=1}^{n} (X_i - \bar{X})^2 \qquad \text{and} \qquad E(MSE) = \sigma^2$$

Both expected values coincide only if $\beta_1 = 0$. If $\beta_1 \neq 0$, the value of $E(MSR)$ will be higher than $E(MSE)$. Therefore, if $\beta_1 \neq 0$, *MSR* will be higher than *MSE*. The decision rule is based on this idea to lead to a conclusion.

It can be shown that if H_0 is true assuming normality and constant variance of the error terms,

$$F = \frac{MSR}{MSE} \sim F_{1;n-2}$$

Then, large values of *F* support H_a and values closer to 1 support H_0, which suggest an upper-tail test. Therefore, the decision rule at a significance level of α is:

If $F > F_{1-\alpha;1;n-2}$ then we conclude H_a, $F_{1-\alpha;1;n-2}$ being the $(1-\alpha)100$ percentile of the *F* distribution with one degree of freedom in the numerator and $n-2$ degrees of freedom in the denominator.

Assuming that the model is valid, we may summarize the analysis of variance in Table 10.3.

Table 10.3. Analysis of variance table using Model (2).

Source of variation	*SS*	*df*	*MS*	*F*
Regression	*SSR*	1	*MSR*	$F = \frac{MSR}{MSE}$
Error	*SSE*	n - 2	*MSE*	
Total	*SST*	n - 1		

```
anova(Model2)
          Analysis of Variance Table

          Response: NO3
                     Df      Sum Sq      Mean Sq      F value      Pr (> F)
     pH               1      65967       65967        46.852       1.947e-07 ***
     Residuals 28            39424        1408
          ---
 Signif. codes:  0  '***'  0.001  '**'  0.01  '*'  0.05  '.'  0.1  ' '  1
```

Figure 5. ANOVA Table of the R output of the linear regression of nitrates and pH.

In analyzing the relationship between nitrates and pH values with computer software, John gets the following table (Fig. 5):

The *F*-statistic value is high enough to provide a strong evidence to conclude that there is a linear relationship between the pH and the nitrate levels ($P < 0.0001$) proving the pH usefulness for nitrate level prediction.

The *t*-Student statistic is an alternative test only for the simple linear regression. If the linear Model (2) is valid and ε_i are independent $N(0,\sigma^2)$, then

$$\frac{b_1 - \beta_1}{S_{b_1}} \sim t_{n-2} \quad \text{where} \quad S_{b_1} = \sqrt{\frac{MSE}{\sum_{i=1}^{n}(X_i - \overline{X})^2}}$$

The Chemaly industrial engineer calculates the statistic for the same hypothesis and has the following result:

$$\frac{b_1}{S_{b_1}} = \frac{314.56}{45.96} = 6.84 \quad \text{then the } P < 0.0001$$

These results can be observed in the computer output in Fig. 3. John obviously reaches the same conclusion with ANOVA. ANOVA results are the same as the *t*-test only for the simple regression and only with bilateral, hypothesis.

It is important to point out that if the null hypothesis that $\beta_1 = 0$ is not rejected, this does not necessarily mean that *X* and *Y* are not related. H_0 non-rejection may be attributed to two situations: i) there is a type II error, or ii) *X* and *Y* have a non-linear relationship.

3.4 Estimation of the Expected value of Y

John Denitri was able to predict the nitrate levels from the pH values from his sample by replacing the pH values in the estimated equation: $\widehat{Y} = b_0 + b_1 X$. But he is interested in extending his conclusions beyond the sample results. Point estimation itself does not provide information about the precision of this estimation. As you may surely remember, for each pH value there is a probability distribution of the nitrate levels whose expected value is given by:

$$E(Y) = \beta_0 + \beta_1 X$$

John wants to estimate this expected value by a confidence interval for each X_p, value using the following expression:

$$\widehat{Y} \pm t_{n-2;1-\alpha/2} S_{\hat{Y}} \quad \text{being} \quad S_{\hat{Y}} = \sqrt{MSE}\sqrt{\frac{1}{n} + \frac{(X_P - \overline{X})^2}{\sum_{i=1}^{n}(X_i - \overline{X})^2}}$$

John notes that not all the intervals have the same width. Intervals are narrower and therefore more precise as X_p is closer to $\overline{X}$, thus when $X_p = \overline{X}$, the interval reaches the minimum width with a radius equal to $t_{n-2;1-\alpha/2}\sqrt{\frac{MSE}{n}}$. As X_p is more distant from $\overline{X}$, intervals have larger radius and therefore less precision in the estimation.

As mentioned, John's point estimation of the nitrate level for the pH value 7.5 was 273.7 ppm. Hence, the boundary values of the 95 percent confidence interval for the same pH value are 247.62 ppm and 299.75 ppm. The information provided by this interval may be useful for calculating the capacity of the denitrification plant. For this reason, John would like that the interval be narrower and therefore, more precise.

Note: We have assumed that X has values that are known constants. This can be easily held if, for example, John Denitri were carrying out an experiment at a pilot plant controlling the pH values. However, especially in observational studies as the one presented here, this does not occur. At studying the nitrate levels as a function of the pH variable, X (pH) is a random variable and John does not have any control on the values for each site. It can be shown that the results obtained by estimation and hypothesis testing provide a good approximation even when X is a random variable, as the statistical procedures based on the regression model when X is fixed are fairly robust.

3.5 Correlation Coefficient

In situations when both variables are random and the objective is to check if these variables are related, instead of using one variable to predict the values of the other, the correlation analysis may be used. In this case, the study variables have a symmetrical role because neither of them is automatically assigned as a response variable. To describe the strength of the linear relationship between the variables the *correlation coefficient* is used as a descriptive measure. This coefficient is obtained with the following expression:

$$r = \frac{\sum_{i=1}^{n}(X_{1i} - \overline{X}_1)(X_{2i} - \overline{X}_2)}{\sqrt{\sum_{i=1}^{n}(X_{1i} - \overline{X}_1)^2 \sum_{i=1}^{n}(X_{2i} - \overline{X}_2)^2}}$$

It is shown that $-1 \leq r \leq 1$

The correlation coefficient measures the strength and direction of the linear relationship between two quantitative variables. A correlation coefficient value of +1 indicates a perfect positive relationship: if a variable increases, the other variable also increases and all the observations belong to a straight line. On the other hand, a correlation coefficient value equal to –1 indicates a perfect negative relationship: if a variable increases, the other variable decreases. A zero correlation indicates that there is no linear relationship between the variables. In real life, correlations do not take on exactly these values, but as they are closer to –1 or to +1, the linear relationship is stronger.

Self-evaluation

10.10 Check whether the following statement is True or False. Justify your answer:
"*MSR* always provides an unbiased estimator of σ^2".

10.11 In simple linear regression, are the hypothesis tested for the β_1 parameter always the same for *t*-and *F*-tests?

10.12 Check whether the following statement is True or False. Justify your answer:
"If β_1 is a value different from zero and negative, the statistic *F* is expected to be negative."

10.13 You are analyzing the advertising costs of the product Premium, in relation with the sales of the product of the company that hired you. Five minutes before your presentation, you accidentally spill coffee all over your preliminary report, just on the ANOVA summary table. You can read only two figures from the table. Considering that your analysis was based on 16 pairs of observations from a sample of the variables advertising costs and sales profits, could you reconstruct the table without having to re-run the analysis and be able to draw conclusions from these results?

Source of variation	*SS*	df	*MS*	*F*
Regression				
Error			10.73	
Total	252.34			

10.14 Refer to Problem 10.4,

a) Check with Analysis of Variance the usefulness of the linear regression by using the copper level in the pasture as a predictor for the SOD levels in the animals.

b) Find a 95 percent confidence interval of the mean SOD when Cu is 30 ppm and when Cu is 10 ppm. Why are the interval widths different?

10.15 A species aptitude to pulping represents its capacity to yield paper having certain characteristics and with a reasonable performance. These characteristics depend on physical and chemical properties, which are intrinsic to raw material and to the operative conditions of pulping. The aim of a study (Monteoliva, S.; Area, M. C.; Felissia, F. E. 2008. "Willow CM Pulps for Newsprint. II Relationships between Woods Characteristics and Pulp Properties" *Cellulose Chem. Technol.*, **42** (1-3), 45-59) was to predict the properties of chemimechanical (CMP) pulp from willows for newsprint, from the viewpoint of wood characteristics. The prediction for the raw material selection with better pulping aptitude saves time and work because it concentrates the selection and improvement on the best individuals for the end use. Some of the variables measured on wood, considered as independent variables, were the following: lumen width (Lu), insoluble lignin (Li), and cellulose content of wood (C). The dependent variable considered in this problem was opacity, an important characteristic of newsprint paper. The article presents the following data:

Opacity (%)	96.5	97.6	96.9	98.3	97.2	98.2	97.7	98.4	95.8	98.5	98.6
Lu (μm)	7.94	8.70	9.11	9.81	9.97	10.40	11.70	10.90	9.20	10.90	9.01
Li (% d.b.)	20.4	20.7	20.5	21.3	20.1	21.0	21.5	22.2	19.0	20.5	21.1
C (% d.b.)	42.3	43.8	42.0	43.8	41.7	44.8	41.2	41.9	42.2	43.1	46.3

a) Consider only insoluble lignin as a predictor variable. Estimate the simple linear regression function.
b) Do you think that insoluble lignin of wood is a useful variable to predict the pulp opacity? Justify your answer with an Analysis of Variance.
c) What proportion of the total variation of opacity remains unexplained in this model?

Answers

10.10 F. Only if the null hypothesis that $\beta_1 = 0$ is true due to

$$E(MSR) = \sigma^2 + \beta_1^2 \sum_{i=1}^{n} (X_i - \overline{X})^2 \cdot$$

10.11 The t-tests are more versatile because they are not only bilateral tests but they allow for unilateral tests and values different from zero. F-test is always bilateral.

10.12 F. The F-statistic is always positive because it is a ratio of mean squares.

10.13

Source of variation	*SS*	*df*	*MS*	*F*
Regression	102.12	1	102.12	9.51
Error	150.22	14	10.73	
Total	252.34	15		

We use the F test: H_0: $\beta_1 = 0$ vs. H_1: $\beta_1 \neq 0$

As the statistic of F test is = 9.51, the P being < 0.001, regression is significant.

10.14

a)

Source of variation	*SS*	*df*	*MS*	*F*	*P*
Regression	5,125,637.853	1	5,125,637.853	45.197	<0.0001
Error	1,474,277.881	13	113,405.991		
Total	6,599,915.733	14			

As $P < 0.0001$ the variable Cu in forage is useful as a predictor for the SOD level in blood.

b) The confidence interval for the expected value of SOD when Cu in forage is 30 ppm is (1,234.80; 1,613.78). The confidence interval for the expected value of SOD when Cu in forage is 10 ppm is (558.43; 1,066.93). The first interval width is 378.98 Ug^{-1} Hb and the second is 508.50 Ug^{-1} Hb. Intervals differ in their width because the more precise prediction is obtained for Cu levels close to the mean of this variable.

10.15

a) $\hat{Y} = 79.936 + 0.852Li$ For each 1 percent increase of the wood insoluble lignin, pulp opacity increases by 0.852 percent.

b)

Source of variation	*SS*	*df*	*MS*	*F*	*P*
Regression	4.994	1	4.994	13.159	0.0055
Error	3.415	9	0.379		
Total	8.409	10			

The wood insoluble lignin is useful to predict the pulp opacity (P= 0.0055)

c) R^2 = 0.594, then 40.6 percent of opacity total variation remains unexplained by this model.

4. Prediction with more than one Independent Variable

John Denitri has proved that by including the pH variable he was able to get a better prediction of the nitrate levels than by using only the arithmetic mean for the nitrate level. He may well think that he could improve the prediction using additional predictor variables. In most cases, a model with a single predictor variable provides a limited description because a number of key variables affect the response variable in a significant and differential way. As mentioned, John is carrying out an exploratory study to determine what variables can be useful for the nitrate level prediction. If he gets a good prediction from a set of variables easy to measure and can find the nature of the relationship of each variable with the response variable, he will be more certain in his decision about the design of the denitrification plants. He uses a model which is an extension of Model (2), that in a generic way and considering p -1 predictor variables, can be written as:

$$Y_i = \beta_0 + \beta_1 X_{1i} + \beta_2 X_{2i} + \ldots + \beta_{p-1} X_{p-1,i} + \varepsilon_i \qquad i = 1,\ldots,n \tag{3}$$

Y_i being the value for the response variable, nitrate level, at the ith observation, $\beta_0, \beta_1, \ldots, \beta_{p-1}$ are the regression coefficients, $X_{1i}, X_{2i}, X_{p-1,i}$ are the possible predictor variables, assumed as known constants, and ε_i is the random error term with $E(\varepsilon_i) = 0$ y $V(\varepsilon_i) = \sigma^2$. Error terms are assumed as not correlated, that is $\sigma(\varepsilon_i, \varepsilon_i) = 0$ if $i \neq j$. For the inference procedures the error terms are assumed as independent $N(0,\sigma^2)$.

John Denitri knows that pure water has low electric conductivity but water in rivers has ions in dissolution and conductivity is proportional to the quantity of ions present. Therefore, high conductivity may indicate high nitrate levels. Thus, he decides to incorporate the variable electric conductivity (EC) in his simple model.

He also uses the least square method to estimate the regression coefficients. To solve the corresponding system of normal equations, as for the rest of the multiple regression analysis, he must use matrix algebra. An alternative option that John chooses is to employ statistical software. From the sample data records at 30 points (see Appendix), he gets the computer output (Fig. 6).

The estimated regression equation for the new linear model with two independent variables is: $\hat{Y} = -1{,}316.784 + 195.588 X_1 + 0.321 X_2$, X_1 being the variable pH and X_2 the variable electric conductivity measured in μL cm^{-1}. The intercept represents an estimated average of the nitrate levels when the pH and the EC variables are simultaneously zero. As mentioned, this coefficient has no meaning for this problem. The other two coefficients, b_1 (= 195. 6) and b_2 (= 0.321), are the *estimated partial regression coefficients*. The adjective *partial* refers here to the estimated change in the average of the response variable for one-unit increase in the corresponding independent variable, holding the other variable constant. Thus he estimates that the mean nitrate level increases by 195.6 ppm for each additional unit pH when EC is held constant. On the other hand, the average nitrate level increases up to 0.321 ppm for each 1 μL cm^{-1} EC increase, holding the pH constant.

```
Model3 <- lm(NO3~pH + EC, data=nitro)
summary(Model3)
Call:
lm(formula = NO3 ~ pH + EC, data = nitro)
anova(Model3)
Coefficients:
              Estimate   Std. Error    t value    Pr(>|t|)
  (Intercept) -1.317e+03  2.701e+02    -4.875     4.26e-05 ***
  pH           1.956e+02  3.843e+01     5.090     2.39e-05 ***
  EC            .321e-01  5.774e-02     5.567     6.68e-06 ***
  ---
  Signif. codes:  0  '***'  0.001  '**'  0.01  '*'  0.05  '.'  0.1  ' '  1

Residual standard error: 26.07 on 27 degrees of freedom
Multiple R-squared: 0.8258,    Adjusted R-squared: 0.8129
F-statistic: 64.01 on 2 and 27 DF,  p-value: 5.668e-11

Analysis of Variance Table

Response: NO3
           Df      Sum Sq      Mean Sq     F value      Pr(>F)
pH          1       65967        65967      97.027   1.961e-10 ***
EC          1       21067        21067      30.986    6.677e-06 ***
Residuals  27       18357          680
                    ---
Signif. codes:  0  '***'  0.001  '**'  0.01  '*'  0.05  '.'  0.1  ' '  1
```

Figure 6. R regression output of the linear regression of nitrate, pH and EC levels.

The existence of a statistical relationship between the response variable and the predictor variables does not necessarily mean that the response depends causally on the predictor variables. The regression model does not imply a cause-effect relationship, no matter how strong the statistical relationship is between variables. In our case, the electrical conductivity increase does not cause an increase in the nitrate levels, but an increase in the nitrate level increases electric conductivity. However, John is interested in predicting the nitrate levels from the EC.

From Figs. 5 and 6 results, he notices that *MSE* decreased from 1,408 when he considers only pH at 680 when adding the variable EC to the model; that is, an important part of the nitrate level unexplained variability can now be explained from incorporating EC to the linear model. On the other hand, R^2 increased from .63 to .83. This means that in this new model there is only 17 percent of the total variation of the nitrate levels that remains unexplained.

The ANOVA applied to the multiple regression model is used to test hypotheses that, extending to p -1 predictor variables, can be written as:

$$H_0 : \beta_1 = \beta_2 = ... = \beta_{p-1} = 0 \qquad H_a : \text{There is at least one } \beta_k \neq 0$$
$$k = 1,2,..., p-1$$

The procedure to decide between these hypotheses is a generalization of the method used in the simple linear regression. The $p-1$ degrees of freedom are associated with SSR because there are $p-1$ independent variables in the model and SSE has $n-p$ degrees of freedom, since p parameters need to be estimated in the regression function for model (3).

According to ANOVA results (Fig. 6), John concludes that at least one of the predictor variables is useful to explain nitrate level variation.

On rejecting H_0 of the *overall* F *test*, the following step is to assess each independent variable contribution to the regression model the rest of the variables being in the model, to determine if any term $\beta_k X_k$ can be eliminated and then use a simpler model. The *partial test criterion* is then used:

$$H_0\text{: } \beta_k = 0 \qquad H_1\text{: } \beta_k \neq 0$$

In a similar way to simple linear regression, even though calculation is more difficult, the test statistic is:

$$t^* = \frac{b_k}{s(b_k)}$$

From Fig. 6 , it may be inferred that both predictor variables, pH and EC, are useful to predict nitrate levels, since P for both tests are lower than 0.001 at a significance level of α =0.05.

Self-evaluation

10.16 Refer to Problem 10.15:

a) State the estimated regression function considering all the predictor variables measured.
b) Set up the ANOVA table for the model including all the predictor variables considered and draw a conclusion.
c) What proportion of the total variation in opacity remains unexplained when the predictors are introduced into the analysis? Is this proportion relatively small or large?

Answers

10.16

a) $\hat{Y} = 68.362 + 0.539Li + 0.349Lu + 0.341C$
For each 1 percent increase in Li holding Lu and C constant in the model, the pulp opacity increases 0.539 percent. For each 1 percent increase in Lu holding Li and C constant in the model, the pulp opacity increases by 0.349 percent. For each 1 percent increase in C, holding Li and Lu constant in the model, the pulp opacity increases by 0.341 percent.

b)

Source of variation	*SS*	*df*	*MS*	*F*	*P*
Regression	7.487	3	2.496	18.939	0.0010
Error	0.922	7	0.132		
Total	8.409	10			

At least one of the wood predictor variables considered is useful to predict the pulp opacity according to the linear model presented.

c) Eleven percent of the total opacity variation is left unexplained. It is a small value. The prediction of this optic characteristic of the pulp can be made from these predictor variables measured in wood. This can be considered when planning a plant breeding program for the forest species considered.

4.1 Looking for a better fit

Encouraged by results, John thinks that if he adds any other predictor variable he will significantly increase the model predictive ability. According to previous information, he knows that oxygen dissolved in water (DO) may be the other possible predictor variable since NO_3^- ion production requires oxygen for denitrification. He expects that increasing quantities of oxygen dissolved in water will be related with decreasing quantities of nitrates.

Figure 7 contains some of the main regression results when the nitrate levels are regressed on pH, EC and DO.

From Fig. 7, John checks that he has improved the precision of his fitted model. The coefficient of determination has increased from 0.83 – corresponding to the model considering the pH and EC – to 0.85 in this model in which DO has been added. However, improvement has been less than expected. How much information does the DO provide to the understanding of the nitrate level variability when the other variables are already in the model? Prediction improvement corresponds to a decrease in the variability of the response variable that has not been explained yet by

```
Model4 <- lm(NO3~pH + EC + DO, data=nitro)
summary(Model4)
Call:
lm(formula = NO3 ~ pH + EC + DO, data = nitro)
anova(Model4)
Coefficients:
              Estimate     Std. Error    t value     Pr(>|t|)
(Intercept) -1.023e+03     2.884e+02     -3.547      0.001504 **
pH           1.709e+02     3.789e+01      4.511      0.000122 ***
EC           1.974e-01     7.929e-02      2.490      0.019509 *
DO          -1.032e+01     4.813e+00     -2.143      0.041609 *
                           ---
Signif. codes:  0  '***'  0.001  '**'  0.01  '*'  0.05  '.'  0.1  ' '  1
Residual standard error: 24.5 on 26 degrees of freedom
Multiple R-squared: 0.852,      Adjusted R-squared: 0.8349
F-statistic: 49.88 on 3 and 26 DF,  p-value: 6.365e-11

Analysis of Variance Table
Response: NO3
           Df       Sum Sq      Mean Sq      F value      Pr(>F)
pH          1        65967        65967     109.9432     7.830e-11 ***
EC          1        21067        21067      35.1109     2.984e-06 ***
DO          1         2757         2757       4.5941     0.04161 *
Residuals  26        15600          600
                     ---
Signif. codes:  0  '***'  0.001  '**'  0.01  '*'  0.05  '.'  0.1  ' '  1
```

Figure 7. R regression output of the linear regression of nitrate, pH, EC and DO levels.

the rest of the incorporated variables, *SSE*(pH, EC). This may be expressed as the sum of squares from the regression corresponding to DO given that pH and EC are included in the model, *SSR*(DO/pH, EC):

$SSR(\text{DO}/\ \text{pH, EC}) = SSE(\text{pH, EC}) - SSE(\text{pH, EC, DO}) = 18{,}357 - 15{,}600 = 2{,}757.$

If this result is divided by *SSE*(pH, EC), we get the *coefficient of partial determination* ($R^2_{YDO/pH,EC}$) between the nitrate levels and DO, given that pH and EC are in the model. We can see that when DO is added to the regression model containing pH y EC the error sum of squares SSE(pH, EC) is reduced by 15 percent, which means a small reduction on the unexplained variation. However, adding the DO, being the rest of the variables in the model, is significant ($P = 0.0416$). On the other hand, the DO as a predictor variable can be considered in the same fashion as the EC, nitrates can affect the oxygen dissolved in water, but not conversely. Therefore, we have to be careful in interpreting causation in this case.

Based on the concepts of the reduction of the error sum of squares with the addition of a new variable, we can obtain a variety of decomposition of the regression sum of squares into extra sums of squares. Consider three predictor variables, X_1, X_2, and X_3, that are incorporated to a model one at a time. We can calculate $SSR(X_1)$, $SSR(X_2/X_1)$ and $SSR(X_3/X_1,X_2)$ in the same way as we have done in the previous paragraph. It can be shown that:

$$SSR = SSR(X_1) + SSR(X_2/X_1) + SSR(X_3/X_1,X_2)$$

Each extra sum of squares is associated with one degree of freedom. Table 10.4 shows the ANOVA table decomposition for the case of three X variables often used in regression software. R software gives this ANOVA table by default, as you can see from Figs. 6 and 7. The partial contribution of each variable according to this decomposition can be tested as usual, by calculating the F statistic as a ratio of the corresponding MSR and the MSE.

Table 10.4. ANOVA table with decomposition of *SSR* for three *X* variables

Source of variation	*SS*	*df*	*MS*
Regression	SSR	3	MSR
X_1	$SSR(X_1)$	1	$MSR(X_1)$
X_2/ X_1	$SSR(X_2/ X_1)$	1	$MSR(X_2/ X_1)$
$X_3/ X_1, X_2$	$SSR(X_3/ X_1, X_2)$	1	$MSR(X_3/ X_1, X_2)$
Error	SSE	$n - 4$	MSE
Total	SST	$n - 1$	

Someone from the Chemaly company suggests that the chlorine quantity (Cl) in the water can also explain the nitrate level variation found in the river. His reasoning is that he has analyzed the nitrate level data at the sites using a simple linear model and considering only the Cl variable. He has obtained the results contained in Fig. 8.

Note that the variable Cl explains 62 percent of the total variability of the nitrate level. Reading these data, John thinks that with the addition of chlorine as an independent variable he can improve much the model predictive ability. However, he is disappointed with the results obtained (Fig. 9).

The coefficient of determination has not almost varied considering the model with variables pH, EC and DO, changing from 0.852 to 0.854. The test for the coefficient of the variable Cl, the rest of the predictor variables being in the model, indicates that this variable is useless for explaining the total variation of the nitrate levels ($P = 0.6032$). Then, why was this result obtained? The contribution of variable Cl for the unexplained variation of nitrate levels, when all the other variables are in the model, is

$SSR(\text{Cl}/\text{pH, EC, DO}) = SSE(\text{pH, EC, DO}) - SSE(\text{pH, EC, DO, Cl}) =$
$15{,}600–15{,}429= 171$

```
Model5 <- lm(NO3~Cl, data=nitro)
summary(Model5)
Call:
lm(formula = NO3 ~ Cl, data = nitro)
   ➤ anova(Model5)
Coefficients:
              Estimate      Std. Error     t value      Pr(>|t|)
(Intercept)  40.55151       24.50177       1.655        0.109
Cl            0.30724        0.04523       6.793        2.23e-07 ***
                                 ---
Signif. codes:  0  '***'  0.001  '**'  0.01  '*'  0.05  '.'  0.1  ' '  1

Residual standard error: 37.7 on 28 degrees of freedom
Multiple R-squared: 0.6224,     Adjusted R-squared: 0.6089
F-statistic: 46.15 on 1 and 28 DF,  p-value: 2.229e-07
Analysis of Variance Table
Response: NO3
            Df        Sum Sq        Mean Sq       F value          Pr(>F)
Cl           1        65592         65592         46.145        2.229e-07 ***
Residuals   28        39800          1421
                                 ---
Signif. codes:  0  '***'  0.001  '**'  0.01  '*'  0.05  '.'  0.1  ' '  1
```

Figure 8. R regression output of the linear regression of nitrate levels and Cl.

```
Model5 <- lm(NO3~pH + EC + DO +Cl, data=nitro)
summary(Model5)
Call:
lm(formula = NO3 ~ pH + EC + DO + Cl, data = nitro)
anova(Model5)
Coefficients:
               Estimate     Std. Error  t value       Pr(>|t|)
(Intercept) -1.074e+03      3.082e+02  -3.485       0.001831 **
pH           1.815e+02      4.338e+01   4.184       0.000308 ***
EC           2.070e-01      8.247e-02    2.510      0.018894 *
DO          -1.182e+01     5.657e+00   -2.090       0.046978 *
Cl          -3.361e-02     6.384e-02   -0.526       0.603183
                                 ---
Signif. codes:  0  '***'  0.001  '**'  0.01  '*'  0.05  '.'  0.1  ' '  1
Residual standard error: 24.84 on 25 degrees of freedom
Multiple R-squared: 0.8536,     Adjusted R-squared: 0.8302
F-statistic: 36.44 on 4 and 25 DF,  p-value: 4.328e-10
Analysis of Variance Table
Response: NO3
            Df       Sum Sq        Mean Sq        F value       Pr(>F)
pH           1        65967         65967        106.8868      1.630e-10 ***
EC           1        21067         21067         34.1348      4.296e-06 ***
DO           1         2757          2757          4.4664      0.04472 *
Cl           1          171           171          0.2772      0.60318
Residuals 25          15429           617
                                 ---
Signif. codes:  0  '***'  0.001  '**'  0.01  '*'  0.05  '.'  0.1  ' '  1
```

Figure 9. R regression output of the linear regression of nitrate, pH, EC, DO and Cl levels.

This implies a coefficient of partial determination between the nitrate levels and Cl, given that pH, EC y DO are already in the model of only 1 percent. The contribution to information provided by Cl levels is very low when the other variables are in the model. The explanation of this apparently puzzling result is that in fact the variable Cl present in the water depends on the pH and, in turn, affects the DO level and EC. Most part of the nitrate level variability explained by the Cl variable has already been explained by variables pH, DO and EC. This correlation phenomenon between independent variables is known as *multicollinearity*.

Multicollinearity can be detected when the F-global test is significant, but the t-tests of significance of the individual parameters are not. In cases of severe multicollinearity, even the signs of the regression coefficients calculated by the least square method can be altered. This suggests that other predictor variable or a set of predictor variables has taken over the information provided by an independent variable. Therefore, it is convenient not to include highly correlated independent variables in the model. On the other hand, the usual interpretation of a regression coefficient measuring the expected change in the response variable when an independent variable has one-unit increase, holding the rest of the predictor variables constant, is not completely applicable when severe multicollinearity exists. As variable Cl is highly related to the other predictor variables, it is not included in this model since the information provided is redundant. For a more detailed treatment of the tests to detect this problem as well as some remedial measures see Neter et al. (1996) and Belsley et al. (2005).

Self-evaluation

10.17 Data below have been collected from sports cars selected at random to determinate whether the prices in thousand of dollars, can be predicted from the weight in kg, and velocity in km/hour reached at 400m.

Price (P)	71	55	36	36	46	56	63	58	35	50
Weight (W)	1,443	1,262	1,337	1,281	1,366	1,380	1,380	1,468	1,322	1,417
Velocity (V)	167	156	144	146	152	158	163	160	143	155
Price (P)	78	45	39	50	37	47	55	50	69	
Weight (W)	1,408	1,409	1,398	1,477	1,357	1,257	1,206	1,223	1,442	
Velocity (V)	172	150	146	153	145	152	158	154	167	

a) Test whether there is a regression relation.
b) Calculate each coefficient of partial determination. Explain what each coefficient measures and interpret your results

c) Do you suspect a problem of multicollinearity? Justify.

10.18 The relationship of rice yield (kg per ha), plant height (cm), and number of tillers per meter is investigated. The results are shown below.

Height in cm (H)	110.5	107.0	105.1	97.0	93.6	88.5	88.5	83.5
Number of tillers (T)	13	15	16	20	22	22	23	22
Yield in kg per ha (Y)	5,755	5,939	6,010	6,545	6,730	6,750	6,899	6,862

a) Estimate the regression coefficients of the multiple linear regression model giving the meaning of them in terms of the problem.
b) Calculate the coefficients of determination of the two simple regression models and compare them with the coefficients of partial determination. Comment.

Answers:

10.17

a) $F = 957.375$ $P < 0.001$
b) $R^2_{YV/W} = 0.9906$ $R^2_{YW/V} = 0.0250$
c) $SSR(W) = 330.170$ and $SSR(W/V) = 0.6$ $SSR(V) = 2{,}802.399$ $SSR(V/W) = 2{,}472.829$

The reason why $SSR(W/V)$ is so small compared with $SSR(W)$ is that W and V are highly correlated with each other and with the response variable. When V is already in the regression model, the marginal contribution of W in reducing the error sum of squares is comparatively small because V contains much of the same information as W.

10.18

a) $b_0 = 5{,}855.801$ kg per ha, this coefficient does not have any particular meaning because the zero setting for the two predictors is a nonsensical combination.
$b_1 = -11.752$ kg per ha. cm, this coefficient indicates the estimated change in yield per centimeter increase in plant height when the number of tillers is held constant.
$b_2 = 89.778$ kg per ha, this coefficient indicates the estimated change in yield per unit of tiller increase per plant when plant height is held constant.
b) $R^2_{H} = 0.942$, $R^2_{T} = 0.990$, $R^2_{YH/T} = 0.578$, $R^2_{YT/H} = 0.925$

The variable height explains 94.2 percent of total yield variability when it is considered as the unique predictor variable, but this percentage is reduced to 57.8 percent once the number of tillers is in the model. This difference may be explained by the relationship between the predictor variables.

4.2 Checking the Fitness of the Model

John is now wondering whether his model is appropriate or if he should add curvature or another variable. He also wants to check whether his assumptions are valid. He wonders if the errors in the prediction are the result of a true lack of relationship between variables or if they are caused by certain data characteristics not considered in the regression model. Residual plots are useful to make decisions about the model. In general, these charts are plotted with the residuals against X, or the residuals against the fitted values, $\widehat{Y}_i$. John draws a residual plot against fitted values in the model with variables pH, EC and DO. He notes that the linear regression function seems appropriate because the residuals fall within a horizontal band centered on zero, displaying no systematic tendencies, as shown in Fig. 10.

If John had not included the pH in his model and had only considered variables EC and DO, he would have obtained the graph shown in Fig. 11 when plotting the residuals against the omitted variable pH. In this plot we can see a linear tendency suggesting that this variable explains part of the unexplained variation after fitting the regression with EC and DO.

What other residual patterns could John have found? Figure 12 shows some of the possible problems he could have detected by analyzing the residual plot.

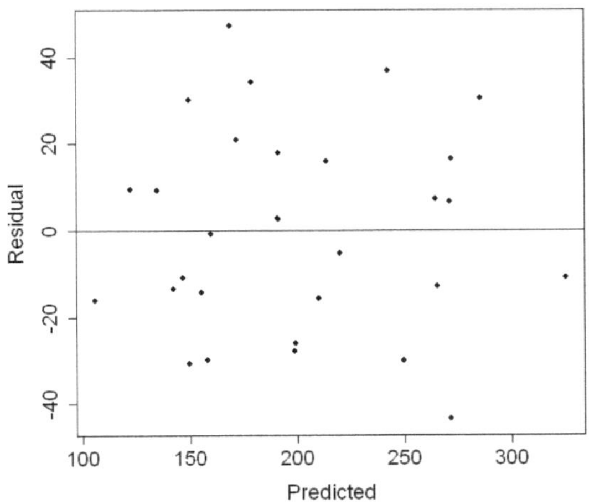

Figure 10. Residual plot of the model with pH, EC and DO versus predicted.

```
Model6<- lm(NO3~pH + EC + DO, data=nitro)
plot(predict(Model6),residuals(Model6), pch=20, xlab="Predicted",
ylab="Residual", col="blue")
abline(h=0)
```

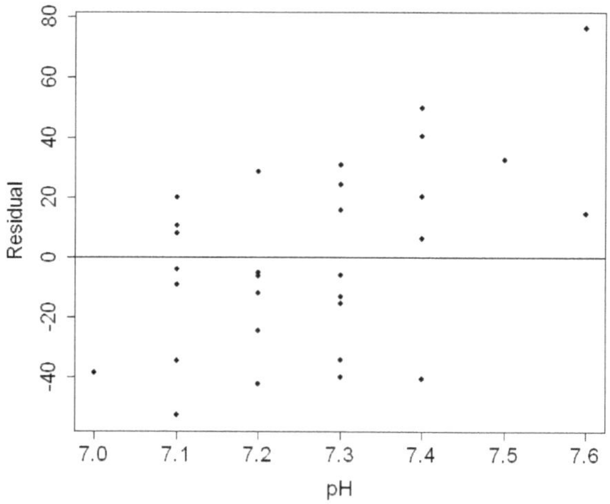

Figure 11. Residuals of the model with EC and DO versus pH.

```
Model7<- lm(NO3~EC + DO, data=nitro)
plot(nitro$pH,residuals(Model7), pch=20, xlab="pH", ylab="Residual")
abline(h=0)
```

Figure 12a shows a clear curvilinear relationship which suggests the need to add a quadratic term to the X_1 variable in the model. Figure 12b points out that as the predicted values increase, the absolute residual values also increase. This indicates different dispersion in error distributions for the different values of the predictor variables, which violates the homogeneity of variance assumption.

Figure 12c shows the presence of an *outlier*. Outliers are extreme observations which do not fit in to the tendency of the rest of the data. In this case, one observation has a great influence on the results. The estimated line has a negative slope. If this observation is eliminated from the data set, the slope of the new line will be positive and the intercept will be much lower. This observation has more influence on the regression results than the rest of the values, since if any other observation is eliminated from the data set, it will not have much influence on the equation obtained. If there is evidence that the outlier comes from an error, it can be discarded and conform to a new regression equation. If the observation is valid, intermediate values for the predictor variable may be obtained to have a better description of the variable relationship between variables. There are more refined procedures for identifying outliers (as the studentized residuals or the Cook distance, among others, which can be looked up in Belsley et al. (2005) and Neter et al. (1996)).

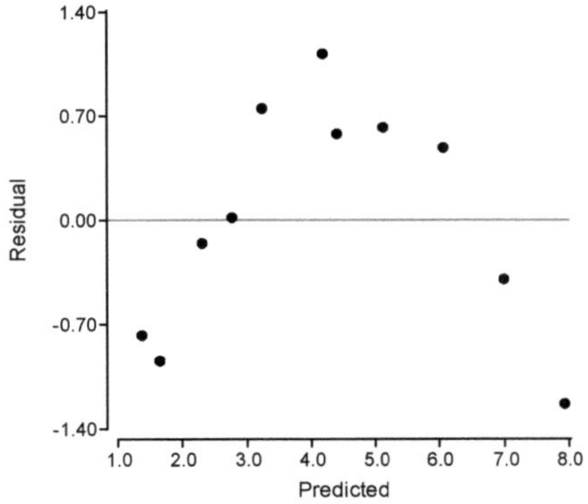

a) Curvilinear relationship

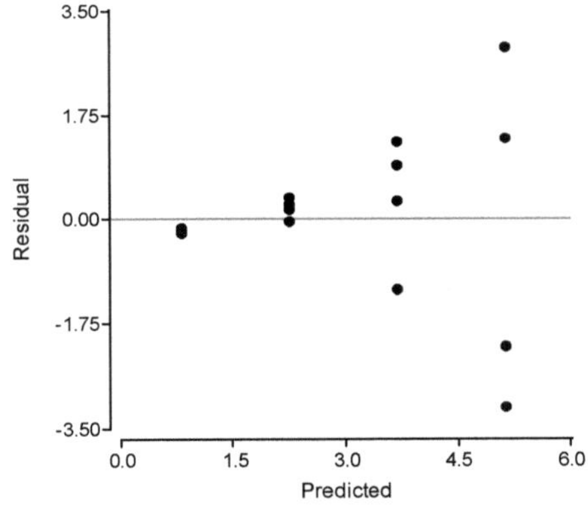

b) Nonconstancy of Error Variance

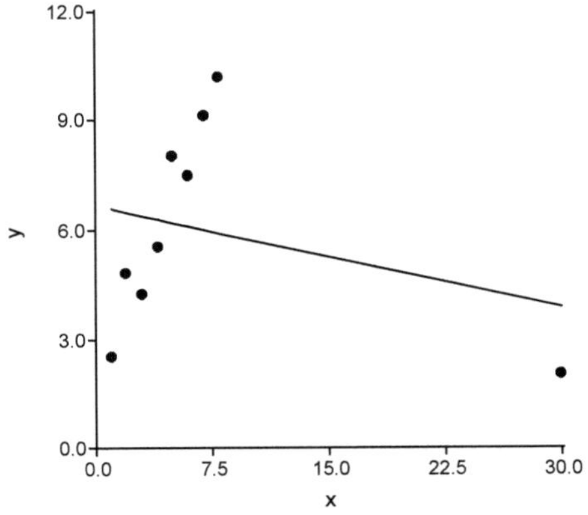

c) Presence of Outliers

Figure 12. Further residual patterns.

Residuals are also useful for detecting departures from the normal distribution of errors in the model. We just need to consider if there is a severe departure from normality. It is common to draw box-plots or residual histograms to detect important deviations. Most statistical software can be used to assess the normality of residuals via *Normal probability plot*. Each residual is plotted against its expected value if distribution is normal. If points lie close to the straight line, the normality assumption is satisfied, but if they lie far from the line, errors may not have a normal distribution. Figure 13 shows the normal probability plot of the residuals corresponding to the nitrate levels, accounting for pH, EC and DO variables. Note that points are close to the line indicating normality; therefore we may consider that this assumption is valid.

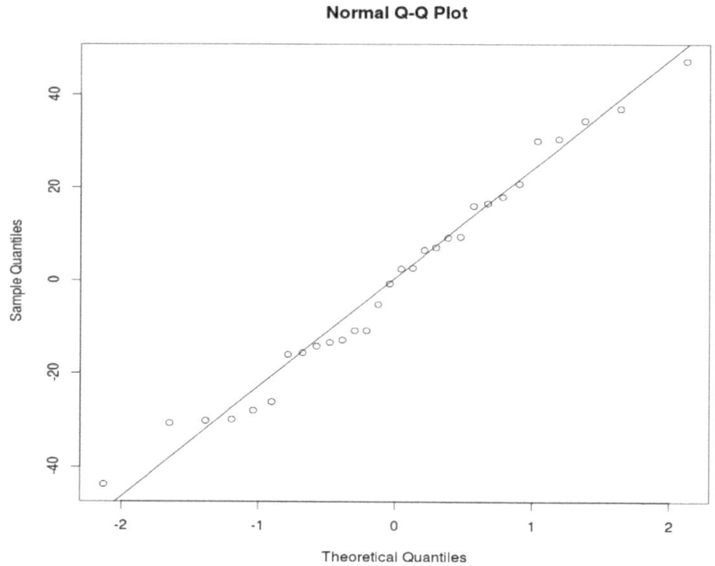

Figure 13. Normal probability plot of the residuals corresponding to the nitrate problem considering the predictor variables pH, EC and DO.

```
qqnorm(residuals(Model6))
qqline(residuals(Model6))
```

Self-evaluation

10.19 Refer to Problem 10.8

Plot the residuals against the fitted values. What are your findings?

10.20 Refer to Problem 10.15

a) Plot the residuals of the estimated regression considering only Lu and C as predictor variables against Li. What does your plot show?

b) Plot the residuals of the complete model against the predicted values. What is your conclusion?

c) Prepare a normal probability plot of the residuals. What do you conclude?

Answers

10.19 The plot shows a curvilinear relationship.

10.20 a) The residuals depart from zero in a systematic fashion suggesting a strong Li effect not considered in the model.

b) This plot shows that the linear regression model considering the three predictor variables is appropriate. The residuals fall within the horizontal band centered on zero displaying no systematic tendencies.

c) The plot suggests that the normal assumption is satisfied.

4.3 Model selection

John is really satisfied with his first-order model with predictor variables pH, EC and DO, but he wishes to check whether this model is his best choice. However, not necessarily there is a unique "best model". He can select one from a set of many "good models", according to his expertise and the easiness of measurement of the independent variables in the model.

The all-possible regressions procedure considers all the possible sets of predictor variables and identifies a subset according to a criterion. One of the criteria for model selection uses the coefficient of determination R^2. In each set of p predictor variables, the model accounting for the highest R^2 is selected. This criterion is equivalent to using the error sum of squares *SSE*.

Table 10.5 shows the all-possible sets of predictor variables for the river nitrate levels. R^2 are plotted against the number of parameters in the regression model, connecting by lines the highest values for each number of parameters in the model. Figure 14 shows that when only one predictor variable is considered, the model with the DO variable is the best, whereas in models with two predictor variables, the model with pH and EC has the highest R^2. It can also be noted that there is little improvement by adding a third predictor variable, DO, and there is practically no improvement in the model with the four variables under analysis. If John is interested in getting a good and parsimonious model, his best choice would be the model including only pH and EC. The little improvement of the model predicting

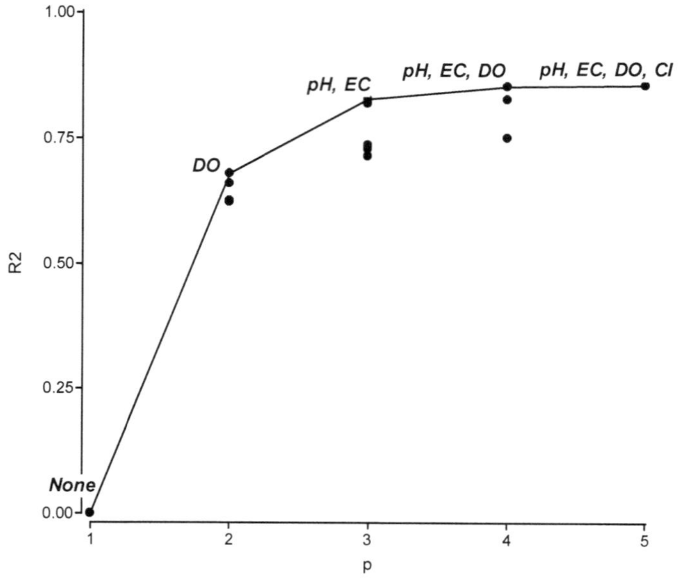

Figure 14. R^2 plot of the nitrate levels.

ability by adding new variables can be explained by the high correlation of the predictor variables, which do not contribute much information when the other variables are in the model.

R^2 value is always increased by adding more predictor variables. This is due to the fact that with more predictor variables, *SSE* decreases and *SST* does not depend on the predictor variables because it always has the same value. As R^2 can be higher because of a great number of predictor variables (even when some variables are not significant), it is more useful to have a measure modified by the number of predictor variables in the model. The *adjusted coefficient of multiple determination* denoted R_a^2 is obtained adjusting R^2 by dividing each sum of squares by its degrees of freedom associated with it.

$$R_a^2 = 1 - \frac{\frac{SSE}{n-p}}{\frac{SST}{n-1}} = 1 - \left(\frac{n-1}{n-p}\right)\frac{SSE}{SST}$$

The adjusted coefficient of multiple determination may actually become smaller when another predictor variable is added to the model, because the decrease in *SSE* may be more than offset by the loss of a degree of freedom in the denominator $n - p$. The R_a^2 value for the four-variable predictor model (Table 10.5) is somewhat smaller than the one corresponding to the three-variable predictor model, pH, EC and DO. The R_a^2 for the different models may be used as another criterion for the selection of the best model, as its equivalent the *MSE*.

Table 10.5. R^2, R_a^2, SSE and MSE values for all possible regression models for the nitrate levels.

Variables in the Model	Number of parameters (p)	SSE	MSE	R^2	R^2_a
None	1	105,391.367	3,634.185	0	0
pH	2	39,423.941	1,407.998	0.626	0.613
EC	2	35,970.072	1,284.645	0.659	0.647
DO	2	33,789.679	1,206.774	0.679	0.668
Cl	2	39,799.614	1,421.415	0.622	0.609
pH, EC	3	18,356.897	679.885	0.826	0.813
pH, DO	3	19,319.465	715.536	0.817	0.803
pH, Cl	3	28,576.702	1,058.396	0.729	0.709
EC, DO	3	27,810.065	1,030.002	0.736	0.717
EC, Cl	3	28,902.592	1,070.466	0.726	0.705
DO, Cl	3	30,374.670	1,124.988	0.712	0.690
pH, EC, DO	4	15,600.358	600.014	0.852	0.835
pH, EC, Cl	4	18,124.389	697.092	0.828	0.808
pH, DO, Cl	4	19,318.909	743.035	0.817	0.796
EC, DO, Cl	4	26,233.703	1,008.989	0.751	0.722
pH, EC, DO, Cl	5	15,429.279	617.171	0.854	0.830

There are several other procedures for deciding whether to add or eliminate variables in the model, but none of them can replace subject-matter, knowledge or judgment.

Common procedures include *stepwise, forward selection, backward selection,* which are automatic search procedures. These procedures are iterative. Variables are included or excluded one by one and the new model is assessed. The process continues until a stopping criterion is satisfied.

Self-evaluation

10.21 Refer to problem 10.15.

Find the best subset of variables according to the R_a^2 criterion

Answers:

10.21 The regression function including the three predictor variables is the best subset according to the criterion specified. The R_a^2 for this model is 0.843, while R_a^2 for the best subset with two predictor variables (Li, C) is 0.714.

4.4 Qualitative predictor variable

At a technical monthly meeting at *Chemaly Inc.*, one of John's colleagues shows interest in the results he obtained in the prediction of the nitrate levels of the Rughi-Rughi River and tells John that he has made some observations of the nitrate and the pH levels in the Bambon River, where there might be other nitrifying bacteria. The Bambon River data are shown at Table 10.6. Both chemicals engineers are interested in comparing the nitrate and pH levels of both rivers because they think that the bacterial species behave differently according to the pH.

Table 10.6. Nitrate levels and pH observed in the Bambon River.

Nitrate levels	544.60	683.20	751.30	794.70	749.90	605.70	903.90	768.10	602.30	749.10
pH	7.00	7.20	7.40	7.50	7.40	7.10	7.60	7.40	7.00	7.50
Nitrate levels	900.60	729.20	676.00	684.90	830.00	849.10	656.50	534.90	785.50	
pH	7.60	7.30	7.20	7.30	7.50	7.60	7.20	7.00	7.50	

This case shows the relationship between a quantitative response variable, Y, nitrate levels, and two predictor variables: a quantitative variable, X_1, pH, and a qualitative variable, X_2, river. We may expect a linear relationship between nitrate and pH levels for both rivers. The qualitative variable may take on only two "values", Rughi-Rughi River and Bambon River. Then it is included in the model using a dummy or indicator variable X_2 denoted:

$$X_2 = \begin{cases} 1 & \text{if it is Rughi-Rughi} \\ 0 & \text{if it is Bambon} \end{cases}$$

As John and his colleague assume that the change of the nitrate level values at a pH increase will be different for each river, they will include an interaction term between the predictor variables, written as the product X_1X_2. Under this condition, the model regression function is:

$$E(Y) = \beta_0 + \beta_1 X_1 + \beta_2 X_2 + \beta_3 X_1 X_2$$

The interpretation of the model parameters can be obtained by assigning values to the qualitative variable X_2:

$$E(Y)=\begin{cases}\beta_0+\beta_1X_1 & \text{if } X_2=0\\(\beta_0+\beta_2)+(\beta_1+\beta_3)X_1 & \text{if } X_2=1\end{cases}$$

Thus, we can see that the parameter β_0 is expected value of the intercept for the Bambon River, and β_1 is the expected change in nitrate levels for each increase unit on X_1, pH, for the Bambon River. On the other hand, the parameter $\beta_0+\beta_2$ indicates the expected value of the intercept for the Rughi-Rughi River. Then, β_2 is the difference between both intercepts. The parameter $\beta_1+\beta_3$ indicates the expected change in nitrate levels for each increase unit on X_1, pH, for Rughi- Rughi River. Therefore, β_3 is the difference in the change of $E(Y)$ for each unit increase on pH between both rivers. The scatter plot for both sets of data is shown in Fig. 15.

The scatter plot shows that the Bambon River has higher nitrate levels than the Rughi-Rughi River at the same pH levels. It can also be supposed that at a pH increase there is a higher nitrate level in the Bambon River. Figure 16 shows the data analysis output.

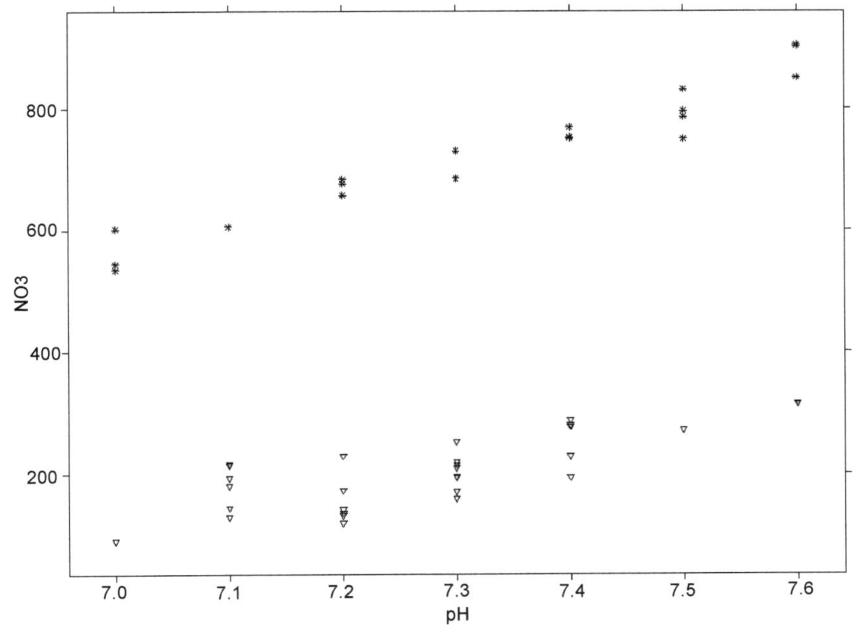

Figure 15. Scatter plot of nitrate and pH levels in two rivers (stars for Bambon River and triangles for Rughi-Rughi River).

xyplot(NO3 ~ pH,group=RIO, col=c("black", "black"),pch=c(8,6))

```
Call:
lm(formula = NO3 ~ RIO + pH + INTERAC)
              Coefficients
                       Estimate Std. Error t value Pr(>|t|)
              (Intercept) -2951.89     284.21 -10.386 1.57e-13 ***
              RIO            866.34     416.79   2.079  0.04338 *
              pH             501.69      38.75  12.947  < 2e-16 ***
              INTERAC       -187.13      57.10  -3.277  0.00203 **
              ---
              Signif. codes:  0 '***' 0.001 '**' 0.01 '*' 0.05 '.'
              0.1 ' ' 1
              Residual standard error: 34.25 on 45 degrees of freedom
              Multiple R-squared: 0.9851,    Adjusted R-squared:
              0.9841
              F-statistic: 989.3 on 3 and 45 DF,  p-value: < 2.2e-16
              anova(Model8)
              Analysis of Variance Table
              Response: NO3
                        Df  Sum Sq Mean Sq  F value     Pr(>F)
              RIO        1 3218469 3218469 2744.163 < 2.2e-16 ***
              pH         1  249959  249959  213.122 < 2.2e-16 ***
              INTERAC    1   12594   12594   10.738  0.002026 **
              Residuals 45   52778    1173
              ---
              Signif. codes:  0 '***' 0.001 '**' 0.01 '*' 0.05 '.'
              0.1 ' ' 1
```

Figure 16. R output of a model with a dummy variable.

The interaction effect is significant ($P = 0.0020$). The change of the nitrate levels for each unitary pH increase is significantly higher in Bambon River. Both colleagues write the following estimated equations for each river:

Bambon River: $\hat{Y}_i = -2,951.89 + 501.69X_{1i}$

Rughi-Rughi River:

$\hat{Y}_i = (-2,951.89 + 866.34) + (501.69 - 187.13)X_1 = -2,085.55 + 314.56X_1$

On some occasions, the qualitative variable can take on different k values. One way of analyzing them is using $k - 1$ dummy variables, each with a value 0 and 1.

Self-evaluation

10.22 A seed company is studying the relationship between employee salaries and seniority. Analysts expect a linear relationship between the average salary and years of seniority for both women and men. Draw a possible scatter plot and indicate the signs for each regression coefficient for the following situations:

a) Starting salary does not depend on gender, but the increase of the mean value of salaries for each year of seniority is higher in men than in women.

b) Starting salary is higher for men than for women, but the increase of the mean value of salaries for each year of seniority is equal for both women and men.
c) Starting salary and the increase of the mean value of salaries for each year of seniority is higher for men than for women.
d) Salary does not depend on gender, but it increases by years of seniority.
e) Salary does not change with years of seniority or gender.

10.23 A Veterinary Lab technician is interested in predicting antibody titres (titre of Ac in UI/mg) against a bovine disease from the vaccine antigen level, expressed as a logarithm of the antigen concentration ($\log_{10}$ Ag). He is also interested in finding out if the relationship between these two variables is the same for the two vaccine formulations (aqueous and oil emulsion). Sixteen bovine animals above 6 months old were vaccinated with different antigen and formulation levels. The results are shown as follows:

Animal	1	2	3	4	5	6	7	8
Vaccine 1= Aqueous 0= oil emulsion	1	0	1	0	1	1	1	0
$\log_{10}$ Ag	5	5	6	6	6	7	8	7
Ac titre	1.05	2.00	1.18	2.86	1.20	2.27	3.55	3.30
Animal	9	10	11	12	13	14	15	16
Vaccine 1= Aqueous 0= oil emulsion	1	1	0	1	0	0	0	0
$\log_{10}$ Ag	5	7	8	8	6	5	7	8
Ac titre	1.00	2.30	3.86	3.48	2.90	2.10	3.35	3.92

a) Propose a model for this problem and interpret the parameters.
b) Plot the data identifying the different vaccines. Which parameters of the model do you think may be significant?
c) Verify your answer to item (b) testing the corresponding hypothesis.
d) Predict the antibody titer of a bovine vaccinated with an oil emulsion vaccine with $\log_{10}$ Ag = 6.5. Repeat this prediction but for an aqueous vaccine.

Answers

10.22

Considering X_2=0 if FEMALE and X_2=1 if MALE

a) $\beta_1 > 0 \quad \beta_2 = 0 \ \beta_3 > 0$
b) $\beta_1 > 0 \quad \beta_2 > 0 \ \beta_3 = 0$
c) $\beta_1 > 0 \quad \beta_2 > 0 \ \beta_3 > 0$

d) $\beta_1 > 0 \quad \beta_2 = 0 \ \beta_3 = 0$
e) $\beta_1 = 0 \quad \beta_2 = 0 \ \beta_3 = 0$

10.23

a) $Y = \beta_0 + \beta_1 X_1 + \beta_2 X_2 + \beta_3 X_1 X_2 + \varepsilon$
β_0 is the expected value of the intercept for the oil emulsion vaccine. β_1 is the expected change in antibody titre for each unit increase on $\log_{10}$ Ag for the oil formulation. β_2 is the difference between both intercepts. β_3 is the difference in the change of $E(Y)$ for each unit increase on $\log_{10}$ Ag between both formulations.
b) $\beta_1, \beta_2, \beta_3$
c) The null hypothesis for the parameters $\beta_1, \beta_2, \beta_3$ are rejected with $P <$ 0.05.
d) Oil emulsion $\hat{Y} = 3.04$
Aqueous $\hat{Y} = 2.00$

4.5 Some final comments on model building for the nitrate example

John Denitri carried out an observational and exploratory study for *Chemaly Inc*. He looked for predictor variables which may be related to the response variable. From his results, he will continue with the model validation to study its stability and determine how robust the regression coefficients are. The model validation can be achieved in three different ways: i) by collecting new data to check its predictive ability; ii) by comparing its actual results with the theoretically expected results or with previous empirical results; iii) by splitting the database in two sets, estimating the model with one and using the other set for cross-validation.

Later on, once he knows the intervening variables in the relationship with the nitrate levels, he may design an experiment in which he will control some of these variables to study their effect on, for example, the velocity in which bacteria transform water toxic nitrates in non-toxic forms of nitrogen. These experiment results will be useful for building the company denitrification plants.

4.6 Expressing curvilinear responses

So far, we have analyzed models in which no curvatures were described. However, in everday activities we find curvilinear relationships. In a study about electricity consumption (kWh) in houses in the function of monthly average temperatures (°C) the following data plotted in Fig. 17. was obtained:

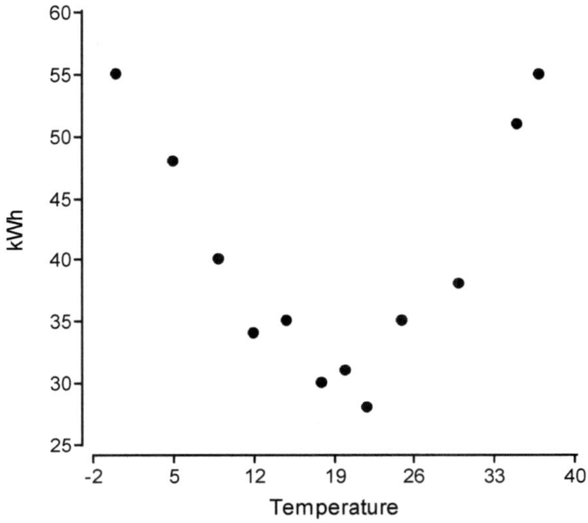

Figure 17. Electricity consumption and temperature.

When we look at the scatter plot for the electricity consumption against temperature, we see that at low temperatures consumption is high. However, as temperature increases, consumption decreases until it reaches a minimum temperature close to 20°C. Then, consumption increases again with temperature. The function accounting for the data is no longer a straight line but a parabola that may be denoted with a second-order polynomial. In this situation, the appropriate model, being X_i temperature and Y_i consumption, is

$$Y_i = \beta_0 + \beta_1 X_i + \beta_2 X_i^2 + \varepsilon_i \quad (4)$$

The estimated equation for this example is:

$$\hat{Y}_i = 57.60 - 2.79 X_i + 0.07 X_i^2$$

It can be of interest to estimate the temperature at which the minimum consumption is reached. By using the derivative calculation, we find that it is 19.9°C and replacing the values in the estimated model, it corresponds to a consumption of 29.8 kWh.

We may consider that (4) is a special case of the linear regression model (3) because if it is stated that $X_{1i} = X_i$ y $X_{2i} = X_i^2$, then you can write model (4) as

$$Y_i = \beta_0 + \beta_1 X_{1i} + \beta_2 X_{2i} + \varepsilon_i$$

that is the general form of the linear regression (3), carrying out the regression analysis in the usual fashion.

Another example of the curvilinear regression may be expressed by an exponential model. It is usual that these models have time as the predictor variable. This model is used in growth studies where the rate of growth at a given time X is proportional to the amount of growth remaining as time increases. Another use is to relate the concentration of a substance to elapsed time. If only one predictor variable is used, the regression model is

$$Y_i = \beta_0 \exp(\beta_1 X_i) * \varepsilon_i .$$

A transformation by logarithms allows the model linearization, transforming it in the form of model (3) with the response variable being transformed.

Self-evaluation

10.24

Refer to Problems 10.8. According to the results obtained, propose a model expected to be appropriate.

a) Plot the new fitted regression function and the data. Does this regression function appear to be a good fit?
b) Test whether or not each regression coefficient is significant.
c) Compare the R^2 calculated with this model with the corresponding R^2 obtained in Problem 10.8. What do you find?
d) Find the phosphate level that results in the maximum yield and estimate which is the maximum yield.

Answers

10.24

a) The data have a good fit to a quadratic curve.
b) All the coefficients are significant with $P < 0.0001$.
c) $R^2 = 0.97$
d) The maximum seed yield is obtained with 21.9 kg/ha phosphate, and it is 155 kg seed/ha.

5. Regression Model Expansions

Expanding to models where errors do not have a normal distribution, there is a type of models, known as *generalized linear models* that, apart from including the regression models already dealt with in this chapter and the ANOVA models seen in other chapters, deal with other models where response variables are categorical, such as the logistic regression and Poisson models. The former are widely used when the response variable is qualitative (in most cases dichotomous) and the predictor variable is quantitative. This model is used to make predictions for an event probability, for example, a student will pass a test according to his hours of study or a person will develop a lung disease due to the number of cigarettes he smokes per day. If only one predictor variable is considered, the expected value for the *i*th observation is $\pi_i = \dfrac{e^{\beta_o+\beta_1 X_i}}{1+e^{\beta_o+\beta_1 X_i}}$, π_i being the probability that the event will take place. The Poisson model is used when the response variable is a count and where large counts are a rare event for example, the number of faulty pieces produced by an employee per day. As predictor variables you may use, among others, the seniority of an employee or years of use of a machine.

In these models, instead of using the Least Square Method, the *Maximun-likelihood* method is employed. This method specifies the probability distribution function of error terms. This method chooses as estimates those values of the parameters that are more consistent with the sample data given the distribution function of the response are variable.

Solved problems

1) For model (1), determine that $\widehat{\beta}_0 = \overline{Y}$.
2) For model (1), determine that $\sum_{i=1}^{n} e_i = 0$
3) The Marketing Department of a cosmetic company wants to predict the sales for their anti-ageing cream. A sample of 20 cities with approximately equal populations is selected during one-month test period. Sales are considered (in million dollars) for the product during the testing period. The table below shows the obtained observations as well as the TV, radio and newspaper advertising.

City	Sales (u$s millions)	TV Advertising costs (u$s thousands)	Radio Advertising costs (u$s thousands)	Newspaper Advertising costs (u$s thousands)
1	7.4	1	0	20
2	6.5	1	1	18
3	5.0	2	3	4
4	6.5	3	3	5
5	8.4	8	3	9
6	7.5	9	2	8
7	9.8	10	6	12
8	10.0	9	6	14

contd....

contd.

City	Sales (u$s millions)	TV Advertising costs (u$s thousands)	Radio Advertising costs (u$s thousands)	Newspaper Advertising costs (u$s thousands)
9	10.3	10	12	5
10	11.2	11	11	6
11	12.0	13	12	20
12	12.1	14	11	20
13	11.8	18	12	4
14	10.9	19	13	5
15	13.0	20	15	0
16	15.1	20	17	3
17	15.0	28	12	8
18	15.8	25	16	9
19	16.0	28	17	18
20	18.1	30	17	17

a) Predict the sales in million dollars of the anti-ageing cream without considering advertising costs.
b) Estimate the sales variance in the cities under analysis (without considering any predictor variable).
c) Make a scatter plot for the sales volume considering the sales as a response variable and the TV advertising costs as predictor. Comment.
d) Estimate the parameters of the simple linear regression model considering the TV advertising costs as predictor. Explain its meaning in terms of the problem.
e) Estimate the error mean square and compare it with the one obtained in item b.
f) Estimate the multiple linear regression function considering all the predictor variables.
g) Explain the meaning of each of the regression coefficients in terms of the problem.
h) What percentage of the total variation of the sales is explained by the fitted model?
i) Set up the ANOVA table and conclude.
j) Check that each variable is significant when the rest of the variables are included in the model.
k) How much has the error mean square decreased by adding the radio and newspaper cost variables compared to the result obtained when only the TV advertising costs were considered?

4) In a simple regression linear model, what values does R^2 take if all the points are aligned on a straight line with a negative slope?
a) 0 b) - 1 c) + 1 d) It is necessary to make calculations

5) Show that the point $(\overline{X},\overline{Y})$ always belongs to the fitted regression line.
6) For a certain item, the following production volume sample and the total operation cost were obtained.

Production volume (units)	200	400	450	550	600	700	750	0
Total cost (u$s)	3,000	4,300	4,500	4,800	5,800	6,400	7,000	1,000

   a) Infer the regression equation that can predict the total cost for a definite production volume.
   b) What is the estimated fixed cost?
   c) What is the estimated additional cost for each production unit?
   d) If 500 units were produced at u$s 5,500, how far is this value from the prediction made using the value of its estimated function?
7) Water spinach (*Ipomoea aquatica*) is a vegetable that grows in wet areas. It has a high potential production and it offers the possibility of improving the nutrition of the inhabitants of the tropic areas. An Intercropping experiment with cassava was carried out to study the effect of nitrogen and defoliation of cassava in water spinach production. The results in ton per ha of fresh matter are shown below:

Defolia	**N (kg per ha)**												
	0	**0**	**15**	**15**	**15**	**30**	**30**	**30**	**30**	**60**	**60**	**60**	**60**
YES	1.50	1.30	2.13	2.16	2.00	3.09	2.80	3.12	3.00	3.45	3.32	3.50	3.90
NO	1.40	1.36	1.46	1.52	1.53	2.15	2.30	2.00	2.25	3.10	3.20	3.16	3.02

   a) Plot the data identifying the use of defoliation. Comment.
   b) Fit the regression model and state the estimated regression function.
   c) Is the effect of defoliation the same for all the levels of nitrogen? Test the significance of the corresponding regression coefficient.
   d) Estimate water spinach fresh matter when using defoliation and 40 kg per ha of nitrogen.

Answer

1) As mentioned, to find $\hat{\beta}_0$ the $\min_{\beta_0} \sum_{i=1} (Y_i - \beta_0)^2$ must be calculated, and this expression must be derived with respect to β_0. The $\hat{\beta}_0$ searched value is that value β_0 annulling that derivative, thus

$$2\sum_{i=1}^{n}(Y_i - \beta_0) = 0$$

Simplifying and expanding, we have

$$\sum_{i=1}^{n} Y_i - \sum_{i=1}^{n} \beta_0 = 0 \Rightarrow \sum_{i=1}^{n} Y_i = n\beta_0 \Rightarrow \overline{Y} = \widehat{\beta}_0$$

2) $\sum_{i=1}^{n} e_i = \sum_{i=1}^{n} (Y_i - \widehat{Y}_i) = \sum_{i=1}^{n} Y_i - \sum_{i=1}^{n} \widehat{Y}_i = n\overline{Y} - n\overline{Y} = 0$

3) a) 11.12 b) 13.12 d) $b_0 = 5.91$; $b_1 = 0.374$ e) $MSE = 1.34$
f) $\hat{Y} = 4.45 + 0.23TV + 0.26RADIO + 0.10NEWS$ h) $R^2 = 0.96$
i)

Source of variation	*SS*	*df*	*MS*	*F*	*P*
Regression	238.719	3	79.573	120.641	<0.0001
Error	10.553	16	0.660		
Total	249.272	19			

k) 0.676
4) c)
5) $\widehat{Y}_i = b_0 + b_1 X_i$. In the case that $X_i = \overline{X}$, then
$\widehat{Y}_i(\overline{X}) = b_0 + b_1\overline{X} = (\overline{Y} - b_1\overline{X}) + b_1\overline{X} = \overline{Y}$
6) a) $\hat{Y} = 1{,}164.41 + 7.53X$
b) 1,164.41
c) 7.53
d) 570.59
7) b) $\hat{Y} = 1.206 + 0.032N + 0.448DEFOLIA + 0.003N * DEFOLIA$
c) As the hypothesis for $\beta_3 = 0$ is not rejected ($P = 0.5644$) the interaction term is not significant, so the effect of defoliation is the same for all the levels of N.
d) $\hat{Y} = 3.054 kg/ha$

Non-solved problems

1) A company producing balanced feed for chicken is interested in partially replacing soya beans and fish meal with dry cassava leaves. Protein content in cassava leaves is relatively high and can replace other costly protein sources. To this purpose, a technician from the company has carried out an experiment with concentrates including cassava meal in different proportions (0%, 10%, 15%, 20%, 25% and

30%). Eight weeks after, the average weight gain of each 10-chicken pen in grams is recorded (see the table below):

CASSAVA	**0**	**0**	**0**	**0**	**10**	**10**	**10**	**10**	**15**	**15**	**15**	**15**
WEIGHT GAIN	1350	1000	1200	1500	1810	1350	2000	1800	2096	1750	2250	2150
CASSAVA	20	20	20	20	25	25	25	25	30	30	30	30
WEIGHT GAI N	2333	1980	2000	2587	2100	1850	2300	1920	1860	1950	1780	1650

a) Plot the data. Does the linear regression function appear to be a good fit here?
b) Find R^2 for the linear regression function. Do you consider that this linear fit is good?

2) A textile engineer is interested in studying the resistance to tension of a new synthetic fiber which will be used in fabric for trouser material. He knows that resistance is affected by the cotton percentage in the fiber, which fluctuates from 10 to 40 percent. He takes 6 samples and tests the 7 cotton percentages in each sample and obtains the following results of resistance to tension (lb per plg^2):

Percent of cotton						
10	**15**	**20**	**25**	**30**	**35**	**40**
9	9	17	17	20	10	8
10	11	14	19	21	11	10
7	9	18	16	22	13	10
9	8	14	17	20	11	8
7	11	13	18	23	12	8
8	8	16	19	22	10	7

For each of the following situations, formulate the model for the analysis, indicate the model assumptions, use the corresponding tests, and draw conclusions.

a) If your objective is to determine that there are significant differences in the resistance to the average tension due to the differences in cotton percentage without considering the nature of the relationship between variables.
b) If your objective is to predict the tear resistance at 22 percent cotton level.

Note: According to different objectives, different analysis can be carried out with the same data set and under the same assumptions. In the first case, the differences in cotton percentage are considered as different treatments, without taking into account the quantitative nature of the independent variable. The objective is to determine if there are significant differences in the mean responses of the tension resistance for the differences in cotton percentages. In the second case, the objective is to find the best curve fit to the model in order to estimate the tension resistance value at an intermediate level of the cotton percentage.

3) A chemist is studying in a chemical process if the reaction time measured in hours is related to the chamber temperature (°F) where the reaction takes place.

temperature (°F)	200	210	220	230	240	250	210	215	235	245
time (hours)	2.7	2.5	2.0	1.9	1.8	1.4	2.2	2.3	1.7	1.6

   a) Estimate the regression coefficients and interpret them in terms of the problem.
   b) The value estimated for β_1 is small; do data support that the β_1 value is significant?
   c) What percentage of the variation observed in the reaction time is explained by this model?
   d) If there are two chambers A and B, A having a higher temperature than B, how can reaction times be expected?

4) Analysis of dam breaching and the resulting floods are essential for identifying and reducing potential for loss of life and damage in the downstream floodplain. The paper Predicting *Peak Outflow from Breached Embankment Dams.* prepared by M. W. Pierce, C. I. Thornton and S. R. Abt for the National Dam Safely Review Board Steering Committee on Dam Breach Equations (2010), available in:

http://www.damsafety.org/media/documents/RESEARCH/ResearchReports/PredictPeakOutflwBrchEmbkDms2010.pdf (27 June 2012)

Document QR Code

combined two data bases containing data describing the height in meters (H) and volume in m^3 (V) of water behind the dam at failure to predict the peak outflow in m^3 per s (Q_p). The data are presented in the file OUTFLOW.xls. Transform the three variables to logarithms before the analysis.

   a) Interpret the parameter estimations of the linear model with two explanatory variables.
   b) Can you determine if any of the variables under study is useful to explain the peak outflow? Justify your answer statistically.
   c) Calculate the determination coefficient and interpret it.

d) Calculate the two coefficients of partial determination. Explain.
e) Estimate the peak flow for a depth of water of 2.20 m and a volume of water of 5,000 m^3.
f) Obtain the residuals and plot them against predicted values. What information is provided by your plot?

Appendix A

River Rughi-Rughi data

NO3	pH	Cl	DO	EC
288.6	7.4	694	5.9	461.8
213.5	7.1	345.9	8.5	386.2
128.3	7.1	362.4	8.1	259.4
128.7	7.1	481.4	7.5	146.5
173	7.2	572.5	7	322.9
219.4	7.3	640.8	6.4	460.2
252.3	7.3	686.1	4.5	440
228	7.4	793.1	5.7	449.1
316.1	7.6	644.1	6.5	388
170.9	7.3	471.3	8.8	329
141.5	7.2	415.8	8.8	196.6
193.8	7.4	537	9.3	229.8
135.9	7.2	344.8	9.1	167
179.9	7.1	398.7	8.3	227
216.3	7.1	495.2	7.7	294.1
192.9	7.1	411.2	7.2	282.8
279.7	7.4	783.8	6.3	333.8
194.5	7.3	672.6	7	291.7
144	7.1	411.4	9	188.2
131.6	7.2	189.5	10.9	137
193.7	7.3	520.6	8.4	267.9
277.6	7.4	718.7	4.8	398.8
209.4	7.3	573.9	8.4	270.6
90.1	7	345.1	10.2	191.8
159.5	7.3	390.5	9.4	164.4
119.2	7.2	361.1	9.4	198.6
314.4	7.6	788.7	3.9	453.2
271.6	7.5	533.5	5.4	310.3
229.5	7.2	509	6.4	364.5
214.8	7.3	504.7	6.8	331.3

Acknowledgements

The authors wish to thank Teresa Boca for her assistance with the R software and Julián López for giving us good ideas for the main example. Our special thanks to our colleague Norberto Bartoloni whose corrections improved the chapter.

Reference

Belsley, D.A., E. Kuh and R.E. Welsch. 2005. Regression Diagnostics: Identifying influential Data and Sources of Collinearity. John Wiley & Sons. New Jersey.

Neter, J., M.H. Kutner, C.J. Nachtsheim and W. Wasserman. 1996. Applied Linear Statistical Models. McGraw-Hill. Boston.

R Development Core Team. 2008. R: A language and environment for statistical computing. R Foundation for Statistical Computing, Vienna, Austria. ISBN 3-900051-07-0, http://www.R-project.org.

Index

Color Plate Section

Chapter 3

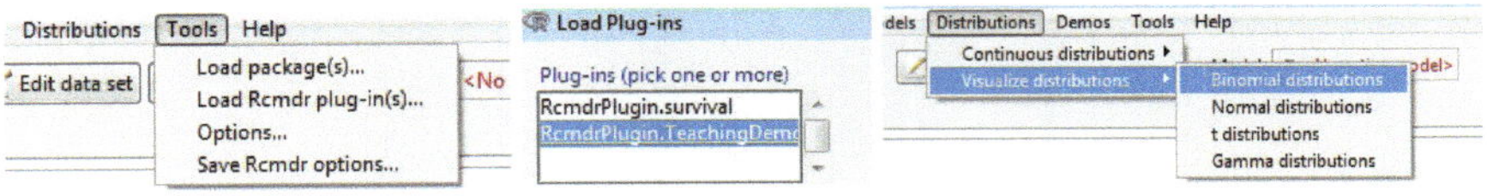

Figure 16. Steps to initialize visualization of Binomial distribution.

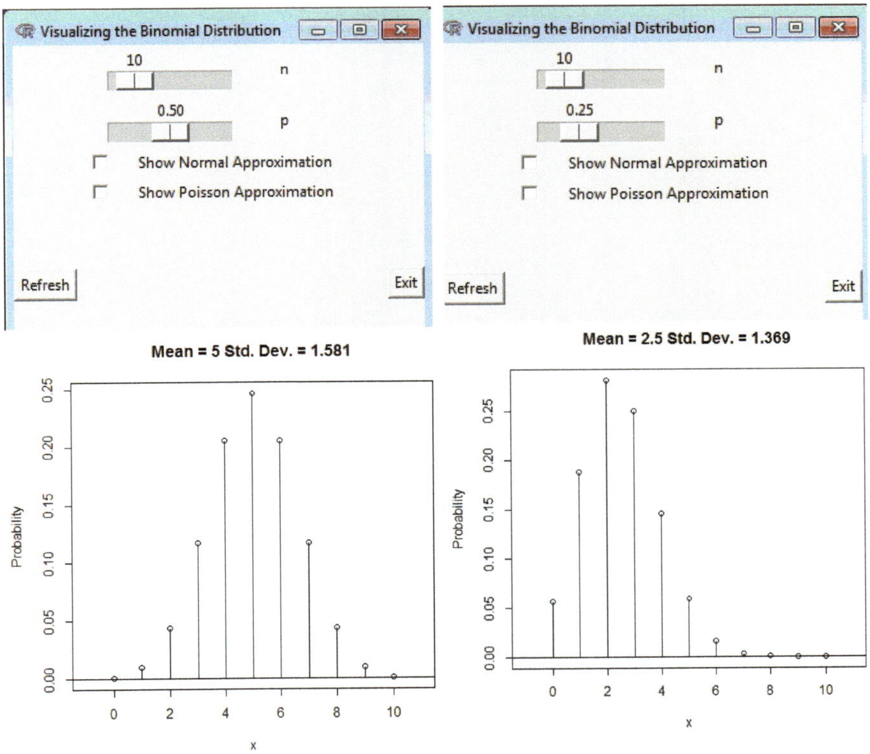

Figure 17. Examples of the graphs generated in the package Rcmdr, R software, for the binomial distributions with parameters n =10, p=0.50 (left graph) and p=0.25 (right graph).

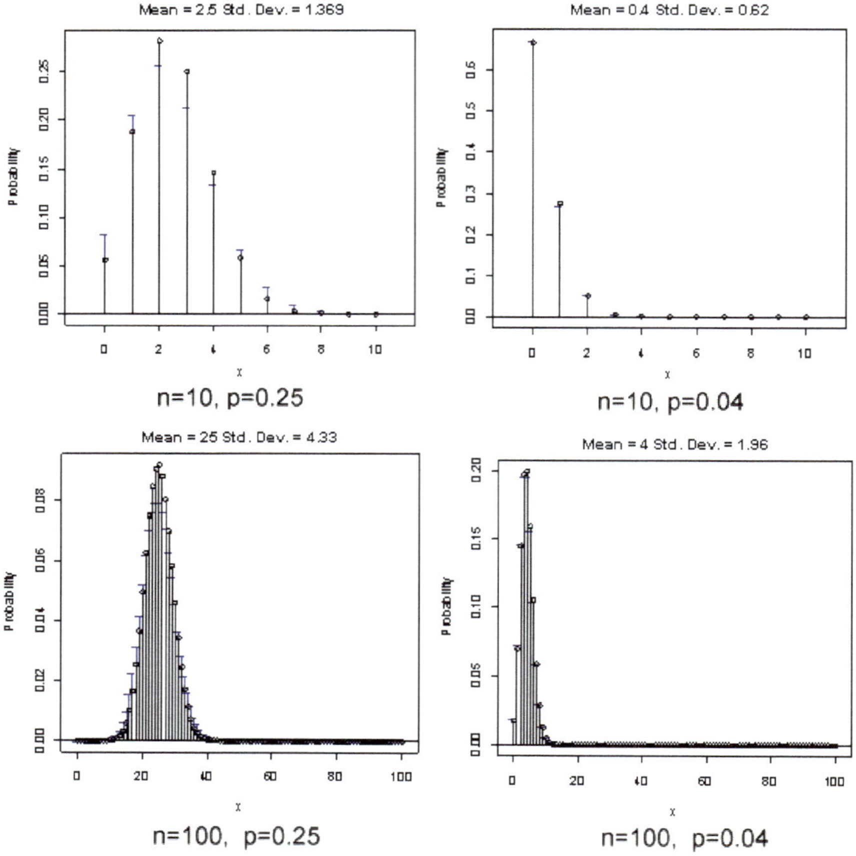

Figure 18. Examples of the graphs generated in the package Rcmdr, R software, for binomial distributions with different parameters and compared to Poisson distribution.

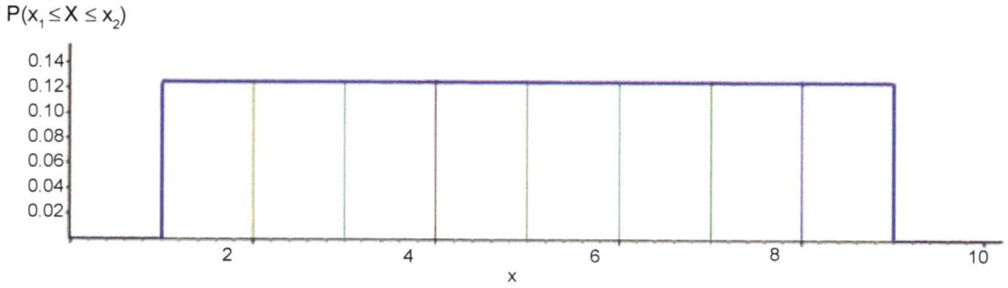

Figure 20. Graph 1 Probability distribution for the experiment of the spinner.

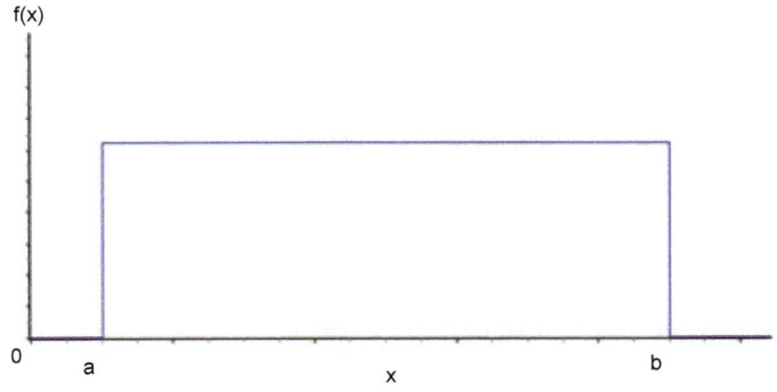

Figure 21. Graph 2: pdf for a spinner with sample space [a, b].

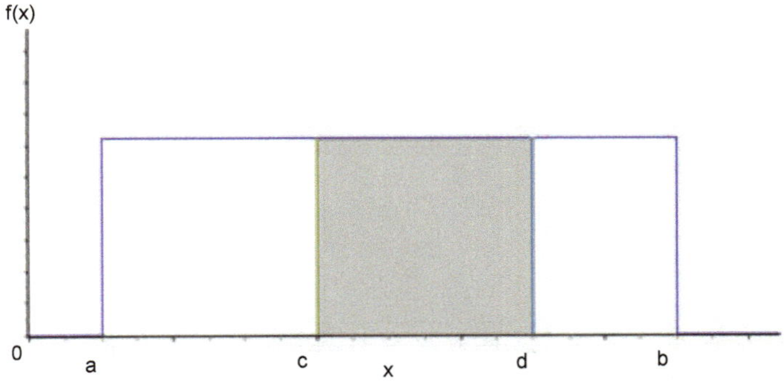

Figure 22. Graph 3: PDF for a spinner with a sample space [a,b] and interval of occurrence [c,d].

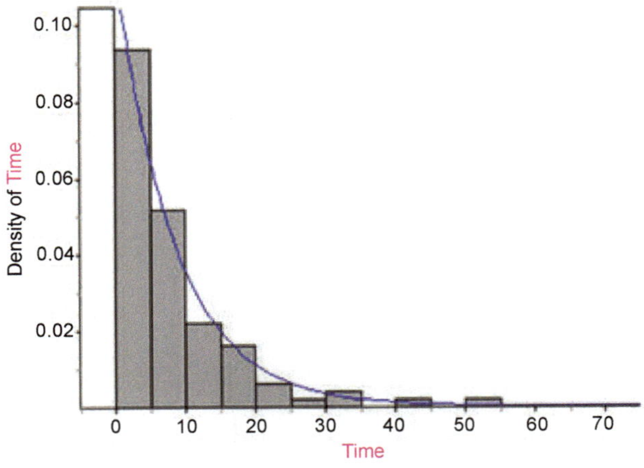

Figure 24. Graph 5: A possible pdf for the time in minutes spent by 100 costumers at a quick market.

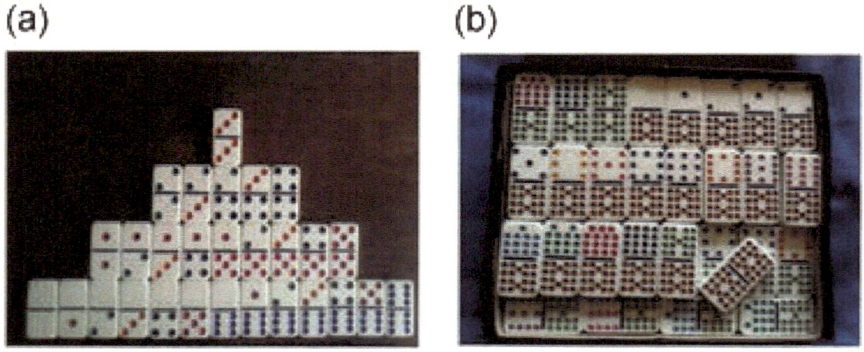

Figure 27. (a) Graph made with a set of dominoes, using the sum of the dots as the variable; (b) Set of dominoes of 190 pieces.

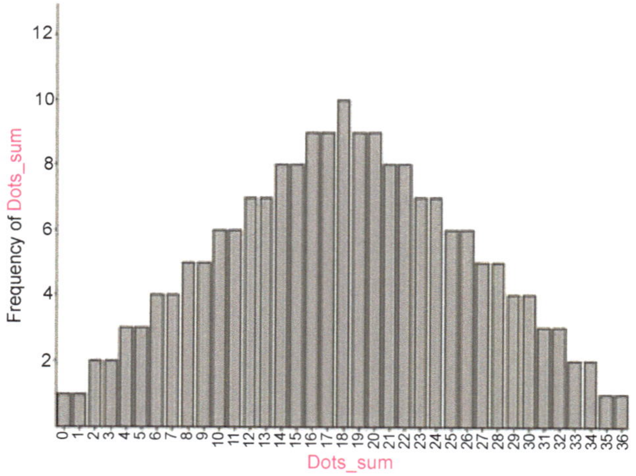

Figure 28. Graph 8: Bar graph for the sum of dots on the dominoes pieces.

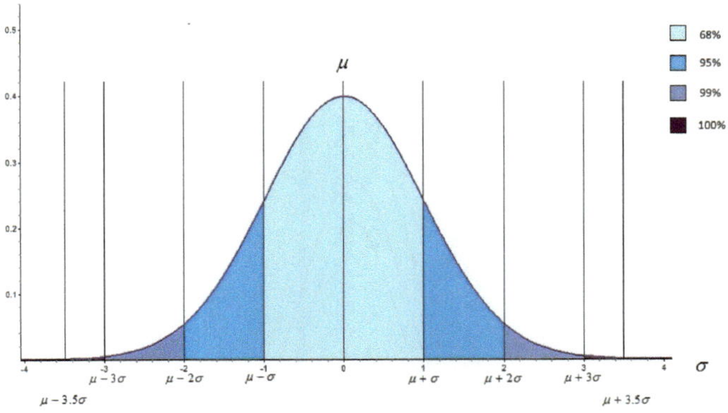

Figure 29. Empirical rule.

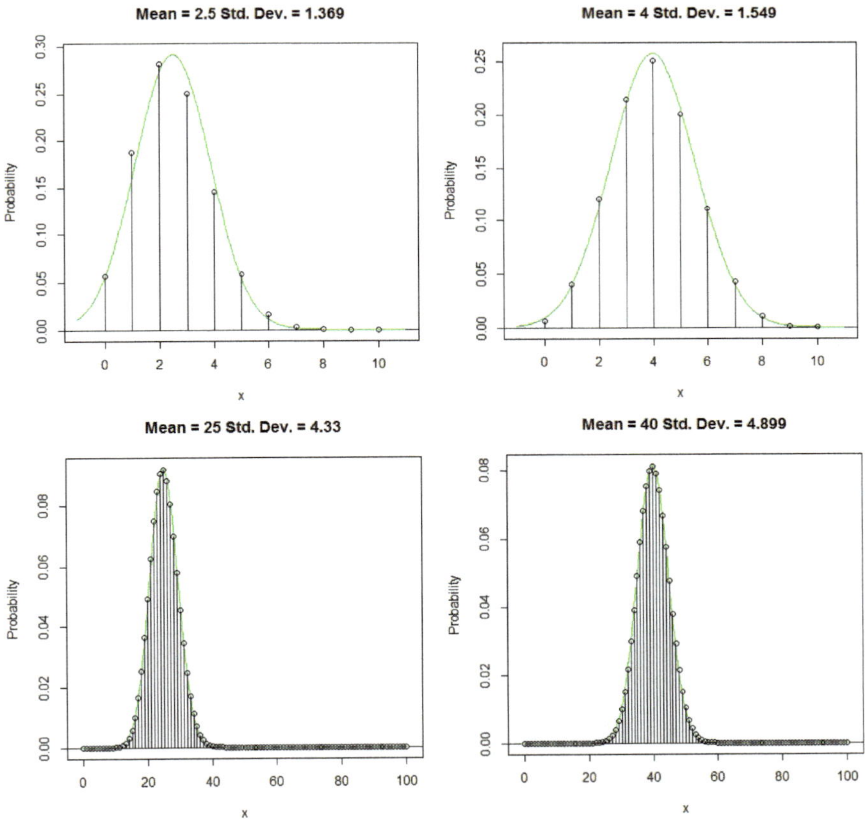

Figure 36. Examples of graphs constructed in package Rcmdr, R software, for binomial distributions with different parameters, compared to normal distributions.

Chapter 5

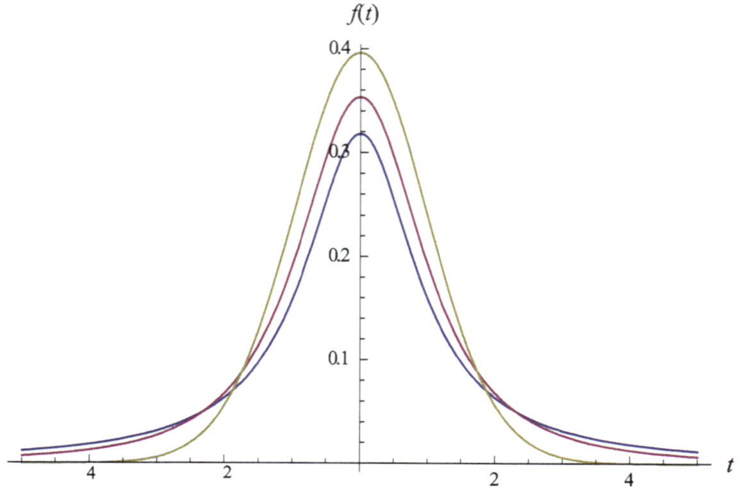

Figure 8. From bottom to top, t distributions having degrees of freedom 1, 2 and 40, respectively.

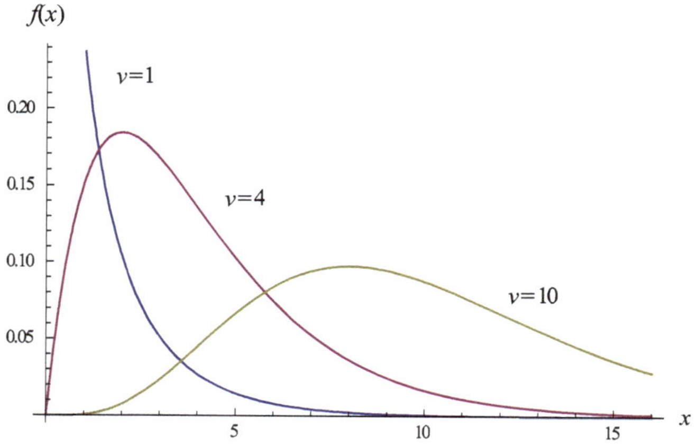

Figure 9. Chi-squared distributions having 1, 4 and 10 degrees of freedom.